Layered Double Hydroxide-Based Catalytic Materials for Sustainable Processes

Layered Double Hydroxide-Based Catalytic Materials for Sustainable Processes

Editors

Ioan-Cezar Marcu
Octavian Dumitru Pavel

MDPI • Basel • Beijing • Wuhan • Barcelona • Belgrade • Manchester • Tokyo • Cluj • Tianjin

Editors
Ioan-Cezar Marcu
University of Bucharest
Romania

Octavian Dumitru Pavel
University of Bucharest
Romania

Editorial Office
MDPI
St. Alban-Anlage 66
4052 Basel, Switzerland

This is a reprint of articles from the Special Issue published online in the open access journal *Catalysts* (ISSN 2073-4344) (available at: https://www.mdpi.com/journal/catalysts/special_issues/Layered_ Double_Hydroxide_Based_Catalytic_Materials).

For citation purposes, cite each article independently as indicated on the article page online and as indicated below:

LastName, A.A.; LastName, B.B.; LastName, C.C. Article Title. *Journal Name* **Year**, *Volume Number*, Page Range.

ISBN 978-3-0365-4979-8 (Hbk)
ISBN 978-3-0365-4980-4 (PDF)

Contents

About the Editors

Ioan-Cezar Marcu

Prof. Ioan-Cezar Marcu received a BSc degree in chemistry and physics in 1995, and an MSc degree in heterogeneous catalysis in 1996 at the University of Bucharest (UB), Romania. He obtained his PhD in catalysis in 2002 at the Institute of Catalysis—University Lyon 1, France, and he was a CNRS postdoctoral researcher from October 2006 to September 2007 at the Institute Charles Gerhardt Montpellier, France. He became a Senior Researcher at the Research Center for Catalysts and Catalytic Processes of the UB starting from 2007. He received his habilitation in catalysis in 2013 and became a Full Professor at the UB in February 2020, in charge of the chemical technology and catalytic materials disciplines. His research interests cover the field of catalysis by metal oxides, including LDH-derived mixed oxides. He has coauthored more than seventy publications, including three book chapters and four encyclopedia articles.

Octavian Dumitru Pavel

Dr. Octavian Dumitru Pavel obtained his PhD diploma in catalysis in 2006 at the University of Bucharest, Romania, while his postdoctoral research was carried out at Laboratoires de Réactivité de Surface, Université Pierre et Marie Curie, Paris, France, in 2007. In 2005 he received a doctoral research internship at the same university in France. Additionally, he was a Visiting Research Associate (2011) and Visiting Research Fellow (2016) at the School of Chemistry and Chemical Engineering, Queen's University Belfast, United Kingdom of Great Britain and Northern Ireland. Starting in 2007, he became a member of the Catalysts and Catalytic Processes Research Center and a Senior Lecturer at the University of Bucharest, Faculty of Chemistry. His research interests are in heterogeneous catalysis, especially in the field of layered materials. Additionally, a secondary area of his interests is ionic liquids. He was a project director for 4 national grants and a member of the work team for 23 projects. Additionally, he was a coauthor of 70 ISI articles, 5 book chapters, and 1 encyclopedia paper.

Editorial

Layered Double Hydroxide-Based Catalytic Materials for Sustainable Processes

Ioan-Cezar Marcu * and Octavian Dumitru Pavel *

Research Center for Catalysts and Catalytic Processes, Faculty of Chemistry, University of Bucharest, 4-12, Blv. Regina Elisabeta, 030018 Bucharest, Romania
* Correspondence: ioancezar.marcu@chimie.unibuc.ro (I.-C.M.); octavian.pavel@chimie.unibuc.ro (O.D.P.)

Citation: Marcu, I.-C.; Pavel, O.D. Layered Double Hydroxide-Based Catalytic Materials for Sustainable Processes. *Catalysts* **2022**, *12*, 816. https://doi.org/10.3390/catal12080816

Received: 20 July 2022
Accepted: 21 July 2022
Published: 25 July 2022

Publisher's Note: MDPI stays neutral with regard to jurisdictional claims in published maps and institutional affiliations.

Layered double hydroxides (LDH) or hydrotalcites (HT), together with their corresponding mixed oxides, continue to arouse a great deal of research interest. Due to their amazing properties [1], they are highly appreciated in various fields such as catalysis [2–7], pharmacy [8,9], medicine [10–12], environmental protection [13–15], etc. All the applications of LDH-based materials are due to their specific properties, which can be tailored by their composition and the synthesis methods employed for their preparation. Traditionally [16], the co-precipitation method provides a fast synthesis path with very high yields, wherein salts containing the targeted cations are contacted with an alkaline solution. Additionally, the traditional preparation methods for LDH include ion-exchange, rehydration using the structural memory effect, the hydrothermal method, and secondary intercalation. Currently, there are alternative methods for obtaining LDH-based materials including electrochemical synthesis, exfoliation in an aqueous solution, dry exfoliation, and the mechano-chemical method. These methods can be used to obtain LDH-based materials with peculiar physicochemical characteristics and, hence, specific properties, including excellent catalytic performances.

This Special Issue, entitled "Layered Double Hydroxide-Based Catalytic Materials for Sustainable Processes" is a collection of 12 articles, including one review paper, presenting the preparation of different LDH-based materials, their physicochemical characterization, and the study of their catalytic performance in various chemical reactions.

In their review paper, Stamate et al. [17] emphasize the most relevant studies related to the large group of polyoxometalate (POM)-intercalated LDH solids, with a focus on their synthesis, characterization, and catalytic applications. Althabaiti et al. [18] bring to bear new information about CoMgAl-LDH used in Michael addition of an aryl halide compound onto activated olefin as a Michael acceptor. The ultrasound was found to have a beneficial effect on this reaction due to the cavitation phenomenon. Zn is a widely used cation in obtaining LDH-type materials with photocatalytic applications. Therefore, by two different methods, Trujillano et al. [19] prepared Zn,Al-layered double hydroxides as well as their corresponding mixed oxides by calcination at 650 °C, to be used in adsorption and photocatalytic degradation of 4-nitrophenol in an aqueous solution. Here, the mixed oxides display a better performance in the adsorption/degradation of the contaminant than ZnO, also showing the memory effect. Zăvoianu et al. [20] present a study of the influence of chemical composition of Zn_xAl- and Mg_xAl-LDH (x = 2–5) and Mg_yZn_zAl-LDH (y + z = 4, y = 1, 2, 3) on the olefin epoxidation with H_2O_2 in the presence of acetonitrile. It was found that the catalytic activity and the basicity of the samples varies in the order LDH (OH^-/CO_3^{2-}) < Reconstructed LDH (majority OH^-) < Calcined LDH (O^{2-}), and the yield to epoxycyclohexane increases almost linearly when the number of weak and medium-strength base sites in the brucite-type layer rises in the range 4.5–8.5 $mmol \cdot g^{-1}$. The same author and coworkers [21] synthetized MgZnAl-LDH by a nontraditional mechano-chemical method in the presence of an organic base, tetramethylammonium hydroxide, with good activity in Claisen–Schmidt condensation between benzaldehyde and cyclohexanone. Indeed, conversions higher than

90% after 2 h of reaction with a total selectivity toward 2,6-dibenzylidenecyclohexanone were observed. Korolova et al. [22] study the effect of Al, Ga, Fe, and In trivalent M^{3+} cation from $Mg_6M^{3+}_2(OH)_{16}CO_3 \cdot 4H_2O$ in aldol condensation between furfural and acetone. The authors conclude that the catalytic performance of the rehydrated mixed oxides is determined by the "host" MgO component, rather than by the nature of M^{3+}. Dib et al. [23] provide further evidence that LDH-type materials can play an important role in environmental protection, using CuAlCe ex-LDH mixed oxides in the total oxidation reactions of both toluene and ethanol. The authors claim that the sample with the highest content of Ce showed the best catalytic properties due to the improvement of the reducibility of the copper species and their good dispersion on the surface. In the same vein of environmental processes, Argote-Fuentes et al. [24] use active hydrotalcites in the degradation of Congo red dye through processes assisted by ultraviolet (UV) irradiation and electric current, where maximum degradation was reached with the photoelectrocatalytic process with active hydrotalcites and a copper anode at 6 h with 95% in a half-life of 0.36 h. Additionally, Nayak et al. [25] degraded visible light-triggered Rhodamine B in the presence of MgCr-LDH nanoplatelets with an efficiency of 95% at 0.80 kW/m^2 solar light intensity in 2 h. Additionally, Wang et al. [26] oxidized the formaldehyde with air at low temperature in presence of Mn-containing mixed-oxide-supported bismuth oxychloride (BiOCl), and showed that the complete removal of formaldehyde could be achieved at 70 °C, the removal efficiency being maintained more than 90% for 21 h. Another application of LDH catalysts is the transesterification reaction. Tajuddin et al. [27] studied the catalytic activity of NiAl-layered double hydroxides in tributyrin transesterification with methanol. The activity of the calcined-rehydrated NiAl-LDH materials was found to increase with Ni content and corresponding base site loadings. Huang et al. [28] obtained metallic Ni by reducing a Ni precursor in a H_2 atmosphere at 500 °C for 3 h dispersed on an ex-LDH LiAl mixed oxide deposited on an Al structured framework, i.e., lathe waste strips. This supported catalyst was studied in ethanol steam reforming (ESR). The relatively low acidity of the ex-LDH LiAl oxide support led to low activity for the dehydration of ethanol and high activity for H_2 generation.

All these studies clearly show that LDH-based materials have a great number of applications as catalysts and catalyst supports for a broad range of chemical reactions which are of interest and importance for the sustainable development of our society. This being the case, future development in the research of LDH-based materials looks set to continue.

Finally, we thank the *Catalysts* journal for this great opportunity to produce this Special Issue. We also thank both the authors and the referees for their hard work and contribution to the success of this Special Issue.

Author Contributions: Conceptualization, I.-C.M. and O.D.P.; writing—original draft preparation, I.-C.M. and O.D.P.; writing—review and editing, I.-C.M. and O.D.P.; visualization, I.-C.M. and O.D.P. All authors have read and agreed to the published version of the manuscript.

Funding: This research received no external funding.

Conflicts of Interest: The authors declare no conflict of interest.

References

1. Cavani, F.; Trifiro, F.; Vaccari, A. Hydrotalcite-type anionic clays: Preparation, properties and applications. *Catal. Today* **1991**, *11*, 173–301. [CrossRef]
2. Sahoo, D.P.; Das, K.K.; Mansingh, S.; Sultana, S.; Parida, K. Recent progress in first row transition metal Layered double hydroxide (LDH) based electrocatalysts towards water splitting: A review with insights on synthesis. *Coord. Chem. Rev.* **2022**, *469*, 214666. [CrossRef]
3. Gabriel, R.; de Carvalho, S.H.V.; da Silva Duarte, J.L.; Oliveira, L.M.T.M.; Giannakoudakis, D.A.; Triantafyllidis, K.S.; Soletti, J.I.; Meili, L. Mixed metal oxides derived from layered double hydroxide as catalysts for biodiesel production. *Appl. Catal. A* **2022**, *630*, 118470. [CrossRef]
4. Heravi, M.M.; Mohammadi, P. Layered double hydroxides as heterogeneous catalyst systems in the cross-coupling reactions: An overview. *Mol. Divers.* **2022**, *26*, 569–587. [CrossRef] [PubMed]

5. Tichit, D.; Álvarez, M.G. Layered Double Hydroxide/Nanocarbon Composites as Heterogeneous Catalysts: A Review. *ChemEngineering* **2022**, *6*, 45. [CrossRef]
6. Álvarez, M.G.; Marcu, I.-C.; Tichit, D. Recent Innovative Developments of Layered Double Hydroxide-Based Hybrids and Nanocomposite Catalysts. In *Progress in Layered Double Hydroxides—From Synthesis to New Applications*; Nocchetti, M., Costantino, U., Eds.; World Scientific: Singapore, 2022; Chapter 4; pp. 189–362.
7. Marcu, I.-C.; Urdă, A.; Popescu, I.; Hulea, V. Layered Double Hydroxides-based Materials as Oxidation Catalysts. In *Sustainable Nanosystems Development, Properties, and Applications*; Putz, M.V., Mirica, M.C., Eds.; IGI Global: Hershey, PA, USA, 2017; Chapter 3; pp. 59–121.
8. Zhang, L.-X.; Hu, J.; Jia, Y.-B.; Liu, R.-T.; Cai, T.; Xu, Z.P. Two-dimensional layered double hydroxide nanoadjuvant: Recent progress and future direction. *Nanoscale* **2021**, *13*, 7533–7549. [CrossRef]
9. De Sousa, A.L.M.D.; dos Santos, W.M.; de Souza, M.L.; Barros Silva, L.C.P.B.; Koo Yun, A.E.H.; Aguilera, C.S.B.; de França Chagas, B.; Araújo Rolim, L.; da Silva, R.M.F.; Rolim Neto, P.J. Layered Double Hydroxides as Promising Excipients for Drug Delivery Purposes. *Eur. J. Pharm. Sci.* **2021**, *165*, 105922. [CrossRef]
10. Kankala, R.K. Nanoarchitectured two-dimensional layered double hydroxides-based nanocomposites for biomedical applications. *Adv. Drug Deliv. Rev.* **2022**, *186*, 114270. [CrossRef]
11. Rojas, R.; Mosconi, G.; Zanin, J.P.; Gil, G.A. Layered double hydroxide applications in biomedical implants. *Appl. Clay Sci.* **2022**, *224*, 106514. [CrossRef]
12. Choi, G.; Choy, J.-H. Recent progress in layered double hydroxides as a cancer theranostic nanoplatform. *WIREs Nanomed. Nanobiotechnol.* **2021**, *13*, e1679. [CrossRef]
13. Dong, Y.; Kong, X.; Luo, X.; Wang, H. Adsorptive removal of heavy metal anions from water by layered double hydroxide: A review. *Chemosphere* **2022**, *303*, 134685. [CrossRef] [PubMed]
14. Ye, H.; Liu, S.; Yu, D.; Zhou, X.; Qin, L.; Lai, C.; Qin, F.; Zhang, M.; Chen, W.; Chen, W.; et al. Regeneration mechanism, modification strategy, and environment application of layered double hydroxides: Insights based on memory effect. *Coord. Chem. Rev.* **2022**, *450*, 214253. [CrossRef]
15. Zhang, G.; Zhang, X.; Meng, Y.; Pan, G.; Ni, Z.; Xia, S. Layered double hydroxides-based photocatalysts and visible-light driven photodegradation of organic pollutants: A review. *Chem. Eng. J.* **2020**, *392*, 123684. [CrossRef]
16. He, J.; Wei, M.; Li, B.; Kang, Y.; Evans, D.G.; Duan, X. Preparation of Layered Double Hydroxides. *Struct. Bond.* **2005**, *119*, 89–119.
17. Stamate, A.-E.; Pavel, O.D.; Zavoianu, R.; Marcu, I.-C. Highlights on the Catalytic Properties of Polyoxometalate-Intercalated Layered Double Hydroxides: A Review. *Catalysts* **2020**, *10*, 57. [CrossRef]
18. Althabaiti, N.S.; Al-Nwaiser, F.M.; Saleh, T.S.; Mokhtar, M. Ultrasonic-Assisted Michael Addition of Arylhalideto Activated Olefins Utilizing Nanosized CoMgAl-Layered Double Hydroxide Catalysts. *Catalysts* **2020**, *10*, 220. [CrossRef]
19. Trujillano, R.; Nájera, C.; Rives, V. Activity in the Photodegradation of 4-Nitrophenol of a Zn,Al Hydrotalcite-Like Solid and the Derived Alumina-Supported ZnO. *Catalysts* **2020**, *10*, 702. [CrossRef]
20. Zăvoianu, R.; Cruceanu, A.; Pavel, O.D.; Bradu, C.; Florea, M.; Bîrjega, R. Green Epoxidation of Olefins with Zn$_x$Al/Mg$_x$Al-LDHCompounds: Influence of the Chemical Composition. *Catalysts* **2022**, *12*, 145. [CrossRef]
21. Zăvoianu, R.; Mihăilă, S.-D.; Cojocaru, B.; Tudorache, M.; Pârvulescu, V.I.; Pavel, O.D.; Oikonomopoulos, S.; Jacobsen, E.E. An Advanced Approach for MgZnAl-LDH Catalysts Synthesis Used in Claisen-Schmidt Condensation. *Catalysts* **2022**, *12*, 759. [CrossRef]
22. Korolova, V.; Kikhtyanin, O.; Veselý, M.; Vrtiška, D.; Paterová, I.; Fíla, V.; Capek, L.; Kubička, D. On the Effect of the M^{3+} Origin on the Properties and Aldol Condensation Performance of MgM^{3+} Hydrotalcites and Mixed Oxides. *Catalysts* **2021**, *11*, 992. [CrossRef]
23. Dib, H.; El Khawaja, R.; Rochard, G.; Poupin, C.; Siffert, S.; Cousin, R. CuAlCe Oxides Issued from Layered Double Hydroxide Precursors for Ethanol and Toluene Total Oxidation. *Catalysts* **2020**, *10*, 870. [CrossRef]
24. Argote-Fuentes, S.; Feria-Reyes, R.; Ramos-Ramírez, E.; Gutiérrez-Ortega, N.; Cruz-Jiménez, G. Photoelectrocatalytic Degradation of Congo Red Dye with Activated Hydrotalcites and Copper Anode. *Catalysts* **2021**, *11*, 211. [CrossRef]
25. Nayak, S.; Parida, K. MgCr-LDH Nanoplatelets as Effective Oxidation Catalysts for Visible Light-Triggered Rhodamine B Degradation. *Catalysts* **2021**, *11*, 1072. [CrossRef]
26. Wang, X.; Wu, G.; Ma, Y. Low-Temperature Oxidation Removal of Formaldehyde Catalyzed by Mn-Containing Mixed-Oxide-Supported Bismuth Oxychloride in Air. *Catalysts* **2022**, *12*, 262. [CrossRef]
27. Tajuddin, N.A.; Manayil, J.C.; Lee, A.F.; Wilson, K. Alkali-Free Hydrothermally Reconstructed NiAl Layered Double Hydroxides for Catalytic Transesterification. *Catalysts* **2022**, *12*, 286. [CrossRef]
28. Huang, S.-H.; Chen, Y.-J.; Huang, W.-F.; Uan, J.-Y. Electrodeposition of a Li-Al Layered Double Hydroxide (LDH) on a Ball-like Aluminum Lathe Waste Strips in Structured Catalytic Applications: Preparation and Characterization of Ni-Based LDH Catalysts for Hydrogen Evolution. *Catalysts* **2022**, *12*, 520. [CrossRef]

 *catalysts*

Review

Highlights on the Catalytic Properties of Polyoxometalate-Intercalated Layered Double Hydroxides: A Review

Alexandra-Elisabeta Stamate, Octavian Dumitru Pavel, Rodica Zavoianu and Ioan-Cezar Marcu *

Department of Organic Chemistry, Biochemistry and Catalysis, Faculty of Chemistry, University of Bucharest, 4-12, Blvd. Regina Elisabeta, 030018 Bucharest, Romania; elisabeta_stamate@yahoo.com (A.-E.S.); octavian.pavel@chimie.unibuc.ro (O.D.P.); rodica.zavoianu@chimie.unibuc.ro (R.Z.)
* Correspondence: ioancezar.marcu@chimie.unibuc.ro; Tel.: +40-213051464

Received: 11 December 2019; Accepted: 26 December 2019; Published: 1 January 2020

Abstract: Layered double hydroxides (LDH) are an extended class of two-dimensional anionic materials that are known for their unique lamellar structure, versatile composition, and tunable properties. The layered architecture allows the intercalation between the positively charged sheets of a vast variety of anionic species, including oxometalates and polyoxometalates (POM). The hybrid composites that were developed using POM and LDH show great advantages when compared to both parent materials causing the appearance of new functionalities, which may lead to remarkable contributions in many areas of application, especially in catalysis. The current review paper emphases all of the crucial works already existing in literature that are related to the large group of POM-LDH solids and their use as catalysts for fine organic synthesis. The new trends in the development of the POM-LDH catalysts are highlighted based on the overview of 121 scientific articles that were published between 1984 and 2019. The main topics are focused primarily on the synthesis, characterization, and the catalytic applications of different LDH systems hosting polyoxometalates with low, medium, and high nuclearity. The intense exploration of the POM-LDH field has led to the obtaining of countless effective catalysts used in various types of reactions, from condensation, esterification, halodecarboxylation, to oxidation and epoxidation.

Keywords: layered double hydroxides (LDH); polyoxometalates (POM); catalytic materials

---

1. Introduction

As the Earth is constantly changing, society has also become driven by ever-increasing requests. The constant need to fulfill the existing demands represents the main priority of the scientists throughout the world. Catalysis has been the key point of our industrialized society, playing a crucial role in the development of the world's economy, allowing for the conversion of raw materials into valuable fine chemicals, in an environmentally friendly and low-cost manner [1]. Catalysts are suitable for applications in a variety of areas (industry, agriculture, environment, ecology, etc.), and it has been estimated that almost 90% of all chemical processes are based on catalysis [1]. This increasing, ongoing need for selective catalysts has enabled the discovery of a series of novel catalytic materials.

Layered double hydroxides (LDH), which are also known as hydrotalcites, are anionic compounds that currently receive increasing attention, owing to their special lamellar structure and other unique properties that make them suitable for applications in several different advanced technological processes: pharmaceutics synthesis, drug delivery, photochemistry, electrochemistry, and many others. Nevertheless, their main application area is in catalysis, either as precursors, catalyst supports, or as actual catalysts [2]. Their variable composition and tunable intrinsic properties clarify the reason why catalytic usage belongs to their main field of applications. As solid base catalysts, LDH exhibit high activities and selectivities for a wide range of reactions, including condensation, alkylation, cyclization, and isomerization, which are commonly carried out in industry while using liquid bases as catalysts. Many of these chemical transformations require stoichiometric amounts of the liquid base for conversion to the desired product, making them economically and environmentally inefficient, so their replacement with hydrotalcite-like catalysts is crucial. From the structural point of view, hydrotalcites can be described as lamellar solids having each layer positively charged with negatively charged interspace species. A wide variety of anionic species can be interlayered either during the formation of the lamellar structure, or by anion exchange [3]. Consequently, layered double hydroxides offer a remarkable opportunity for synthesizing new intercalated composite materials, having improved chemical properties due to a synergistic effect.

Polyoxometalates (POMs) are a versatile group of compounds that contain different transition metal cations, such as Mo, W, V, Nb, etc. at their highest oxidation states. Their properties have been widely exploited in a variety of areas, including catalysis, medicine, pharmaceutical industry, and electronics, due to their variable composition, sizes, rich redox chemistry, and charge distribution [4]. Hybrid materials that are developed using POM and layered double hydroxides have invoked a great deal of interest lately due to their unique properties, also exhibiting advantages over both parent materials, especially in catalytic applications. Several intermolecular interactions, such as electrostatic and hydrogen bond networks, can be established between the brucite-like layers of the hydrotalcites and the intercalated polyoxometalates anions [4]. Additionally, the confinement effect that is displayed by the hydrotalcite interlayer gallery prevents the polyoxometalate anionic species from leaching into the reaction mixture, while also increasing the selectivity of the catalytic transformation [4].

Herein, a complete survey of the scientific literature related to the expanded family of POM-LDH composite materials and their use as catalysts for fine organic synthesis is presented. The main topics will be focused on the synthetic approaches that were developed until nowadays for obtaining different polyoxometalates (POM) that were intercalated within layered double hydroxides. The characterization of these materials, the changes induced in their properties (as compared to those displayed by the same anions in the bulk form), and their catalytic applications are also scrutinized.

2. Layered Double Hydroxides (LDH)-Generalities

Layered double hydroxides (LDH), which are the most known representatives of the anionic clays, are a large family of two-dimensional (2D) materials intensively studied due to their unique catalytic properties conferred by their variable composition [5]. Structurally, layered double hydroxides can be described as lamellar compounds that are similar to brucite, in which every Mg^{2+} ion is octahedrally surrounded by six hydroxyl ions and shares edges to form infinite sheets of $Mg(OH)_2$ [6]. These sheets are overlapping, yielding a layered architecture that is held together by hydrogen bonds or by van der Waals forces [6]. The octahedral coordination of the metals in the layers can be modified if the radius of the cations becomes larger. This can happen when one side of the octahedron is open towards the interlamellar domain with the obtaining of a supplementary coordination of one interlamellar water molecule [4]. If a part of the Mg^{2+} ions is replaced by trivalent cations having a similar ionic radius, such as Al^{3+} or Fe^{3+}, then the entire layer will have a positive charge density. The electrical neutrality of the LDH type solids is maintained by the anions that occupy the interlayer positions (Figure 1).

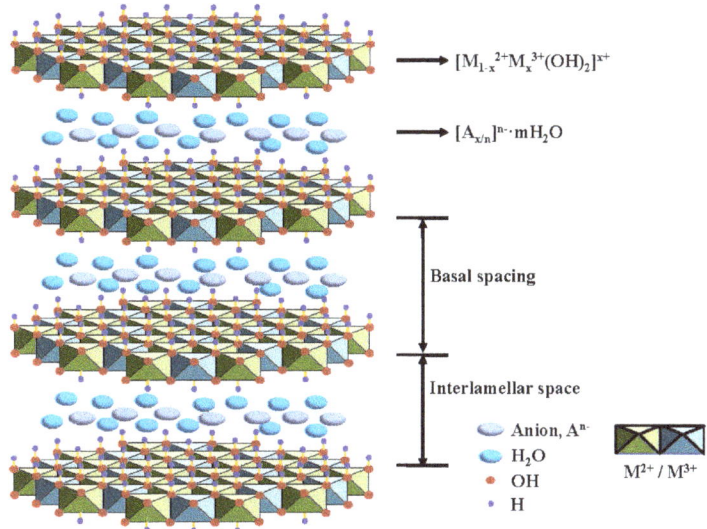

Figure 1. Structural representation of layered double hydroxides (LDH). Adapted from Ref. [4], Copyright 2017, MDPI.

There are two possibilities to stack the layers, either with a rhombohedral (3R symmetry) or a hexagonal cell (2H symmetry). The 3R symmetry is attributed to the hydrotalcite compounds, while the 2H is known as manasseite [4]. Layered double hydroxides are described by the following general molecular formulas: $[M(II)_{1-x}M(III)_x(OH)_2][A^{m-}_{x/m}\cdot nH_2O]$ or $[M(I)_{1-x}M(III)_x(OH)_2][A^{m-}_{(2x+1)/m}\cdot nH_2O]$ [7], where M^+ (M: Li), M^{2+} (M: Mg, Ni, Co, Zn), and M^{3+} (M: Al, Cr, Fe, In, etc.) are metallic cations that are connected with each other through the hydroxyl groups, in a cationic layer, A^{m-} represent the anions, which, along with m molecules of water, form the anionic layer and x is the charge density or the anionic exchange capacity and it takes values between 0.15–0.5 [8]. It can be calculated while using the formula: x = M(III)/(M(III) + M(II)) [3].

The anionic species that can be introduced between the LDH layers can be grouped in the following families of compounds [9]:

- halides;
- non-metal oxoanions;
- oxo- and polyoxometallate anions;
- anionic complexes of transition metals;
- organic anions; and,
- anionic polymers.

It is possible to obtain a large variety of synthetic hydrotalcites-like compounds by simply adjusting their chemical composition in what concerns the nature and proportion of the cations, the nature of the counter ion, or the synthesis conditions that may influence the amount of water in the interlayer region or the crystal morphology and size [10]. Therefore, there is a great possibility to synthesize the materials that fulfill the demands that are specific for different catalytic applications.

2.1. LDH-Preparation Methods

Layered double hydroxides can be synthesized in aqueous medium employing a variety of methods, depending on the specific requirements and the desired properties. The most frequently used is the co-precipitation, which implies mixing the aqueous solutions of precursors (divalent and trivalent metal salts, in well-defined proportions) with an alkaline solution containing the precipitating agent (Na_2CO_3, NH_4Cl, NaOH), which has the role of maintaining a constant pH value to enable the co-precipitation of the mixed hydroxide [11]. A variety of techniques can also be applied for the preparation of the layered double hydroxides, including urea hydrolysis, hydrothermal synthesis, the sol-gel method, and microwave synthesis [12]. These are the most common and widely used solution-phase methods for hydrotalcites synthesis.

Another technique that can be used for both the preparation of LDH and for intercalation purposes is the mechanochemical method. The principle of this method is based on the idea that the precursor salts under the conditions of the mechanical treatment undergo a chemical reaction that results in the formation of the LDH [11]. Given that the formation of the LDH takes place in the solid state, there is no need for aging, which implies higher energy consumption and a longer duration process. Manual grinding or ball milling can be applied as pretreatment as a function of the nature of the precursor (metallic salts, oxides, hydroxides) to obtain uniform mixtures [13]. It has been demonstrated that it is better to utilize as raw materials metallic salts instead of stable hydroxides or oxides in manual grinding [9]. Zhang et al. [11,13] performed an investigation on the formation mechanism of LDH by manually grinding different metallic salts. They have observed that the raw materials disappeared immediately and a considerable amount of LDH phase appeared during milling when hydrated magnesium and aluminum nitrate salts are added into a mortar [13]. Three successive stages are involved in the process of grinding magnesium and aluminum nitrate hydrated salts with Na_2CO_3 and NaOH to form LDH: (i) absorption of water from the ambient air, (ii) formation of $Al(OH)_3$ and $Mg(OH)_2$, and (iii) the formation of LDH phases [13].

This method combines the advantages of both mechanochemistry and hydrothermal treatment, leading to high crystallinity of LDH at a relatively lower temperature or shorter reaction time when compared with the direct hydrothermal process, although the modality of producing is complex when compared with single solution operation [9].

The production of homogenous hybrid structures is a laborious process, therefore it is necessary to employ synthesis methods that present high applicability in terms of cost and environmental protection.

2.2. LDH-Applications

The LDH materials are mainly applied because of their base and redox properties that are present applications in various fields. Over the years, their properties and functionalities have been adjusted by employing different synthesis methods, design principles, and a wide range of metallic cations and organic or inorganic anions. They are often applied as supercapacitors, additives in polymers, or as adsorbents for water remediation [14], in the pharmaceuticals industry, in photochemistry [14], and in electrochemistry. In catalysis, hydrotalcites may be used as such or as mixed oxides obtained by their thermal decomposition, [5,12].

Their variable composition and their tunable intrinsic properties clarify the reason why catalytic usage belongs to their main field of applications. As solid base catalysts, LDHs exhibit high activities and selectivities for different types of reactions, mainly organic transformations (Figure 2), such as alkylation, isomerization, hydroxylation, transesterification, hydroformylation, redox reactions, condensation, and environmentally friendly reactions. Moreover, the mixed oxides that were obtained after calcination of the LDH usually have good catalytic activities and they have been applied to the following organic reactions: aldol and Knoevenagel condensations; epoxidation of olefins; halide exchanges; phenol hydroxylation; and, Michael additions [9].

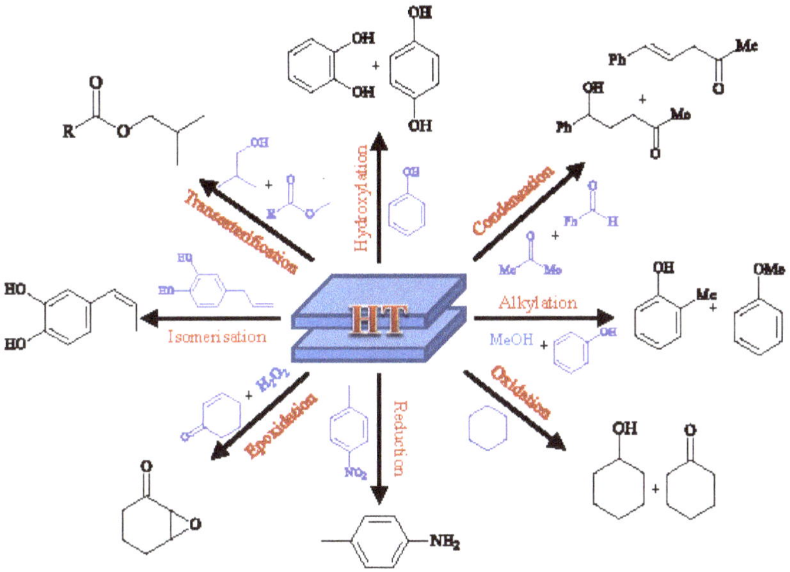

Figure 2. Schematic representation of different organic transformations. Adapted from Ref. [3], Copyright 2015, Royal Society of Chemistry.

3. Synthesis of POM-LDH Nanocomposites

Polyoxometalates (POM) are an extended family of compounds that are obtained from the condensation of metal oxide polyhedra (MO_x, M = W^{VI}, Mo^{VI}, V^V, Nb^V, Ta^V etc., and x = 4 – 7) with each other through corner-, edge-, or, rarely, face-sharing manner [15]. Their synthesis is based on the similar charge to ionic radius ratio and charge density, which: (i) inhibits infinite polymerization; (ii) enables the formation of π-bonds with the O^{2-} ligands; and, (iii) leads to a diversity of coordination geometries [4]. The metal atoms are considered to be addenda atoms, because they can change their coordination with oxygen from four to six, while the MO_x polyhedra condense in solution upon acidification [15]. The nature of the ligand (water, hydroxo, or oxo species) dictates the limit of the polymerization. The nature and the number of the ligands depend on the solution pH and on the metal M^{n+} charge value. There were some POM clusters reported in the literature [15] that contain other atoms or groups, such as sulfur, bromine, nitrosyl, and alkoxy, although oxygen is the most known ligand that is capable of coordinating with the addenda atoms.

The cluster is called isopolymetalate, the Lindqvist type anion $[M_6O_{19}]^{2-}$ being representative when the POM framework consists of addenda metals (from groups 5 and/or 6) and oxygen (Figure 3). If the POM contains additional elements besides addenda metals and oxygen, it is known as a heteropoly complex, being formed by the condensation of MO_x polyhedra around a central heteroatom in acidic medium [15]. A large variety of atoms can act as heteroatoms, such as Be, B, Al, Si, Ge, Sn, P, Te, and the whole first row of transition elements. The atomic ratio between the surrounding atoms (referred to the addenda atoms) and the central heteroatom (arranged in tetrahedral or octahedral coordination) can be 6, 9, 11, or 12 [16]. These ratios come from the most common heteropoly compound structures. The most well-known and studied structure is the so-called Keggin with the general formula $[X^{n+}M_{12}O_{40}]^{(8-n)-}$ (atomic ratio M/X = 12) [16]. POM compounds present high Brønsted acidity due to their large size, which leads to the delocalization of the surface charge density throughout the polyanion, creating a weak interaction between H^+ and the anion. POM display not only high acidity, but also an efficient redox behavior, which conducts fast chemical transformations under soft conditions [16]. This makes POM highly efficient catalysts, especially in the oxidation reactions.

The large diversity of polyoxometalates compounds with well-defined structures offer huge adjustable possibilities for a variety of novel nanocomposite systems [4].

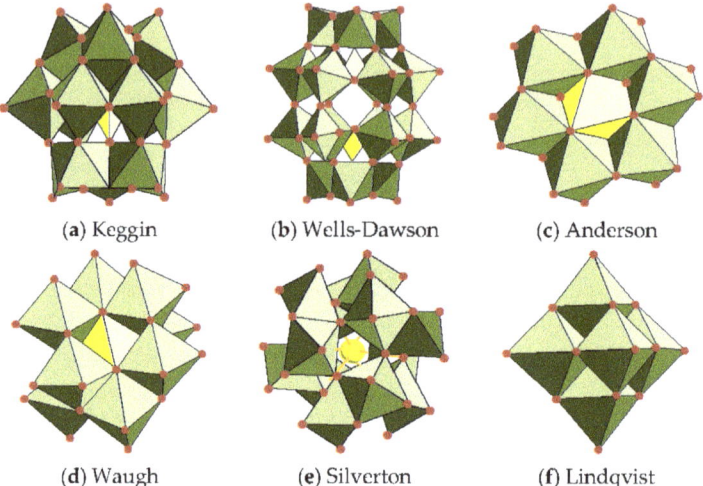

(**a**) Keggin   (**b**) Wells-Dawson   (**c**) Anderson

(**d**) Waugh   (**e**) Silverton   (**f**) Lindqvist

Figure 3. Polyoxometalates (POM) structures in polyhedral representations. Adapted from Ref. [15], Copyright 2013, Elsevier.

The POM/LDH materials have attracted wide interest in catalysis, as they have great advantages over other LDH-based compounds [4] that are attributed to the synergistic effects between the hydrotalcite and the polyoxometalates. The LDH solids offer a remarkable opportunity to modify their functionality without compromising their structural features due to the relatively weak interlayer interactions [4]. The incorporation of catalytically active species, such as inorganic anions, ligands, or biomolecules into the interlayer space of LDHs, has been proven to enhance the catalytic stability and recyclability when compared to their homogeneous counterparts [4]. A series of intermolecular interactions, including electrostatic and hydrogen bonding, can be found between the brucite-like layers of LDHs and the intercalated POM anions [15]. Figure 4 illustrates the general structure of the POM-LDH composition. Although POM-intercalated LDH exhibits superior application capabilities when compared to individual components, the intercalation of POMs into LDH implies some serious complications. First of all, under the neutral to slightly acidic reaction conditions, there is the possibility that M^{2+}/M^{3+} cations leach out during the anionic exchange reactions between LDH precursor and the POM anions [15]. Secondly, the M^{2+}/M^{3+} ratio in the POM pillared LDH must be rigorously controlled, since it helps to determine the charge density of the LDH layers, affecting the basal spacing between the brucite-type layers [10,15]. It is mandatory to maintain, as much as possible, the desired M^{2+}/M^{3+} ratio from the initial step of the preparation towards the final one in order to synthesize POM/LDH solids with different pore size distributions [15].

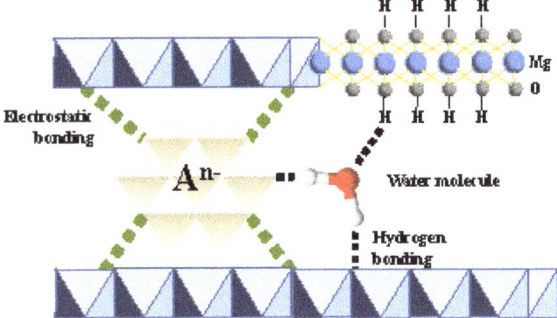

Figure 4. Structural representation of a POM/LDH. Adapted from Ref. [15], Copyright 2013, Elsevier.

It should be mentioned that, according to IUPAC rules [17], the main characteristic of intercalation compounds is the swelling of layered materials that appears when guest species are inserted in the interlayer region, while the term "pillared" should only be applied to the solids that were obtained by thermal or chemical transformation of a layered compound, which shows microporosity or mesoporosity and it is considered to be stable, both chemically and thermally [18]. This is not often the case of POM–LDH hybrids, which are thermally less stable than the compounds that were obtained, starting from cationic clays [18].

Moreover, some POM anions were proven to be unstable at weakly acidic to basic pH [15], explaining why the pillaring reactions are often accompanied by the co-formation of an impurity phase, consisting of the M^{2+} rich salt of the POM that deposits on the surface of the layered double hydroxide [15]. Its deposition can block the micropores of LDH, which leads to the obtaining of low surface areas [15]. It is difficult to prepare POM/LDH nanocomposites in a crystalline form, mostly because LDH hosts act like a base, while most polyoxometalates present acidic properties. Several synthesis methods can be used for the preparation of POM-LDH composites, the co-precipitation, the reconstruction, and the ionic exchange being among the most developed ones [4]. There are also some relatively new preparation techniques that still need improvements, such as electrochemical reduction, ultrasound treatment, and delamination technology [15].

The variety in what concerns the size and features of polyoxometalate compounds can be effectively used in the structural construction of new composite materials that display specific gallery height. Additionally, it is possible to access the gallery region by enlarging the lateral anion spacing if POMs with relatively high-charge density will be used in the construction of the hybrid materials. Therefore, POM can be viewed as ideal functional guest anions for intercalation offering the possibility of enhancing and fine-tuning the properties of the composite. Moreover, the synergistic effect between the constituents of the POM-intercalated LDH composite material might cause the appearance of new functionalities, leading to improved performance and offering the opportunity to increase many areas of application [4].

The POM/LDH hybrids have been intensively applied in catalytic reactions, such as epoxidation of alkenes, molecular oxygen generation, esterification of acetic acid, oxidation of thioether and thiophene, etc. [15]. The prepared POM-LDH composites show superior catalytic performance when compared to the relevant performance of their parent components. The active components are well dispersed in the two-dimensional enclosed space that is defined by the hydrotalcite layers. Moreover, the display of the robust interlamelar zone combined with the intermolecular interactions leads to an increased stability of the system, by preventing the leaching of the POM species into the reaction mixture [15]. The POM/LDH hybrid materials have also been used to enhance the photo-luminescent properties of lanthanide anions and promote the adsorption of harmful components, like dyes, from wastewater [15]. Recently, the delamination of LDH nanosheets has been employed before using them in the development of novel POM/LDH nanocomposite materials. This modern technology represents the ideal solution for the above-stated problems while also retaining the original composition of POMs, apart from creating a new domain of application as photo-luminescent materials based on ultra-thin films [4].

Further investigations and the expansion of the family of POM-LDH composite materials will bring great opportunities in the development of economical and versatile methods for designing specific structural features that could lead reactions towards the desired path and will offer better control over their electron transfer properties [15]. Finally, the exploration of alternative design approaches will enable the construction of POM-LDH materials that present new chemical and physical properties, new types of functionalities, as well as a wider view of catalytic applications [15].

4. Systems Hosting Polyoxometalates

The main topic of this chapter is focused on the synthesis and catalytic activity of layered double hydroxides with a wide variety of cations in the layers, in relation to the nature of the intercalated anion; in this case, polyoxometalates with different nuclearity.

4.1. LDH Intercalated with Low-Nuclearity Polyoxometalates

There are only a few studies of chromate-containing LDH in the literature, and they are resumed in Tables 1 and 2. Cr(VI) in the forms CrO_4^{2-} and $Cr_2O_7^{2-}$ is among the most insidious anionic contaminants in wastewater, so the incorporation of chromates into layered double hydroxides deserves tremendous interest, not only because of their ability to protect the aquatic environment, but also for the valuable application in catalysis of the products obtained [18]. The reported methods used to synthesize LDH with intercalated chromate containing different pairs of divalent/trivalent cations in the brucite-like layers are co-precipitation, anion exchange, and reconstruction. Chromium can be introduced either as a cation, Cr(III) in the brucite-type layers [19–21] or it can be incorporated in its anionic forms CrO_4^{2-} and $Cr_2O_7^{2-}$ in the interlayer [21–30], mostly depending on the nature of the brucite-like cation. In most of the cases, the pairs of cations that were used for the construction of the novel composite materials were Mg–Al, Mg–Fe, Ni–Al, Ni–Fe, Zn–Al, Co–Al, and Ca–Al. A complete study on the use of calcined Mg–Al LDH containing chromium in three different structural sites: as Cr^{3+} in the brucite-like layers (A), as an intercalated Cr^{3+}–EDTA complex (B) and as intercalated chromate (C), in non-oxidative dehydrogenation of ethane to ethylene (EDH), has been reported by Tsyganok et al. [21]. The synthetic strategies that were applied for preparing LDH precursors of chromium catalysts can be seen in Figure 5 and they consist of: (i) conventional co-precipitation of metal cations with carbonate counter-ion under basic conditions; (ii) co-precipitation of Mg(II) and Al(III) with a pre-synthesized EDTA chelate of Cr(III) (method based on the ability of Cr(III) to form a chelate of 1:1 stoichiometry with $(EDTA)^{4-}$ that is much more thermodynamically stable than that of Mg(II) or Al(III) [21]); and, (iii) co-precipitation of Mg(II) and Al(III) with CrO_4^{2-}, chromate anions being present in the solution in slight excess to the amount that is required by stoichiometry. The powder XRD patterns of the materials synthesized by Tsyganok et al. [21] indicated that all of the samples had a layered structure specific to hydrotalcite compounds demonstrated by a set of characteristic diffraction lines at $2\theta = 11.3°, 22.8°, 34.4°$, and $60.5°$ corresponding to reflections from (003), (006), (009), and (110) planes [21].

Table 1. Summary of LDH with intercalated chromium-based anions.

Layer Cations	Molar Ratio (M^{II}:M^{III})	Intercalated Species	Synthesis Method	Reaction	Substrate	Yield (%)	Reference
Mg,Al	2	CrO_4^{2-}	Reconstruction	-	-	-	[22]
Mg,Al	3	CrO_4^{2-}	Co-precipitation	-	-	-	[23,24]
Mg,Al	3	CrO_4^{2-}	Co-precipitation	EDH	Ethane	20	[21]
Mg,Al	3	$Cr^{III}(EDTA)^-$	Co-precipitation	EDH	Ethane	21	[21]
Mg,Al	3	CrO_4^{2-}	Co-precipitation	-	-	-	[25]
Mg,Al	3	$[Cr^{III}(SO_3\text{-salen})]^{2-}$	Anion exchange	Oxidation	Benzyl alcohol	67–88	[28]
Mg,Al	3	$[Cr^{III}(SO_3\text{-salen})]^{2-}$	Anion exchange	Oxidation	Glycerol	30–32	[30]
Mg,Fe	3	CrO_4^{2-}	Co-precipitation	-	-	-	[23,24]
Mg,Fe	2	CrO_4^{2-}	Co-precipitation	-	-	-	[26]
Ca,Al	3	CrO_4^{2-}	Co-precipitation	-	-	-	[23,24]
Co,Al	3	CrO_4^{2-}	Co-precipitation	-	-	-	[23,24]
Co,Al	2	CrO_4^{2-}	Co-precipitation	Oxidation	Benzyl alcohol	23–34	[29]
Co,Fe	3	CrO_4^{2-}	Co-precipitation	-	-	-	[23,24]
Ni,Al	2.88	CrO_4^{2-}	Anion exchange	-	-	-	[27]
Ni,Al	2.83	$Cr_2O_7^{2-}$	Anion exchange	-	-	-	[27]
Ni,Fe	3	CrO_4^{2-}	Co-precipitation	-	-	-	[23,24]
Zn,Al	2	CrO_4^{2-}	Anion exchange	-	-	-	[22]
Zn,Al	3	CrO_4^{2-}	Co-precipitation	-	-	-	[23,24]

Table 2. Summary of CrM-LDH (M-metal) with intercalated chromium-based anions.

Layer Cations	Molar Ratio (M^{II}:M^{III})	Intercalated Species	Synthesis Method	Reaction	Substrate	Yield (%)	Reference
Cu,Cr	2	CrO_4^{2-}	Anion exchange	-	-	-	[19]
Cu,Cr	2	$Cr_2O_7^{2-}$	Anion exchange	-	-	-	[19]
Cu,Cr	2	CrO_4^{2-}	Anion exchange	Alkoxylation	n-Butanol	43–49	[20]
Cu,Cr	2	$Cr_2O_7^{2-}$	Anion exchange	Alkoxylation	n-Butanol	72–83	[20]
Mg,Al,Cr	2;3	CO_3^{2-}	Co-precipitation	EDH	Ethane	20–21	[21]
Mg,Cr	2;3	CO_3^{2-}	Co-precipitation	EDH	Ethane	19–20	[21]

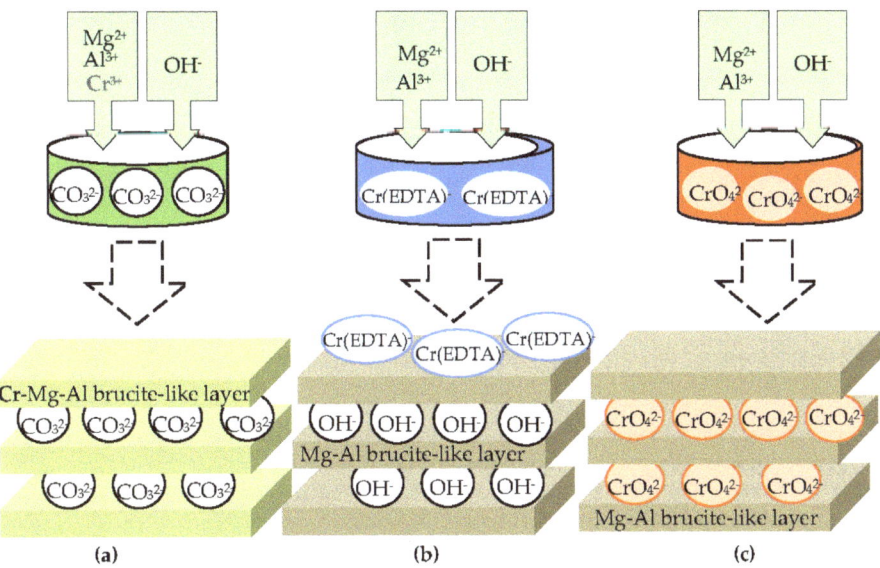

Figure 5. Synthetic approaches for the preparation of LDH precursors of chromium catalysts. Adapted from Ref. [21], Copyright 2007, Elsevier.

For the compound containing chromium pre-chelated with EDTA, the diffraction pattern obtained revealed that the d-spacing was similar with the reference hydrotalcite (i.e. 7 Å) containing only Mg and Al as cation, indicating that Cr(EDTA)$^-$ species were located at the external surface of the LDH crystallites, not in the interlamellar space. The comparison of the XRD pattern of the classical hydrotalcite with that of the LDH containing CrO_4^{2-} species in the interlayers showed that a phase with a larger d-spacing (i.e. 8.4 Å) was formed upon the co-precipitation of Mg(II) and Al(III) with chromate counter-ion, which suggested that the intercalation of the LDH with chromate anion led to an increase of the interlayer distance from 7.8 Å corresponding to the classical hydrotalcite to 8.4 Å for the intercalated LDH, which is in agreement with values that were reported in the literature [21]. On calcination at 500 °C partial oxidation to Cr^{6+} species was observed, and XPS indicated that the molar Cr^{6+}/Cr^{3+} ratio was always larger than 1.0. Ethane dehydrogenation (EDH) was studied at 700 °C during 5 h tests. The activity of all these catalysts significantly decreased, especially during the first hour, due to both the sintering and coking phenomena.

The solids that were synthesized by method (A) displayed the highest selectivity to ethylene (79–84%), but only with 12–15% ethane conversion, while larger conversions (26–30%) with lower selectivity (71–74%) were obtained for samples that were prepared by methods B and C [18]. The deposition of coke was the main problem during the catalytic tests, and its amount was quantified after tests by temperature-programmed oxidation (TPO) while using CO_2 as oxidant. The carbon content was very low (ca. 5%) for the Al-free catalysts [18,21]. After exposure to the EDH reaction and TPO experiments, it was observed that the specific surface areas of the solid sharply decreased, especially for the Al-free samples (ca. 75% decrease).

Wu et al. [28] used a different strategy, which consists first in the incorporation of the anionic salen-like ligand in the interlamellar space and then forming the complex in situ by reaction with Cr^{3+} chloride in order to incorporate chromium complexes in the interlayer of a Mg-Al LDH [28,30], as can be seen in Figure 6.

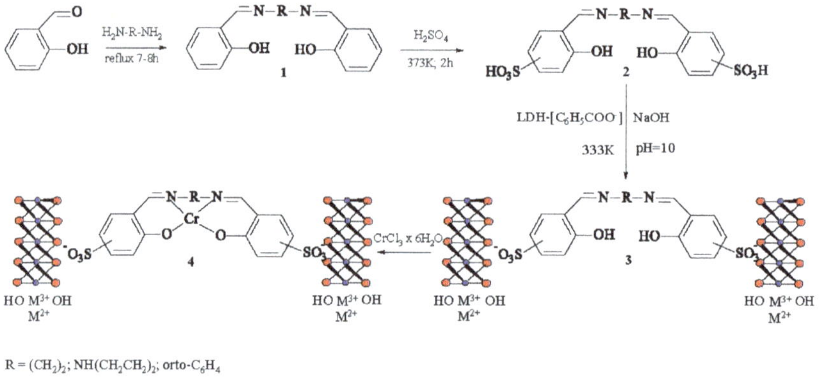

R = (CH$_2$)$_2$; NH(CH$_2$CH$_2$)$_2$; orto-C$_6$H$_4$

Figure 6. The synthesis of MgAl-LDH with intercalated [Cr^{III}(SO$_3$-salen)]$^{2-}$. Adapted from Ref. [28], Copyright 2007, Elsevier.

The reported synthesis method applied for obtaining the sulphonato-salen chromium (III) ligand is the condensation reaction between salicylaldehyde and diamine. For the preparation, Wu et al. [28,30] have tried to use three different diamines (ethylenediamine, diethylenetriamine, and o-phenylenediamine), maintaining the molar ratio salicylaldehyde/amine around 2. The condensation products were further reacted with sulfuric acid to enable the formation of their corresponding phenyl sulfonated compounds, as seen in Step 2, Figure 6, and then with MgAl-LDH, which had the layers swelled by intercalation of benzoate anions.

Finally, the solid obtained was soaked with a $CrCl_3$ aqueous solution, leading to the formation of intercalated $[Cr(SO_3\text{-salen})]^{2-}$ complexes. The FTIR spectra of the three LDH-hosted chromium complexes indicated that the characteristic bands of $C_6H_5COO^-$ anions (1550 cm^{-1}) no longer appeared in the lamellar space of the LDH, which implied that they were completely exchanged with the $[Cr(SO_3\text{-salen})]^{2-}$ anions. The powder X-ray diffraction patterns are characteristic of a veritable layered double hydroxide, also showing the formation of intensively swelled structures, with d_{003} spacing of approximately 20 Å [28]. Wu and his group have published two articles [28,30], in which they have used the calcined solids as catalysts for the oxidation of different substrates, namely benzyl alcohol and glycerin while using H_2O_2 as oxidant. In the first case, the oxidation of benzyl alcohol led to selectivity for benzaldehyde lower than 80% for all of the prepared catalysts. Moreover, benzoic acid (ca. 15%) and benzyl benzoate (ca. 8%) were also formed. It was demonstrated that the catalytic performance of the three catalysts mostly depends on the diamine used, and it decreases depending the diamine backbones in the following sequence: o-C_6H_4 > $NH(CH_2CH_2)_2$ > $(CH_2)_2$ [30]. This decrease can be attributed to the different electronic structures and the steric effect of the three backbones [28,30]. Furthermore, o-C_6H_4 facilitated the contact of H_2O_2 with the active sites due to its planar structure. On the contrary, the other backbones ($NH(CH_2CH_2)_2$ and CH_2CH_2) that are both electron-donor groups led to an increase of the electronic density around the central metal ion hindering the formation of the oxochromium intermediate. Manayil et al. [29] have also studied the oxidation of benzyl alcohol at room temperature while using as catalyst a $CoAl_2$-LDH intercalated this time with chromate anions and tert-butyl hydroperoxide as oxidant. The results showed that an increase in the substrate/oxidant mole ratio will lead to 100% selectivity of benzaldehyde [29].

The LDH-intercalated sulphonato-salen-chromium(III) complex was also a good catalyst for the selective oxidation of glycerin to secondary alcohol-dihydroxyacetone (DHA) with 3% H_2O_2 probably due to the complementary effect that exists between the chromium-Schiff base complex and the weak base LDH host. The highest glycerin conversion reached 73.1%, with 43.5% of selectivity to DHA, when the reaction ran at 60°C for 4 h over 0.2 g catalyst with 25 mL 3% H_2O_2. Moreover, the catalytic performance was maintained even after the catalyst was recycled six times [30].

Malherbe et al. [20] were able to prepare a series of POM/LDH nanocomposites by using Cr and Cu as the cations needed for the construction of the brucite-type layers and CrO_4^{2-} or $Cr_2O_7^{2-}$ as anions intercalated between the layers. The compounds have been tested in the catalytic ethoxylation of butanol by ethylene oxide to produce glycol ethers. The catalytic tests that were carried on in batch reactors have indicated that the catalytic activity of the solids is proportional to the concentration of chromium (VI), which should be theoretically present twice as much in the dichromate intercalated LDH. At maximum conversion, the selectivity is almost the same for both anions [20]. A detailed study of PXRD patterns has led to a better understanding of the ways the layers stack. MgAl–CrO_4 and MgFe–CrO_4 both showed a basal spacing of 10.8 Å and displayed turbostratic disorder [18]. Structural ordering of the first compound was induced by exposing it to a hydrothermal treatment. ZnAl–CrO_4 LDH also present stacking faults with local rhombohedral symmetry, but, in this case, the interlayer spacing was only 7.3 Å, which suggested that the chromate units were grafted to the hydroxyl layers. The MgAl–CrO_4 LDHs have proven to be more stable during calcination (up to 400 °C) than the ZnAl solids (300 °C) due to a fast formation of crystalline ZnO in the last case [25]. Spontaneous contraction of the layers can appear for ZnAl and ZnCr–LDHs that were intercalated with CrO_4^{2-} and/or $Cr_2O_7^{2-}$ after ageing, or even thermal treatment [31]. During ageing, for ZnAl–CrO_4 and ZnCr–Cr_2O_7, a decrease of the basal spacings from 8.61 to 7.86 Å and from 9.20 to 7.34 Å, respectively, appeared. Moreover, it was observed that the Cr(VI) species are more easily reduced to Cr(III) species in the ZnAl systems than in MgAl–CrO_4 ones [25]. Hydrotalcites containing Cu or Zn as cations in the brucite-type layers are not easily synthesized by the usual methods that are described in the literature [31], because their hydroxides do not have the brucite-like structure; instead, they are prepared by adding an aqueous solution of the trivalent cation salt to a suspension of the divalent cation oxide, followed by anionic exchange [31]. Further, NiAl–CrO_4, NiFe–CrO_4, CoAl–CrO_4, and CoFe–CrO_4 LDHs also show turbostratic disorder as the examples aforementioned, maintained, even after exposure to a hydrothermal treatment where the purpose is to increase the crystallite size and favor grafting [18].

The environmental safety, ease of production, relatively low cost of precursors, and enhanced stability make it possible to suggest that layered double hydroxides intercalated with chromate, dichromate, and other different chromium-based anions, such as $Cr^{III}(EDTA)^-$ and $[Cr^{III}(SO_3\text{-salen})]^{2-}$ are promising materials, which are still under development and can be successfully used as catalysts, especially for oxidation reactions.

4.2. LDH Intercalated with Medium-Nuclearity Polyoxometalates: Vanadates and Molybdates

4.2.1. Vanadates

The vanadium-based anions were among the very first polyoxometalates to be successfully intercalated into layered double hydroxides structure. A wide range of polyoxovanadate ions can be inserted in the interlayer space, from discrete to polymeric anions. Although decavanadate is the most used, there are studies that are also devoted to the intercalation of lower oligovanadates. The pH influences the polymerization degree of the oxovanadates. A high nuclearity is associated to a decreased pH value, as seen in Table 3.

Table 3. The polymerization degree of oxovanadates versus pH. Information taken from Ref. [18].

pH	Oxovanadate Species Formed
1–3	Decavanadate $[V_{10}O_{26}(OH)_2]^{4-}$, $[V_{10}O_{27}(OH)]^{5-}$, $[V_{10}O_{28}]^{6-}$
4–6	Metavandate $VO(OH)_3$, $VO_2(OH)$, $[V_3O_9]^{3-}$, $[V_4O_{12}]^{4-}$
8–11	Pyrovanadate $[VO_3(OH)]^{2-}$, $HV_2O_7^{3-}$, $[V_2O_7]^{4-}$
>12	Vanadate $[VO_4]^{3-}$

Similar to the chromate containing LDH, the reported methods used to synthesize hydrotalcites with intercalated vanadium-based anions containing different pairs of divalent/trivalent cations in the brucite-like layers are co-precipitation, anion exchange, reconstruction, and hydrothermal method. Vanadium can be incorporated either in one of its anionic forms in the interlayer, mostly depending on the nature of the brucite-like cations and pH value (Table 4), or can be introduced as a cation, V(III) in the brucite-type layers (Table 5).

Suitable interlamellar distances are required to favor the intercalation of vanadates with similar hindrance. A good solution to increase the interlayer zone is to use proper-sized anions beforehand to swell the layers of the hydrotalcites, such as terephthalate or dodecyl sulphate. The size of the anions, as well as their charge density and orientation within the interlayer, has proven to have a considerable influence on the catalytic activity of the POM-LDH composites. Layered double hydroxides are known to be bifunctional catalysts, exhibiting neither purely acid nor basic character [18]. The Brønsted basicity of hydrotalcites is attributed to the existing surface hydroxyls and their combination with vanadium-based anions is believed to provide some Lewis acidity to the LDH. Polyoxometalates that are based on vanadium have often been applied in heterogeneous catalysis because of their interesting redox properties.

Decavanadate-intercalated LDH, having magnesium and aluminum as cations in the brucite-type layers, can be synthesized with great ease by direct ion exchange of a nitrate precursor with an aqueous $NaVO_3$ solution at an adequate acidic pH value with or without preswelling with terephthalate while following the method that was reported by Villa et al. [32]. The Raman data are particularly useful for distinguishing between intercalated species, such as $[VO_4]^{3-}$, $[V_2O_7]^{4-}$, $[V_4O_{12}]^{4-}$, and $[V_{10}O_{28}]^{6-}$ [33]. The Raman spectrum that was reported by Twu et al. of the LDH containing decavanadate, prepared by direct exchange of an LDH with an acidified V solution, indicates that decavanadate is the dominant anionic species found intercalated, as evidenced by the peaks at 320, 454, 534, 595, 834, 975, and 998 cm^{-1} [34]. Although the signal to noise ratio of the spectrum for the LDH is not as good as that for the $[V_{10}O_{28}]^{6-}$ ions in solution, it is quite clear that all of the characteristic peaks are attributed to decavanadate ions. For the sample that was heated to 160 °C, changes appear in both the XRD and Raman spectrum. These include a new reflection at 4.76 Å and the appearance of a broad band centered at 880 cm^{-1} [34]. The characteristic Raman bands of the decavanadate are still present, as are the XRD reflections that are characteristic of the decavanadate-hydrotalcite complex. However, it was observed that the decavanadate species are transforming into other vanadate species at temperatures above 100 °C [34]. The XRD analysis of the decavanadate-LDH sample indicated a gallery height of 7.1 Å, which also suggested that the $[V_{10}O_{28}]^{6-}$ orientation with the C_2 axis is parallel to the layers [34]. The novel $[V_{10}O_{28}]^{6-}$-MgAl-LDH composites are used as catalysts in a variety of electron transfer reactions.

Maciuca et al. [35] have successfully synthesized a solid containing decavanadate and pyrovanadate anions simultaneously intercalated between the brucite-type layers of a hydrotalcite. The UV–VIS spectrum of the sample exhibits a band at 300–320 nm, which is specific for V^{4+} having a pseudo-tetrahedral geometry, $[V_2O_7]^{4-}$ (representing 30% of V species), and a band around 420–450 nm, characteristic to V^{5+} ions with an octahedral arrangement, $[V_{10}O_{28}]^{6-}$ (70% of V species) [35]. The catalytic activity of the prepared LDH material was evaluated in the oxidation reaction of the tetrahydrothiophene to sulfolane with dilute H_2O_2 aqueous solution as oxidant. It was observed that low conversions were obtained for the vanadium-containing material, probably due to the formation of some complex species, other than $[V_{10}O_{28}]^{6-}$ or $[V_2O_7]^{4-}$, which probably do not possess any catalytic activity. Dobrea et al. [36] have also tried to intercalate decavanadate anions between the brucite-type layers of a MgAl-LDH. In the Raman spectra recorded, they have observed some bands in the regions 800–1000 cm^{-1} and 200–370 cm^{-1} that belong to two types of anions, namely $(VO_3)_n^{n-}$ and $[V_{10}O_{28}]^{6-}$. The ratio between these anionic species strongly depends on the vanadium content in the LDH structure. When the V loading increases, the bands that are specific to the $(VO_3)_n^{n-}$ ions decrease until complete disappearance, while the $[V_{10}O_{28}]^{6-}$ anions prevail [36]. Moreover, the XRD patterns of V-exchanged LDH samples display the same reflections characteristic to the layered materials, but at smaller values of 2θ. There are also some changes in the d_{003}-value that suggest the increase of the basal spacing and, thus, prove the replacement of smaller nitrate anions by larger species. At low concentrations of vanadate, only a partial exchange has occurred and reflections specific for both nitrate and oxyanions coexist in the XRD patterns. For higher V/Al ratios, the exchange was almost complete and the d_{003}-value increased up to 10.8 Å for V-LDH. The XRD patterns of the exchanged materials presented broader and lower intensity reflections when compared to the parent LDH accounting for the intercalation of various oxyanions species, which produce a certain level of disorder in the lamellar structure [36].

Table 4. Summary of LDH with intercalated vanadium-based anions.

Layer Cations	Molar Ratio ($M^{II}:M^{III}$)	Intercalated Species	Synthesis Method	Reaction	Substrate	Yield (%)	Reference
Mg,Al	1	$[V_{10}O_{28}]^{6-}$, $[V_2O_7]^{4-}$	Anion exchange	Oxidation	Tetrahydrothiophene	4.6	[35]
Mg,Al	1	$[V_{10}O_{28}]^{6-}$	Co-precipitation	Epoxidation	Geraniol	90	[32]
Mg,Al	1	$[V_{10}O_{28}]^{6-}$	Co-precipitation	Epoxidation	Nerol	95	[32]
Mg,Al	1	$[V_{10}O_{28}]^{6-}$	Co-precipitation	Epoxidation	Perillyl alcohol	99	[32]
Mg,Al	1	$[V_{10}O_{28}]^{6-}$	Co-precipitation	Epoxidation	Myrtenol	88	[32]
Mg,Al	1	$[V_{10}O_{28}]^{6-}$	Co-precipitation	Epoxidation	Trans-Pinocarveol	79	[32]
Mg,Al	1	$[V_{10}O_{28}]^{6-}$	Co-precipitation	Epoxidation	Isopulegol	76.1	[32]
Mg,Al	1	$[V_{10}O_{28}]^{6-}$	Co-precipitation	Epoxidation	Nopol	54.9	[32]
Mg,Al	1	$[V_{10}O_{28}]^{6-}$	Co-precipitation	Epoxidation	β-Citronellol	13.7	[32]
Mg,Al	1	$[V_{10}O_{28}]^{6-}$	Co-precipitation	Epoxidation	Verbenol	1.4	[32]
Mg,Al	1	$[V_{10}O_{28}]^{6-}$	Co-precipitation	Epoxidation	Crotyl alcohol	65	[32]
Mg,Al	1	$[V_{10}O_{28}]^{6-}$	Co-precipitation	Epoxidation	Cinnamyl alcohol	85	[32]
Mg,Al	1	$[V_{10}O_{28}]^{6-}$	Co-precipitation	Epoxidation	2-Cyclohexen-1-ol	21	[32]
Mg,Al	1	$[V_{10}O_{28}]^{6-}$	Co-precipitation	Epoxidation	Carveol	56.3	[32]
Mg,Al	1	$[V_{10}O_{28}]^{6-}$	Reconstruction	Alkoxylation	n-Butane	14.4	[37]
Mg,Al	1.5	$[V_{10}O_{28}]^{6-}$	Anion exchange	ODH	Propane	25.7	[38]
Mg,Al	2	$[V_{10}O_{28}]^{6-}$	Anion exchange	-	-	-	[34]
Mg,Al	2	$[V_{10}O_{28}]^{6-}$	Anion exchange	-	-	-	[39]
Mg,Al	2	$[V_{10}O_{28}]^{6-}$	Reconstruction	-	-	-	[39]
Mg,Al	2	$[V_{10}O_{28}]^{6-}$	Anion exchange	Oxidation	Cyclohexene	23.2	[40]
Mg,Al	2	$[V_{10}O_{28}]^{6-}$	Anion exchange	ODH	Propane	9.3	[41]
Mg,Al	2	$[V_{10}O_{28}]^{6-}$, $[VO_3]_n^{n-}$	Anion exchange	ODH	n-Butane	-	[42]
Mg,Al	2	$[V_2O_7]^{4-}$	Anion exchange	Oxidation	Dibenzothiophene	85.7	[36]
Mg,Al	2	$[V_2O_7]^{4-}$	Anion exchange	ODH	Propane	7.8	[38]
Mg,Al	2	$[VO_4]^{3-}$	Anion exchange	Oxidation	Cyclohexene	<1	[40]
Mg,Al	2.3;3.2;4.8	$[VO_4]^{3-}$	Co-precipitation	-	-	-	[43]
Mg,Al	2.5	$[V_{10}O_{28}]^{6-}$	Anion exchange	-	-	-	[44]
Mg,Al	3	$[V_{10}O_{28}]^{6-}$	Anion exchange	ODH	Propane	6.47	[45]
Mg,Al	3	$[VO_4]^{3-}$, CO_3^{2-}	Co-precipitation	-	-	-	[46]
Mg,Al	3	$[VO_4]^{3-}$	Co-precipitation	-	-	-	[46]
Mg,Al	3	VO_x^{-}	Anion exchange	Anticorrosion	-	-	[47]

Table 4. *Cont.*

Layer Cations	Molar Ratio (M^{II}:M^{III})	Intercalated Species	Synthesis Method	Reaction	Substrate	Yield (%)	Reference
Mg,Al	3	VO^{2+}	Anion exchange	-	-	-	[48]
Ni,Al	2.03	$[V_{10}O_{28}]^{6-}$	Reconstruction	-	-	-	[49]
Ni,Al	2.31	$[V_{10}O_{28}]^{6-}$	Anion exchange	-	-	-	[49]
Ni,Al	3	$[V_{10}O_{28}]^{6-}$	Anion exchange	ODH	Propane	5.8	[50]
Cu,Cr	2	$[V_{10}O_{28}]^{6-}$	Anion exchange	Alkoxylation	n–Butanol	53	[27]
Cu,Cr	2	$[V_2O_7]^{4-}$	Anion exchange	Alkoxylation	n–Butanol	30	[27]
Zn,Al	1	$[V_{10}O_{28}]^{6-}$	Anion exchange	Oxidation	Cyclohexene	8.3	[40]
Zn,Al	2	$[V_{10}O_{28}]^{6-}$	Anion exchange	Oxidation	Cyclohexene	5.9	[40]
Zn,Al	2	$[V_{10}O_{28}]^{6-}$	Co-precipitation	Anticorrosion	-	-	[51]
Zn,Al	2,3	$[V_{10}O_{28}]^{6-}$	Anion exchange	ODH	Propane	6.1,6.5	[52]
Zn,Al	2	$[V_{10}O_{28}]^{6-}$,$[V_2O_7]^{4-}$,$[VO_3]_n^{n-}$	Anion exchange	-	-	-	[53]
Zn,Al	2	$[V_2O_7]^{4-}$	Anion exchange	Anticorrosion	-	-	[54]
Zn,Al	2	$[V_2O_7]^{4-}$	Anion exchange	-	-	-	[55]
Zn,Al	2	$[V_2O_7]^{4-}$	Anion exchange	-	-	-	[56]
Zn,Al	2	VO_x^{-}	Anion exchange	Anticorrosion	-	-	[47]
Zn,Al	3	$[V_{10}O_{28}]^{6-}$,$[V_2O_7]^{4-}$,$[VO_3]_n^{n-}$	Anion exchange	-	-	-	[53]
Zn,Al,Ce	0.29	$[V_2O_7]^{4-}$	Anion exchange	-	-	-	[57]
Zn,Cr	1.75	$[V_{10}O_{28}]^{6-}$	Anion exchange	-	-	-	[58]
Zn,Cr	2	$[V_{10}O_{28}]^{6-}$	Anion exchange	Oxidation	Cyclohexene	2.9	[40]
Zn,Fe	2	$[VO_4]^{3-}$	Hydrothermally	-	-	-	[59]

Table 5. Summary of $M^{II}M^{III}$V-LDH.

Layer Cations	Molar Ratio (M^{II}:M^{III})	Intercalated Species	Synthesis Method	Reaction	Substrate	Yield (%)	Reference
Mg,Al,V	3	CO_3^{2-}	Hydrothermally	-	-	-	[14]
Mg,Al,V	3	CO_3^{2-}	Hydrothermally	ODH	Propane	1.8	[41]

As previously reported [18], the overlapping and broadening of some reflections at higher 2θ values can be attributed to the turbostratic effects that are caused by the new intercalated anionic species [36]. The samples that were prepared by Dobrea et al. were also tested as catalysts for the oxidation of dibenzothiophene (DBT) to dibenzothiophene-sulfone with H_2O_2. It was observed that an increased amount of vanadium-based anions will produce a change of the oxyanion species from monomers (vanadate) to polymeric species (decavanadate). Moreover, the V-LDH samples suffered major modifications during the oxidation reaction, leading to a decreased catalytic activity that is attributed to their low stability [36].

Villa et al. have reported the obtaining of MgAl-layered double hydroxide pillared with vanadium under mildly acidic conditions, but this time the XRD, Raman, and IR characterization certified the presence in the interlayer region of pure decavanadate ($[V_{10}O_{28}]^{6-}$) anions [32]. The compound was used as a catalyst for the selective epoxidation of different allylic and homoallylic alcohols, mostly of terpene origin. The epoxidation of geraniol while using decavanadate-containing hydrotalcite (without or with preswelling) as catalyst, toluene as solvent, and t-BuOOH dissolved in decane as oxidant, led to high conversions of the substrate with minimal formation of geranial, which indicated that an increased effectiveness is obtained by working under anhydrous conditions that are able to prevent the decavanadate anions hydrolysis and stop any possible leaching. Additionally, it was demonstrated that low conversions and selectivities are obtained when vanadium-based anions are introduced in the LDH structure without rigorous pH control, which might lead to the formation of low-nuclearity polymers [32]. The decavanadate-containing LDH was used as catalyst for the epoxidation of various terpenic allylic alcohols in toluene with t-BuOOH/decane as an oxidant high conversions and epoxide selectivities being obtained for geraniol, nerol, perollyl alcohol, myrtenol, and trans-pinocarveol [32]. Even homoallylic alcohols (C=C–C–COH), such as isopulegol and nopol [32], were successfully epoxidized with high selectivity, but a decreased yield. Additionally, a rather low conversion and epoxide selectivity were obtained when β-citronellol was used as a substrate. A possible explanation for these unsatisfactory results can be attributed to the large distance between the alcohol group and the double bond. The allylic alcohols were smoothly epoxidized, except verbenol, whose bulky carbon skeleton did not allow for the coordination of the alcohol to the V [32]. High conversions with excellent selectivities were also obtained for crotyl alcohol and cinnamyl alcohol [32]. The results of the epoxidation strongly suggest that the allylic alcohol group of the substrate coordinates on the peroxo-V center [32] before oxygen transfer takes place. Therefore, it might be said that the active vanadium species participate in the epoxidation reaction with two coordination sites, one that is needed for the peroxide activation, and the other one for the alcohol coordination [32]. The vanadium atom is initially bound to the pillar with at least five V–O–V bonds [32], so it is mandatory to break one or two of these bonds to be reassured that the reaction takes place. However, there are still enough V–O–V bonds intact to keep all vanadium within the polyanions [32].

The alkoxylation of n-butanol with ethylene oxide (EO) to produce butoxy mono-ethylene glycol ether (BMGE) has been also studied on this type of LDH [37]. Malherbe et al. [37] used a commercial MgAl-LDH, which was reconstructed with intercalated decavanadate (sample MAV) after calcination, calcined at 450 °C (sample MAVCAL), and then rehydrated in water (MAVHYD) or in a 1 M KOH solution (MAVKOH). The calcination–rehydration procedure was used to observe the influence of rehydration on the reactivity of the mixed oxides. The XRD patterns of the three catalysts used in this study were also analyzed. The parent material used, a calcined hydrotalcite, presents the formation of a periclase MgO phase, while aluminum oxide could not be detected because of its amorphous structure. After the anion exchange and another calcination step, the periclase phase is significantly reduced [37]. A possible explanation is that a part of the magnesium participates in the formation of a ternary oxide (Mg–Al–V)–O [37]. The rehydration step decreases, even more, the proportion of the periclase phase in the final material and the XRPD patterns clearly indicate that most of the MgO phase will be transformed into an amorphous hydroxide form [37]. The catalytic performances of the new catalysts that were obtained after calcination or calcination–rehydration were tested in the synthesis of

2-butoxyethanol and compared to that of the untreated MAV. The conversion of ethylene oxide (EO) in a batch reactor increases with the reaction time in the order MAV < MAVHYD < MAVKOH [18]. The calcined solid, MAVCAL shows better conversion for low reaction times but it has the huge disadvantage of being deactivated after 20 h reaction. A likely cause of deactivation is associated to the appearance of a significant amount of a polymer type material in the product. The gelatinous solid is thought to contain polyethylene glycols (PEGs) and polyethoxylates, which strongly adsorb on the surface of the catalyst. A good solution is to rehydrate the calcined LDH in a KOH solution to eliminate the formation of unwanted polyethoxylates [37]. During rehydration, superficial O^{2-} sites are transformed into hydroxyl groups with lower basicity and high catalytic activity [37], while the use of KOH might either lead to an increase in the amount of OH groups or might ensure the formation of stronger ones.

Malherbe et al. [20] reported the synthesis of two systems, namely [CuCr–V_2O_7] and [CuCr–$V_{10}O_{28}$], which were also used as catalysts for the alkoxylation of n-butanol with ethylene oxide (EO) [20]. The parent material, [CuCr–Cl]-LDH was obtained by the co-precipitation method, and then through the appropriate *chimie douce* exchange reactions, the original chloride anions were replaced by decavanadate and pyrovanadate species [20]. From the catalytic tests, it was observed that higher conversions are obtained for the solid modified with decavanadate anions. The increased activity was attributed to a partial reduction of the vanadium species, with the active sites being not V^{5+} but V^{4+} [20].

The vanadate-exchanged layered double hydroxides with cations, such as Mg-Al [38,41,42,45], Ni-Al [50], or Zn-Al [52] in the brucite-like layers provide materials that differ in catalytic performance during the oxidative dehydrogenation (ODH) of various substrates (propane, n-butane), mostly depending on the nature of intercalated vanadate and the vanadium content [52], which can be modified during preparation. Consequently, the applied synthesis methods play a crucial role in the development of compounds with increased catalytic activities for ODH. The most active in the oxidative dehydrogenation of propane were the samples that were derived from the decavanadate-exchanged LDH under their calcined forms, namely their corresponding mixed oxides, while, for n-butane, the highest selectivity to ODH products and lowest selectivity to CO_x were observed for the V-poor solids [42]. These observations can be correlated with the increased nucleophilic character (basicity) of the oxygen atoms present in Mg–O–V units existing in orthovanadate than in V–O–V units (that lead to total oxidation) existing in α-$Mg_2V_2O_7$ and MgV_2O_6. At high vanadium loading, the presence of highly crystalline α-$Mg_2V_2O_7$ accounted for the observed decrease in 1,3-butadiene selectivity [42]. For ODH, acid and basic sites are both necessary for the reaction to take place, but it was proven that the total selectivity for dehydrogenated products was higher on the catalysts with increased base sites concentration [52].

Bahranowski et al. [40] have prepared different decavanadate-exchanged hydrotalcite-like compounds that were catalytically active in the selective oxidation of cyclohexene with dioxygen. Their activities mostly depended on the composition of the brucite layers, and decreased in the order MgAl > ZnAl > ZnCr [40]. The lowering of the M(II)/M(III) ratio in the brucite layer allowed for an increased loading of decavanadate anions, which eventually led to higher activity. Moreover, a pyrovanadate-exchanged-MgAl-LDH was also synthesized and used as a catalyst in the same reaction. It presented a lack of activity, which was probably due to the small interlayer distance that made the catalytic sites inaccessible. A considerable amount of epoxide was still observed in this case, which indicated the presence of decavanadate group in the interlamellar space [40]. Recently, Nejati et al. [59] obtained ZnFe–VO_4-LDH (Figure 7), which was used as an efficient electrocatalyst for water oxidation in alkali solution. In electrochemical water splitting, ZnFe–VO_4-LDH exhibits a superior OER performance, being expressed as lower onset overpotential, smaller Tafel slope, and larger exchange current density [59] when compared to the bare glassy carbon electrode. Layered double hydroxides that were intercalated with vanadium-based species were also used as anticorrosive materials. Tedim et al. [54] successfully reported the preparation of ZnAl–V_2O_7-LDH, which was able to provide efficient active corrosion protection of an aluminum alloy. It was found that the inhibition by vanadates mainly occurs in alkaline solutions, where metavanadates and pyrovanadates are the most abundant. Vanadium can be introduced not only in the interlayer space, but also as a cation in the brucite-like layers (Table 5). Kooli et al. [44] obtained hydrotalcite-like materials containing vanadium, as V(III), together with Mg(II) and Al(III), in the layers, and carbonate in the interlayer. It was found that the nature of the precursor influences the type of the Mg-V-O phases formed; thus, when starting from $Mg_3(VO_4)_2$, the isolated $[VO_4]$ units are formed. Dula et al. [41] have tried to use this solid as a catalyst for oxidative dehydrogenation of propane (ODH), obtaining high conversions of the substrate, but with low selectivities.

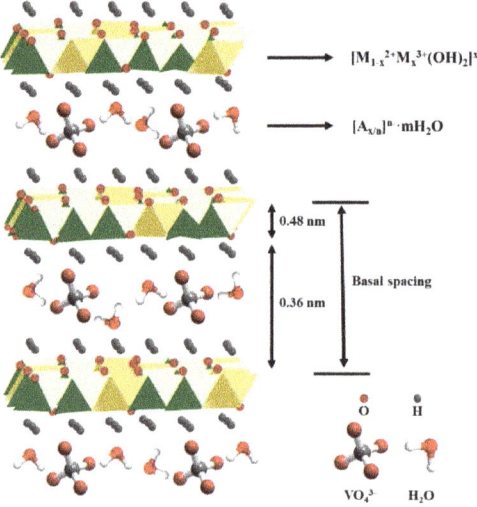

Figure 7. Schematic structure of ZnFe–VO$_4$-LDH. Adapted from Ref. [59], Copyright 2018, Royal Society of Chemistry.

4.2.2. Molybdates

The intercalation of molybdenum-based anions in the structure of layered double hydroxides is a laborious process and it mostly depends on the pH and the Mo(VI) concentration. Moreover, the evolution of Mo(VI) in solutions is enriched, due to the existence of polymers that may lead to the obtaining of a variety of anions, the predominant ones being $[MoO_4]^{2-}$, $[HMoO_4]^-$, $[Mo_6O_{19}]^{2-}$, $[Mo_7O_{24}]^{6-}$, $[HMo_7O_{24}]^{5-}$, $[H_2Mo_7O_{24}]^{4-}$, and $[Mo_8O_{26}]^{2-}$ [60]. The geometry of smaller anions, such as $[MoO_4]^{2-}$, is tetrahedral, while larger anions, like $[Mo_6O_{19}]^{2-}$ and $[Mo_7O_{24}]^{6-}$, possess octahedral or distorted octahedral configuration. The intercalation of vanadates in the interlayer space of layered double hydroxides has been intensively studied, while that of molybdates has been restricted to the heptamolybdate $[Mo_7O_{24}]^{6-}$ ions that can be stabilized at a lower pH and can act as pillars that are used to expand the distance between the brucite-type sheets [61]. Additionally, Van Laar et al. [62] discovered that the $[MoO_4]^{2-}$ hydrolysis at a pH value of 10 inhibits the intercalation in the LDH structure and polymerization of molybdate does not occur until pH < 7 (Figure 8). The molybdate-containing LDH with slightly different chemical composition could be useful as selective catalysts for different processes. The first studies that were recorded for the characterization of oxomolybdate species at different acidic pH values were based on potentiometric titrations [63]. The main conclusion was that $[MoO_4]^{2-}$ ions are stable at pH 6.5, while the heptamolybdate $[Mo_7O_{24}]^{6-}$ exists in equilibrium with $[MoO_4]^{2-}$ at a pH between 4 and 6.5. Further acidification, 1.5 < pH < 2.9, enabled the obtaining of octamolybdate anion $[Mo_8O_{44}]^{2-}$. Larger ionic aggregates are expected to appear at lower pH values, but there are no studies that are related to this statement [64]. Similar to the chromate- and vanadate-containing LDH, the reported methods that are used to prepare hydrotalcites with intercalated molybdenum-based anions containing different pairs of divalent/trivalent cations in the brucite-like layers are co-precipitation, anion exchange, and hydrothermal method. The studies showed that molybdenum can be introduced either as one of its anionic forms in the interlayer (Table 6) or as a cation in the brucite-type layers (Table 7).

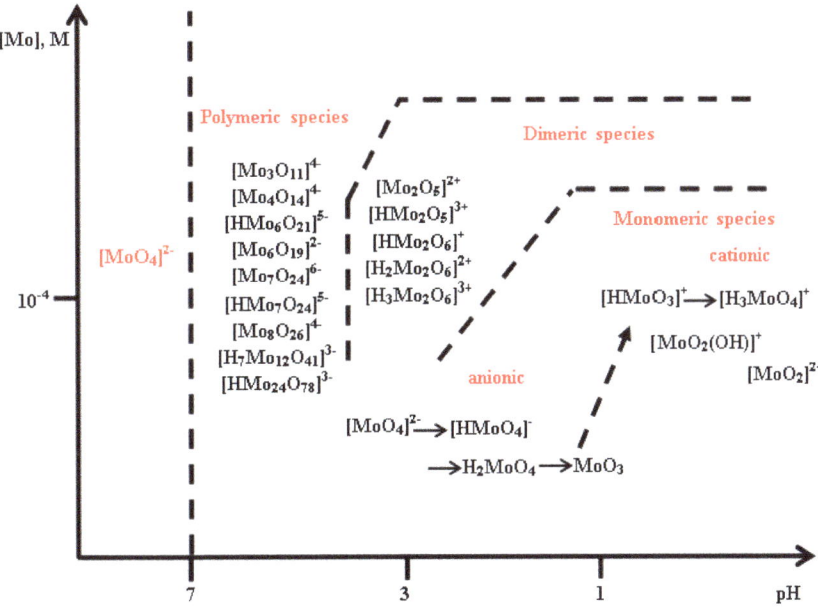

Figure 8. The polymerization degree of oxomolybdate versus pH. Adapted from Ref. [65], Copyright 2004, American Chemical Society.

The most used method for molybdate incorporation in MgAl-LDH structure is the anionic exchange. However, Drezdon et al. [66] reported a new technique derived from the classical ionic exchange that primarily involves the synthesis of an organic anion-pillared precursor, also called swelling agent, which is subsequently exchanged with an adequate polyoxometalate anion under acidic conditions [66]. In this case, the organic anion needs to have slightly larger dimensions than the polyoxometalate to be incorporated. By creating an acidic medium, the organic anion already incorporated in the LDH will be protonated, which weakens the electrostatic interaction between the cationic layers and the intercalated organic species and, thus, allowing for the partial exchange of the organically-pillared hydrotalcite with the polyoxometalate [66]. Moreover, the anion affinity in LDH interlayers depends on the size of the ion and its charge [67]. The monovalent anions have lower affinities and, hence, their participation in anion exchange reactions is favored when compared to divalent anions [67]. Based on the XRD results, Davantes et al. [68] were able to create a representative scheme that describes the interlayer exchange of different anions (Figure 9). The affinity series decreases in the order $[Mo_7O_{24}]^{6-} > CO_3^{2-} > [MoO_4]^{2-} > SO_4^{2-}$ [60]. The diffraction patterns of the Mo-intercalated samples that were obtained by Carriazo et al. [69] are characteristic to hydrotalcite-like structure, indicating a rhombohedral packing of the layers. The powder XRD patterns of layered double hydroxides usually present a decrease in the intensities of the (00*l*) lines, while the value of *l* increases, but, in this situation, it was observed that the intensity of the (006) and (009) lines are larger than that of the (003) line.

Table 6. Summary of LDH with intercalated molybdenum-based anions.

Layer Cations	Molar Ratio (M^{II}:M^{III})	Intercalated Species	Synthesis Method	Reaction	Substrate	Yield (%)	Reference
Mg,Al	1	$[Mo_4^V Mo_8^{VI} O_{40}]^{7-}$	Anion exchange	Epoxidation	Cyclohexene	<1	[70]
Mg,Al	1.46	$[Mo_7O_{24}]^{6-}$	Anion exchange	Oxidation	Dibenzothiophene	94	[71]
Mg,Al	1.46	$[Mo_7O_{24}]^{6-}$	Anion exchange	Oxidation	Anthracene	18.7	[71]
Mg,Al	1.5	$[Mo_7O_{24}]^{6-}$	Anion exchange	Epoxidation	Norbornadiene	49.6	[72]
Mg,Al	1.5	$[Mo_7O_{24}]^{6-}$	Anion exchange	Epoxidation	Benzonorbornadiene	76.1	[72]
Mg,Al	1.5	$[Mo_7O_{24}]^{6-}$	Anion exchange	Epoxidation	Cyclohexene	55.4	[72]
Mg,Al	1.59	$[Mo_7O_{24}]^{6-}$	Hydrothermal	Oxidation	Dibenzothiophene	98	[71]
Mg,Al	1.59	$[Mo_7O_{24}]^{6-}$	Hydrothermal	Oxidation	Anthracene	52.4	[71]
Mg,Al	1.5	$[MoO_4]^{2-}$	Anion exchange	Peroxidation	β-Citronellol	30	[73]
Mg,Al	1.6	$[Mo_7O_{24}]^{6-}$	Co-precipitation	Oxidation	tert-Butanethiol	30	[74]
Mg,Al	1.6	$[Mo_7O_{24}]^{6-}$	Anion exchange	Oxidation	Tetrahydrothiophene	76.4	[35]
Mg,Al	2	$[Mo_2O_7]^{2-}$	Anion exchange	Dehydrogenation	Propane	<30	[75]
Mg,Al	2	$[Mo_7O_{24}]^{6-}$	Anion exchange	Dehydrogenation	Propane	<30	[75]
Mg,Al	2	$[Mo_7O_{24}]^{6-}$	Anion exchange	-	-	-	[69]
Mg,Al	2	$[Mo_7O_{24}]^{6-}$	In situ hydrolysis	Epoxidation	Cyclohexene	<1	[68]
Mg,Al	2	$[Mo_7O_{24}]^{6-}$	Anion exchange	Oxidation	Dibenzothiophene	90	[36]
Mg,Al	2	$[MoO_4]^{2-}$	Anion exchange	Dehydrogenation	Propane	<30	[75]
Mg,Al	2	$[MoO_4]^{2-}$	Anion exchange	Peroxidation	β-Citronellol	33	[73]
Mg,Al	2	$[MoO_4]^{2-}$	Anion exchange	Peroxidation	β-Citronellol	>95	[76]
Mg,Al	2	$[MoO_4]^{2-}$	Co-precipitation	Condensation	Adamantanone	94	[77]
Mg,Al	2	$[MoO_4]^{2-}$	Co-precipitation	Anticorrosion	-	-	[78]
Mg,Al	2;3;4	$[MoO_4]^{2-}$	Co-precipitation	-	-	-	[43]
Mg,Al	2.25	$[MoO_4]^{2-}$, NO_3^-, CO_3^{2-}	Co-precipitation	Oxidation	Cyclohexene	13.9	[79]
Mg,Al	2.3;4.9	$[MoO_4]^{2-}$	Anion exchange	Oxidation	1-Methyl-1-Cyclohexene	11;12;19	[62]
Mg,Al	2.3	$[MoO_4]^{2-}$ 12.5%, pTOS 87.5%	Anion exchange	Oxidation	1-Methyl-1-Cyclohexene	22	[62]
Mg,Al	2.3	$[MoO_4]^{2-}$ 25%, pTOS 75%	Anion exchange	Oxidation	1-Methyl-1-Cyclohexene	21.5	[62]
Mg,Al	2.45	$[MoO_4]^{2-}$	Anion exchange	Peroxidation	β-Citronellol	34	[73]
Mg,Al	2.5	$[Mo_7O_{24}]^{6-}$, pTOS, CO_3^{2-}	Anion exchange	Oxidation	Cyclohexene	16.7	[79]
Mg,Al	2.64	$[MoO_4]^{2-}$	Co-precipitation	Oxidation	tert-Butanethiol	75	[74]
Mg,Al	2.7	$[MoO_4]^{2-}$, NO_3^-, CO_3^{2-}	Co-precipitation	Oxidation	Cyclohexene	12.5	[79]
Mg,Al	2.7	$[MoO_4]^{2-}$, pTOS, CO_3^{2-}	Anion exchange	Oxidation	Cyclohexene	15.3	[79]
Mg,Al	2.75	$[MoO_4]^{2-}$	Anion exchange	Oxidation	tert-Butanethiol	80	[74]
Mg,Al	2.75	$[MoO_4]^{2-}$, CO_3^{2-}	Anion exchange	Oxidation	Cyclohexene	12.5	[79]

Table 6. Cont.

Layer Cations	Molar Ratio (M^{II}:M^{III})	Intercalated Species	Synthesis Method	Reaction	Substrate	Yield (%)	Reference
Mg,Al	3	$[MoO_4]^{2-}$	Anion exchange	Halodecarboxylation	Cinnamic acid	80	[80]
Mg,Al	3	$[MoO_4]^{2-}$	Co-precipitation				[81]
Mg,Al	3	$[MoO_4]^{2-}$	Anion exchange	Peroxidation	β-Citronellol	35	[73]
Mg,Al	3	$[MoO_4]^{2-}$	Co-precipitation	Oxidation	Cyclohexene	14.4	[79]
Mg,Al	3	$[MoO_4]^{2-}$	Co-precipitation				[67,82]
Mg,Al	3	$[MoO_4]^{2-}$	Anion exchange				[83]
Mg,Al	3	$[MoO_4]^{2-}$, CO_3^{2-}	Anion exchange	Oxidation	Cyclohexene	13.1	[79]
Mg,Al	3	$[MoO_4]^{2-}$, NO_3^-	Co-precipitation	Oxidation	Cyclohexene	13.3	[79]
Mg,Al	3	$[MoO_4]^{2-}$, pTOS	Anion exchange	Oxidation	Cyclohexene	15.9	[79]
Mg,Al	3	$[Mo_7O_{24}]^{6-}$, pTOS	Anion exchange	Oxidation	Cyclohexene	17.5	[79]
Mg,Al	4	$[MoO_4]^{2-}$	Anion exchange	Peroxidation	β-Citronellol	35	[73]
Mg,Al	5	$[MoO_4]^{2-}$	Anion exchange	Peroxidation	β-Citronellol		[73]
Ni,Fe	9	$[MoO_4]^{2-}$	Hydrothermal	OER	Water		[84]
Ni,Zn	1.5	$[MoO_4]^{2-}$	Anion exchange	Transesterification	Soybean oil	20	[85]
Ni,Al	2.1	$[Mo_7O_{24}]^{6-}$	Anion exchange	Epoxidation	Norbornadiene	83.1	[72]
Ni,Al	2.1	$[Mo_7O_{24}]^{6-}$	Anion exchange	Epoxidation	Benzonorbornadiene	100	[72]
Ni,Al	2.1	$[Mo_7O_{24}]^{6-}$	Anion exchange	Epoxidation	Cyclohexene	81	[72]
Zn,Al	1	$[Mo_4^{V}Mo_8^{VI}O_{40}]^{7-}$	Anion exchange	Epoxidation	Cyclohexene	<1	[70]
Zn,Al	1	$PMo_{12}O_{40}^{3-}$	Anion exchange	Esterification	Acetic acid	77.2	[86]
Zn,Al	1.56	$[MoO_4]^{2-}$	Co-precipitation	Anticorrosion			[87]
Zn,Al	1.9	$[Mo_7O_{24}]^{6-}$	Anion exchange	Epoxidation	Norbornadiene	64.5	[72]
Zn,Al	1.9	$[Mo_7O_{24}]^{6-}$	Anion exchange	Epoxidation	Benzonorbornadiene	100	[72]
Zn,Al	1.9	$[Mo_7O_{24}]^{6-}$	Anion exchange	Epoxidation	Cyclohexene	77.27	[72]
Zn,Al	2	$[MoO_4]^{2-}$	Anion exchange	Anticorrosion			[88]
Zn,Al	2	$[Mo_7O_{24}]^{6-}$	Co-precipitation	Epoxidation	Cyclohexene	<1	[68]
Zn,Al	2	$[Mo_7O_{24}]^{6-}$	Anion exchange				[69]
Zn,Al	2.06	$[Mo_7O_{24}]^{6-}$	Anion exchange				[60]
Zn,Al	2.1	$[MoO_4]^{2-}$	Chimie douce				[89]

Table 7. Summary of MMo-LDH (M-metal).

Layer Cations	Molar Ratio (M^{II}:M^{III})	Intercalated Species	Synthesis Method	Reaction	Substrate	Yield (%)	Reference
Co,Mo	3	CO_3^{2-}	Co-precipitation	Adsorbent			[90–92]
Zn,Mo	2.3;4,9	CO_3^{2-}	Co-precipitation				[93]

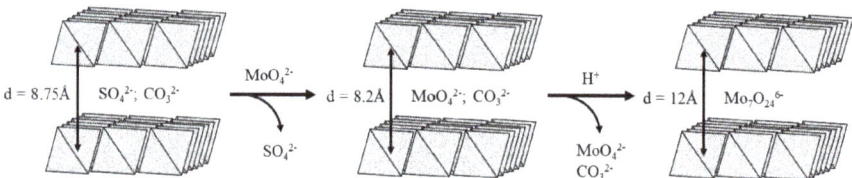

Figure 9. Schematic representation of molybdate exchange in LDH. Adapted from Ref. [60], Copyright 2013, American Chemical Society.

This abnormal behavior is specific for LDH intercalated with different POM and it appears to be due to the large atomic scattering factor of the interlamellar species [69]. The ZnAl-LDH containing heptamolybdate in the interlayer presents sharp and very symmetric lines probably due to its ordered packing of the layers. Additionally, the value of the interlayer distance was similar to the height of the heptamolybdate (7.2 Å) with its C_2 axis being perpendicular to the layers. This orientation favors the formation of a strong interaction between the brucite-like layers and the intercalated anion [69]. For the samples MgAl–Mo and NiAl–Mo, a broad reflection that is close to 11 Å is observed, which suggests the presence of a low-ordered phase of Mg-POM [94] or a novel phase with the anion grafted to the hydroxyl layer [69]. Raman studies [95] indicated that this is due to a small amount of $[MoO_4]^{2-}$ anions that resulted from the polyanion hydrolysis [69]. Dobrea et al. [36] have also tried to intercalate molybdenum-based anions between the brucite-type layers of a MgAl-LDH with different Mg/Al ratios. In the Raman spectra that were recorded for the Mo-LDH samples, it was observed that the most important bands are present in the regions 890–950 and 300–360 cm^{-1}, and they are most likely attributed to the main types of molybdenum-based anions that can appear within the hydrotalcite structure: $[MoO_4]^{2-}$, $[Mo_2O_7]^{2-}$, and $[Mo_7O_{24}]^{6-}$ [36].

At low Mo loading, the Raman spectra display bands at about 895 and 320 cm^{-1}, which clearly belong to the monomer $[MoO_4]^{2-}$ species. In these spectra, there are also some bands at 1050 and 709 cm^{-1} that are probably attributed to the residual NO_3^- anions [36]. By increasing the Mo loading, a more complex Raman spectra was obtained, due to the formation of octahedrally coordinated polymolybdate species (i.e., $[Mo_2O_7]^{2-}$ with band at 920 cm^{-1} and $[Mo_7O_{24}]^{6-}$ with bands at 947 and 358 cm^{-1}), while the bands characteristic for the nitrate and molybdate ions decreased in intensity until their complete disappearance. The sample with the highest Mo concentration mainly contained heptamolybdate anions [36]. It can be stated that different oxyanions are inserted in the structure of hydrotalcites after anion exchange, and their nature mostly depended on the molybdate species loading and, of course, on the pH used by analyzing the wavenumbers that are specific for various Mo species found in literature (Table 8) and the Raman spectra of the compounds that were prepared by Dobrea et al. [36]. Infrared spectroscopy appears to be a better technique for analyzing the structure of Mo(VI) species, but only in aqueous solutions, not necessary intercalated within the hydrotalcite materials, even though the Raman measurements have been widely used for the characterization of polymolybdate anions intercalated between the cationic layers of LDH.

Table 8. Raman frequencies (cm^{-1}) for Mo oxyanions species. Data taken from Refs. [12,36,71].

Vibration Type *	Anion Species			
	$[MoO4]^{2-}$	$[Mo_2O_7]^{2-}$	$[Mo_7O_{24}]^{6-}$	$[Mo_8O_{26}]^{4-}$
v_s (Mo=O)	895–898	920–930	937–0945	965
v_{as} (Mo=O)	837–846	-	903	925
δ (Mo=O)	300–320	355	355-365	370
υ (Mo-O-Mo)	-	-	564	860

* v_s = symmetric stretching mode attributed to Mo=O bond vibrations; v_{as} = antisymmetric stretching mode; δ = bending mode.

However, Yu et al. [88] were able to record the FT-IR spectrum of $[MoO_4]^{2-}$-pillared ZnAl hydrotalcite, which consists of bands attributed to hydroxyl stretching vibrations (3430 cm^{-1}) and bending modes of interlayer water (1630 cm^{-1}). There is also a band at 1368 cm^{-1} that is specific for CO_3^{2-} stretching mode that might indicate the contamination of the material by atmospheric CO_2 [88]. The band corresponding to the antisymmetric mode of Mo–O–Mo in $[MoO_4]^{2-}$ is observed at 834 cm^{-1}. Moreover, the bands at 620, 559, and 428 cm^{-1} can be related to the vibrations modes of the oxygen atoms at the layer crystal lattice [88]. Polyoxometalates that are based on molybdenum have been often applied in heterogeneous catalysis due to their interesting redox properties.

Layered double hydroxides containing heptamolybdate $[Mo_7O_{24}]^{6-}$ anions in the interlamellar space that were prepared by anion exchange starting from suitable precursors proved to be effective catalysts in the epoxidation of different substrates.

Carriazo et al. [72] have prepared, by the ion exchange method, starting from suitable precursors containing terephthalate, three different LDH, combining Al with different divalent cations (Mg, Zn, and Ni) to obtain the brucite-like layers and heptamolybdate in the interlayer [72]. The compounds were further used as catalysts for the epoxidation of norbornadiene, benzonorbornadiene, and cyclohexene. The hydrotalcite composition and the reaction solvents, i.e. hydrogen peroxide, dioxane, or dioxane–butyl maleate influenced the selectivity to epoxides [72]. For the hydrotalcites containing NiAl–Mo and ZnAl–Mo, the selectivity in epoxide was very good, approximately 85% for norbornadiene and almost 100% for benzonorbornadiene [72]. When cyclohexene was used as a substrate, the same tendency as in previously mentioned cases was observed: NiAl–Mo and ZnAl–Mo catalysts led to the highest selectivities in epoxide. As for the solvent, the best results were obtained while using dioxane or acetone. The differences between these two solvents were quite small. However, the replacement of dioxane with dioxane–butyl maleate mixture led to a decrease in the selectivity to epoxide [72].

Cyclohexene was also used as a substrate for the epoxidation reaction in the work that was reported by Sels et al. [70]. The catalysts, in this case, were two hydrotalcites containing MgAl or ZnAl as cations in the layers. Both LDH were obtained by co-precipitation and intercalated by the anion exchange method with molybdenum blue, a mixed-valency isopolyacid containing Mo^V and Mo^{VI}, namely $[Mo_4^VMo_8^{VI}O_{40}]^{7-}$ [70]. The exchange within the LDH is rapid because of its high negative charge. However, it was observed that the nature of the cations from the brucite-type layers of the LDH considerably influenced the long-term stability of Mo blue. With Mo blue on MgAl-LDH, the material has changed its color from deep blue to colorless within a few weeks [70]. This indicates that Mo blue, which is synthesized in a solution at pH 3, is decomposed in time at the surface of MgAl-LDH, which has pronounced basic properties [70]. Such decomposition problems do not appear with Mo blue that is exchanged on ZnAl-LDH or on DS-ZnAl-LDH (ZnAl-LDH pre-exchanged with dodecyl sulfate) [70]. The results on epoxidation of cyclohexene with H_2O_2 and with Mo blue exchanged on MgAl-LDH as catalyst indicate a low olefin conversion. Before the addition of H_2O_2 in the reaction medium, the catalyst presents the yellow shade that is specific to the Mo^{VI} form of the isopolyacid [70]. The color changed rapidly into red, suggesting the presence of tetraperoxomolybdate $[Mo(O_2)_4]^{2-}$ indicating the isopolyacid structure destruction [70]. A slightly higher olefin conversion is obtained with Mo blue that is exchanged on ZnAl-LDH.

Gardner et al. [68] have also prepared two catalysts that are derived from Mg$_2$Al and Zn$_2$Al layered double hydroxides intercalated by heptamolybdate species and then used them for epoxidation of cyclohexene. The catalysts were synthesized by different methods, in situ hydrolysis in the case of Mg$_2$Al-LDH and anion exchange for the material containing Zn. In the epoxidation reaction, it was observed that all of the heptametallate-intercalated LDH are not stable under the reaction conditions being converted into lower nuclearity species, which migrate to the LDH surface and lead to low conversions and selectivities [68].

Zavoianu et al. [79] have studied the effect of the synthesis methods on the physicochemical properties and catalytic performance of MgAl-molybdate-LDH in cyclohexene oxidation with hydrogen peroxide [79]. Five samples have been prepared by (i) ionic exchange procedure, as described by Kwon et al. [63], between a hydrotalcite containing carbonate anions previously obtained by co-precipitation at pH = 10 and low supersaturation with aqueous Na_2MoO_4; (ii) competitive ionic exchange at pH 10 while using a hydrotalcite containing carbonate anions, aqueous Na_2MoO_4, and p-toluene sulfonic acid (pTOS) as swelling agent, based on Van Laar studies [62]; (iii) exchange of the carbonate precursor LDH with pTOS at pH 4.5, followed by the exchange with molybdate at pH 4.5 under a constant stirring for 24 h and a readjustment of the pH to 10 using NaOH; (iv) direct synthesis at high supersaturation and pH 10 [18]; and, (v) direct synthesis at low supersaturation and pH 10 [18,79]. The characterization techniques used (FTIR, Raman, and UV–Vis spectroscopies) indicate the presence of $[MoO_4]^{2-}$ and smaller amounts of $[Mo_7O_{24}]^{6-}$ species in all of the samples. The catalytic activity of the solids is correlated with the basicity, which decreases with the increase in Mo loading [18,79]. The samples containing a low Mo amount (about 2.5–2.8%) in the form of $[MoO_4]^{2-}$ in the interlayer and increased magnesium concentration in the brucite layers have proven to favor the hydroperoxidation reaction of cyclohexene, while the catalysts also containing $[Mo_7O_{24}]^{6-}$ species in their structure encouraged the appearance of epoxides [79].

Van Laar et al. [62] reported a molybdate-exchanged layered double hydroxide heterogeneous catalyst for the conversion of H_2O_2 into singlet molecular dioxygen ($^1O_2$). The distinction between epoxidation and $^1O_2$ oxygenation was studied by using 1-methyl-1-cyclohexene as a substrate. The highest product yield at complete H_2O_2 consumption was obtained by the LDH catalyst containing less Mo amount [62]. Whalen et al. [73] have also studied the disproportionation of H_2O_2 into singlet oxygen by using molybdate-exchanged MgAl-LDH with Mg/Al ratio between 1.5 and 5. The chemical trapping (CT) of $^1O_2$ with β-citronellol performed the production of singlet oxygen. The efficiency of H_2O_2 has been enhanced with the increase of Mg/Al ratio. The yield of the generated $^1O_2$ was around 30–35%. The rate of H_2O_2 disproportionation sharply decreases when the Mo loading is increased due to the high Mo amount, which reduces the access of β-citronellol to the $[MoO_4]^{2-}$ anions in the interlayer space, leading to the loss of a part of the $^1O_2$ produced within the LDH. A few years later, Whalen et al. [76] discovered a way of increasing the oxidant efficiency of molybdate-containing MgAl-LDHs, which consists in pretreating the catalyst with 1,3-propanediol, glycerol, ethylene glycol, or 1,2-propanediol [76]. The peroxidation of different olefins with this glycol-modified catalyst led to conversions above 88% and selectivities up to 99% [76]. Mild oxidation of tetrahydrothiophene (THT) to sulfolane has been studied on molybdenum-containing LDH that was obtained by anionic exchange of a nitrate-LDH precursor with an aqueous solution of $Na_2MoO_4 \cdot 2H_2O$ at pH 4.5 by Hulea et al. [35]. The oxidation of tetrahydrothiophene was highly selective towards the sulfoxidation reaction, exclusively resulting in sulfoxide and sulfolane. A conversion of 98% was reached after 45 min. reaction, with a selectivity to sulfolane of 78%.

Ciocan et al. [71] have synthesized two series of Mo-containing hydrotalcites by different techniques: (i) anion exchange at atmospheric pressure, after complete synthesis of LDH, and (ii) anion exchange under hydrothermal conditions during the aging step of LDH synthesis. The influence of the preparation technique on the catalytic properties was investigated in the oxidation of dibenzothiophene (DBT) and anthracene with hydrogen peroxide [71]. The catalytic tests that were performed under moderate conditions (40–70 °C) indicated that both types of catalysts are efficient, but the solids that were prepared under hydrothermal conditions are more active, probably due to their larger specific surface area, which makes the access to active sites easier [71]. Dobrea et al. studied the oxidation of dibenzothiophene with H_2O_2 while using as catalyst a MgAl-LDH intercalated with heptamolybdate species [45]. The catalyst was characterized by Raman spectroscopy before and after the reaction to verify the stability. The DBT conversion strongly increased with the metal content in the Mo-LDH [45], which was obviously due to the increase in the number of active sites. Large polyoxymolybdate species were formed at high metal loadings, leading to an important increase in the distance between the LDH layers and, hence, to increased accessibility of the large DBT molecule to the catalytic centers [45]. The Raman spectra recorded confirmed the increased stability of Mo-oxoanions contained in layered double hydroxides, in contrast with the V-oxoanions [45].

Zavoianu at al. [74] have studied the oxidation of tert-butanethiol (t-BuSH) while using as catalysts Mo-LDH samples that were obtained from two different molybdenum sources, e.g. Na_2MoO_4 or $(NH_4)_6Mo_7O_{24}$ and prepared by two methods: (i) ionic exchange and (ii) co-precipitation at pH 10 under high supersaturation [74]. When Na_2MoO_4 was used as a molybdenum source, crystalline materials with relatively high basicity and fine dispersion of molybdate species were obtained. These samples showed the best catalytic activities, leading to high conversions of t-BuSH of about 80%. By changing the molybdenum source to $(NH_4)_6Mo_7O_{24}$, solids with a lower crystallinity, smaller surface area, lower basicity, and catalytic activity were obtained [88]. Mitchell and Wass [75] have synthesized the Mo-LDH catalysts while applying different preparation methods based on the anion exchange technique to be used in propane ODH. The most active and selective catalyst in this reaction was that prepared by classical anion exchange between a calcined commercial hydrotalcite and ammonium dimolybdate with prior impregnation with terephthalic acid and pH adjusted to 4.5 with nitric acid [18,75].

Choudary et al. [80] have reported the use of molybdate-containing MgAl-LDH catalyst for the synthesis of β-bromostyrenes in aqueous medium via the halodecarboxylation reaction of cinnamic acid. The high catalytic activity of the LDH system was attributed to its basic character and the large positive charge of brucite layers which may further lead to the enrichment of bromide ion near the solid surface [80]. Moreover, the excess of positive charge on the LDH surface shields the negative charge of the peroxomolybdate and bromide, ensuring high halide oxidation rates in this way [80]. Das and Parida [86] reported a parallel study of the esterification of acetic acid with n-butanol on ZnAl-LDH intercalated with tungstophosphoric (TPA) and molybdophosphoric (MPA) acid. The conversion of acetic acid was higher for the solid that was prepared with MPA (84.15%), although the selectivity to the ester was 100% for both of the samples. This behavior might be ascribed to the increased surface Brønsted acidity and the specific surface area shown by the Mo-containing sample [18,86]. Colombo et al. [85] prepared at two different pH values a NiZn LDH initially intercalated with acetate ions, which were subsequently replaced with molybdate anions via an ion exchange reaction. The basal spacing in the NiZn-LDH decreased from 13.08 Å to ca. 9.5 Å regardless of the working pH, [85], which might suggest the intercalation of hydrated molybdate anions. The as-synthesized material and the solid that was obtained by thermal treatment at 250 °C (basal spacing reduced to 7.35 Å due to dehydration) were tested as catalysts for the methyl transesterification of soybean oil [85]. All of the materials were catalytically active and led to high conversions of the substrate, but poor selectivities [85].

Klemkaite-Ramanauske et al. [77] prepared molybdate-containing MgAl layered double hydroxides by using different variations of the co-precipitation method and used them as catalysts for the synthesis of 2-adamantylidene(phenyl)amine Schiff base. All of the synthesized molybdate-containing layered double hydroxides presented similar catalytic activity for the studied reaction, no matter what co-precipitation route was used for preparation [77]. LDH is usually prepared from divalent and trivalent cations, but Mostafa et al. [90–92] succeeded to synthesize by co-precipitation method while using ammonium hydroxide and ammonium carbonate as precipitating agents, a highly crystalline CoMo-LDH, as a new type of $M^{2+}M^{6+}$-LDH. This new type of LDH had shown a highly energetic surface due to the formation of +4 charges between Co^{2+} and Mo^{6+}, as identified by XPS analysis. Muramatsu et al. were also successful in the preparation of a ZnMo-LDH belonging to this new generation of hydrotalcites.

Apart from the crucial role in the conventional catalysis, Mo-containing LDH can also be used as photocatalyst for electrocatalytic water oxidation [84], as adsorbent [60,69,89], or as an anti-corrosion material [78,88].

4.3. Layered Double Hydroxides Intercalated with High-Nuclearity Oxometalates: Iso and Hetero-Polyoxometalates

Polyoxotungstates (POW) are a large family of metal-oxo anionic clusters that have a variety of structures with a definite size and shape and a wide range of applications [96,97]. The preparation of tungstate-based anions intercalated within the hydrotalcite-like structure is a difficult task, as these materials are usually unstable under basic conditions and the synthesis process is strongly dependent on pH and other reaction conditions (temperature, contact time between the LDH and POM, etc.) [65]. Polyoxotungstates that are incorporated into the interlayer of layered double hydroxides present different structures: Keggin, Dawson, or Finke type [65]. Each type of POW can impose a particular gallery pore structure upon pillaring reaction, which results in new members of the class of LDH-POW derivatives. Figure 10 summarizes the pH ranges that correspond to the stability of the different W-POM.

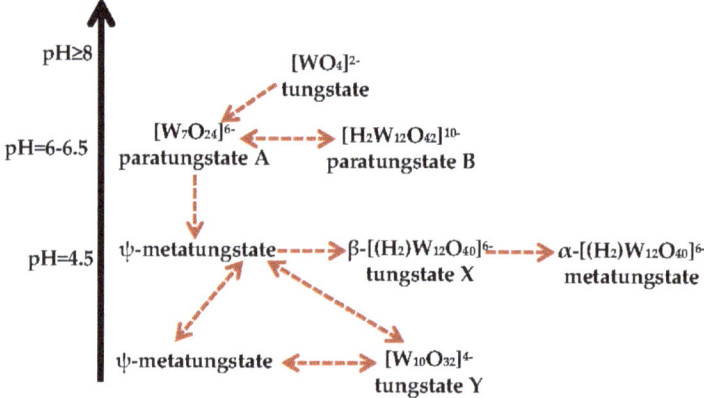

Figure 10. pH ranges corresponding to the stability of different polyoxotungstates. Adapted from Ref. [65], Copyright 2004, American Chemical Society.

Having a nearly spherical shape due to its tetrahedral symmetry, the α-Keggin ion $[\alpha\text{-}H_2W_{12}O_{40}]^{6-}$ will limit the possible gallery heights upon intercalation in LDH structure [97]. Moreover, this particular Keggin structure presents a diameter of approximately 10 Å, which is two-times larger than the brucite-like layer thickness (ca. 4.8 Å). However, LDH materials intercalated with larger POW than Keggin ions are more interesting, due to the increased access to the interlayer space [97]. The other two POW with lower molecular symmetries, as seen in Figure 11, can lead to gallery micropores of different sizes, depending on their interlayer orientations [97]. For example, the parallel and perpendicular gallery orientations of $[\alpha\text{-}P_2W_{18}O_{62}]^{6-}$ having the D_{3h} symmetry attributed to the Dawson structure, and of $Co_4(H_2O)_2(PW_9O_{34})_2{}^{10-}$ with the C_{2h} symmetry specific for the Finke structure, can result in 2.1 and 5.0 Å differences, respectively, in the pore size [97]. The reported methods used to prepare hydrotalcites with intercalated tungsten-based anions containing different pairs of divalent/trivalent cations in the brucite-like layers are anion exchange, co-precipitation, and reconstruction, similar to the chromate-, vanadate-, and molybdate-containing LDH (Figure 12). Table 9 summarizes all of the reported studies.

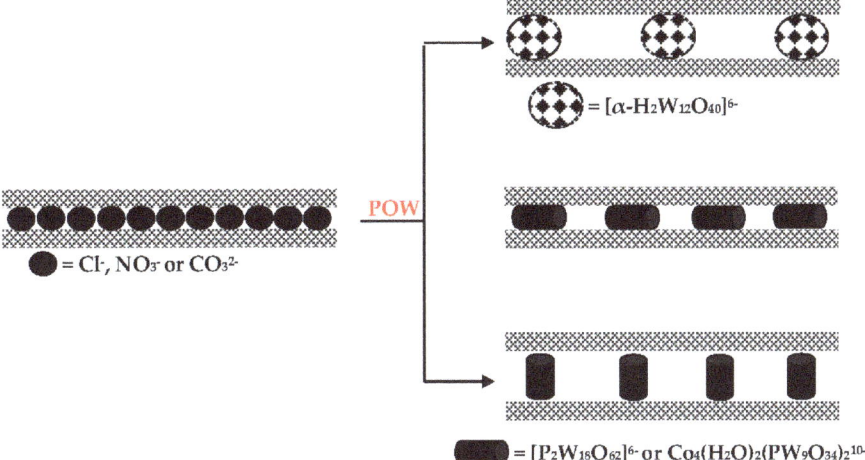

$= [\alpha\text{-}H_2W_{12}O_{40}]^{6-}$

$= Cl^-,\ NO_3^-$ or CO_3^{2-}

POW

$= [P_2W_{18}O_{62}]^{6-}$ or $Co_4(H_2O)_2(PW_9O_{34})_2^{10-}$

Figure 11. Schematic representation of LDH pillared by Keggin POW ions with nearly spherical symmetry and by Dawson and Finke POW ions with cylindrical symmetry. Adapted from Ref. [97], Copyright 1996, American Chemical Society.

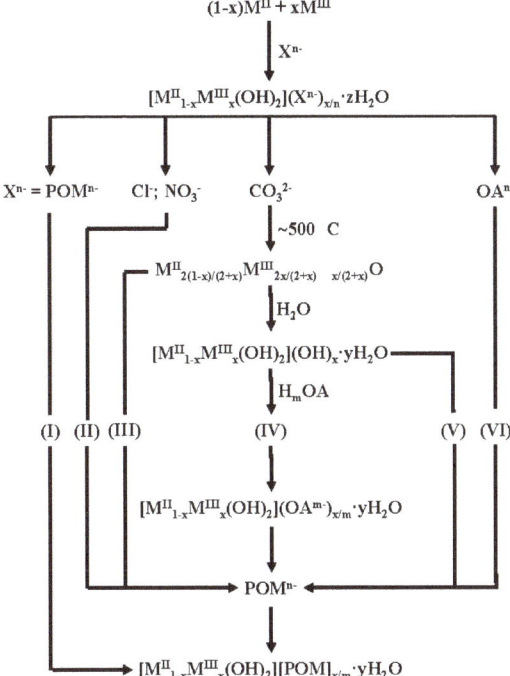

Figure 12. Representative methods used to prepare POW-LDH: (I) co-precipitation; (II) anion exchange; (III) anion exchange with calcined metal oxide; IV and VI, organic precursor methods, consisting in adding a preswelling agent; (V) anion exchange between an aqueous synthetic meixnerite and a POW solution. Adapted from Ref. [97], Copyright 1996, American Chemical Society.

The acid–base interaction that exists between POW anions (hydrolytically unstable at weakly acid to basic pH) and LDH (with a basic character) induces a partial dissolution of the brucite-like layers influencing the final cation ratio, also leading to the formation of another phase, probably a Mg-rich salt of the POW that deposits on the surface of LDH. Therefore, special attention must be given to the method that is used to synthesize POW-LDH and the working conditions. Choudary at al. [98] have reported the obtaining of a tungstate-intercalated LDH, which was further used as biomimetic catalyst for mild oxidative bromination and bromide-assisted epoxidation reactions of different aromatic and olefinic substrates. While taking that conventional bromination reactions typically use elemental bromine, a pollutant and health hazard, into consideration, this study is also important from environmental point of view [18]. The NiAl-LDH-WO$_4^{2-}$ catalyst prepared enabled the electrophilic bromination of the substrates in mild and well-controlled conditions. The majority of the reactions presented a high selectivity of halonium ions with complete stereoselectivity [98]. The methoxybromination of some aliphatic olefins showed moderate chemoselectivity due to the formation of a considerable amount of dibromides, while, in all other cases, e.g. in the methoxybromination of aromatic olefins, or in the bromohydroxylation of aromatic or aliphatic olefins, chemo-, regio-, and stereoselectivity were high [98]. Suitable substrates for bromide-assisted epoxidation have proven to be geminally di-, tri-, and tetra-substituted olefins. Additionally, the reaction can be easily switched from bromohydroxylation to epoxidation by changing the solvent [98]. The bromide-assisted epoxidation shows much higher turnover frequencies than classical W-catalyzed epoxidations [98].

Jacobs et al. [99] have also used MAl–Cl-LDH (M = Mg, Ni) exchanged with tungstate as catalysts for mild oxidative bromination. The XRD data of the prepared compounds did not show any swelling upon exchange, despite Raman spectroscopy indicating the typical features of tetrahedral [WO$_4$]$^{2-}$ units, which suggested that [WO$_4$]$^{2-}$ should be mainly located in edge positions. Moreover, the solids were recycled after each bromination reaction and then reused in the same chemical transformation maintaining their oxidative stability and unaffected catalytic activity [99]. The prepared catalyst allowed for the electrophilic bromination of a wide range of nucleophilic substrates (olefins, 1,3-diketones, aromatics) with high selectivity under mild conditions [99]. Choudary et al. [100] have prepared by anion exchange two MgAl-LDH intercalated with [WO$_4$]$^{2-}$ and PO$_4$[WO(O$_2$)]$_4$, respectively, and used them as catalysts for the oxidation of thioanisole with H$_2$O$_2$. The LDH-WO$_4$ exhibited the highest turnover frequency of 21.7 h^{-1} when compared to the other LDH catalyst.

Usually, when a reaction that is catalyzed by POM–LDH solids occurs inside the interlamellar region, a decrease in hydrophobicity would lead to an intensified diffusion of the substrate and/or the final products [18]. Following this statment, Palomeque et al. [101] have studied the epoxidation of cyclohexene with MgAl-hydrotalcite-intercalated organotungstic complexes that were synthesized by different methods. The tungsten-based anions were incorporated within the LDH structure either by anionic exchange or by complexation with phosphonic acids previously incorporated between the sheets [101]. Phenyl-/dodecylphosphonic acids and peroxotungstic salts were intercalated into the LDH and they led to the expansion of the lamellar lattice and the complexation of phosphonic acids with W-peroxo species [101]. These catalysts were tested in the epoxidation of cyclohexene with hydrogen peroxide or tertbutyl hydroperoxide as oxidants. The materials that were prepared by anionic exchange only gave epoxycyclohexane, but with moderate yields, whereas those that were prepared from previously incorporated phosphonic acids gave high selectivities (50–72%) towards allylic oxidation products [101]. The efficiency of the phosphonato-peroxotungstatic species is the result of the intrinsic reactivity of the WO$_5$–O=P subunit [101]. Carriazo et al. [102] have synthesized polyoxometalates [H$_2$W$_{12}$O$_{40}$]$^{6-}$ and [W$_4$Nb$_2$O$_{19}$]$^{4-}$ intercalated between the brucite-like layers of MgAl and ZnAl hydrotalcites by anion exchange, starting from the corresponding nitrate LDH precursors. The samples were also tested in the epoxidation reaction as in the previous case, but the substrate was cyclooctene and the oxidants were H$_2$O$_2$ or t-BuOOH, this time. The results showed that both of the anions are effectively located in the interlayer space, maintaining their pristine structures without depolymerization [102].

Table 9. Summary of LDH with intercalated tungsten-based anions.

Layer Cations	Molar Ratio ($M^{II}:M^{III}$)	Intercalated Species	Synthesis Method	Reaction	Substrate	Yield (%)	Reference
Mg,Al	1	$[WO_4]^{2-}$	Anion exchange	Mild oxidative bromination	Olefins	84	[99]
Mg,Al	1	$[WO_4]^{2-}$	Anion exchange	Mild oxidative bromination	Olefins	61.89	[99]
Mg,Al	1	$[WO_4]^{2-}$	Anion exchange	Oxidation	Thioanisole	94	[100]
Mg,Al	1	$PO_4[WO(O_2)]_4$	Anion exchange	Oxidation	Thioanisole	90	[100]
Mg,Al	1	$C_6H_5PO_3(WO_5)_2$	Anion exchange	Epoxidation	Cyclohexene	12.4	[101]
Mg,Al	1.92	$[CoW_{12}O_{40}]^{5-}$	Anion exchange	Oxidation	Benzaldehyde	98.8	[103]
Mg,Al	1.99	$[WO_4]^{2-}, [W_7O_{24}]^{6-}$	Anion exchange	Oxidation	Benzothiophene	60	[35]
Mg,Al	1.99	$[WO_4]^{2-}, [W_7O_{24}]^{6-}$	Anion exchange	Oxidation	Dibenzothiophene	80	[35]
Mg,Al	1.99	$[WO_4]^{2-}, [W_7O_{24}]^{6-}$	Anion exchange	Oxidation	Diphenyl sulfide	90	[35]
Mg,Al	1.99	$[WO_4]^{2-}, [W_7O_{24}]^{6-}$	Anion exchange	Oxidation	Benzyl-phenyl sulfide	100	[35]
Mg,Al	1.99	$[WO_4]^{2-}, [W_7O_{24}]^{6-}$	Anion exchange	Oxidation	Methyl-phenyl sulfide	100	[35]
Mg,Al	2	$[W_7O_{24}]^{6-}$	Anion exchange	Photodegradation	Hexachlorocyclohexene	-	[104]
Mg,Al	2	$[W_7O_{24}]^{6-}$	Anion exchange	-		-	[105]
Mg,Al	2	$[W_7O_{24}]^{6-}$	Anion exchange	Epoxidation	Cyclohexene	30.6	[106]
Mg,Al	2	$[W_7O_{24}]^{6-}$	Anion exchange	Peroxidation	Cyclohexene	17	[68]
Mg,Al	2	$[H_2W_{12}O_{40}]^{6-}$	Anion exchange	Epoxidation	Cyclooctene	7	[102]
Mg,Al	2	$[W_4Nb_2O_{19}]^{4-}$	Anion exchange	Epoxidation	Cyclooctene	7	[102]
Mg,Al	2	$[W_7O_{24}]^{6-}$	Anion exchange	Sulfoxidation	Dimethylsulfoxide	80	[107]
Mg,Al	2.05	$[W_7O_{24}]^{6-}$	Anion exchange	Oxidation	Methyl-phenyl-sulfide	18.6	[108]
Mg,Al	2.05	$[W_7O_{24}]^{6-}$	Anion exchange	Oxidation	Dibenzothiophene	<5	[108]
Mg,Al	2.1	$[WO_4]^{2-}$	Anion exchange	Oxidation	Methyl-phenyl-sulfide	17.64	[108]
Mg,Al	2.1	$[WO_4]^{2-}$	Anion exchange	Oxidation	Dibenzothiophene	<5	[108]
Mg,Al	2.1	$[WO_4]^{2-}$	Anion exchange	Oxidation	Sec-butyl-sulfide	9.52	[108]
Mg,Al	2.1	$[WO_4]^{2-}$	Anion exchange	Oxidation	Benzothiophene	<5	[108]
Mg,Al	2.57	$[WCo_3(CoW_9O_{34})_2]^{12-}$	*Chimie douce*	Epoxidation	Prenol	93.1	[109]
Mg,Al	3	$[WO_4]^{2-}$	Anion exchange	Mild oxidative bromination	Aromatics	-	[98]
Ni,Al	3	$[WO_4]^{2-}$	Anion exchange	Mild oxidative bromination	Aromatics	-	[98]
Mg,Al	3	$[WO_4]^{2-}$	Anion exchange	Sulfoxidation	Dimethylsulfoxide	90	[107]
Mg,Al	3	$[PW_{11}O_{39}]^{7-}$	Anion exchange	Photodegradation	Methyl orange	-	[110]
Mg,Al	3	$\alpha\text{-}[SiW_9O_{37}[Co(H_2O)]_3]^{10-}$	Reconstruction	Oxidation	Cyclohexanol	67	[111]
Mg,Al	3	$[WZnMn_2(ZnW_9O_{34})_2]^{12-}$	Anion exchange	Epoxidation	Prenol	87.3	[109]
Mg,Al	3	$[PW_{11}O_{39}]^{7-}$	Anion exchange	Epoxidation	Prenol	84.6	[109]

Catalysts **2020**, 10, 57

Table 9. Cont.

Layer Cations	Molar Ratio ($M^{II}:M^{III}$)	Intercalated Species	Synthesis Method	Reaction	Substrate	Yield (%)	Reference
Ni,Co	2.57	$[W_2O_7]^{2-}$	*Chimie douce*	-	-	-	[112]
Ni,Al	2	$[WO_4]^{2-}$	Anion exchange	Mild oxidative bromination	Olefins	21–96	[113]
Ni,Al	2	$[WO_4]^{2-}$	Anion exchange	Mild oxidative bromination	Cyclic enol ethers		[106]
Ni,Al	2	$[SiW_{12}O_{40}]^{4-}$	Anion exchange	Epoxidation	Cyclohexene	<30	[114]
Zn,Al	1	$[H_2W_{12}O_{40}]^{6-}$	Anion exchange	Oxidation	Isopropanol	10.3	[115]
Zn,Al	1	$[SiV_3W_9O_{40}]^{7-}$	Anion exchange	Oxidation	Isopropanol	8	[115]
Zn,Al	2	$[W_{12}O_{41}]^{10-}$	Anion exchange	Photodegradation	Glycerol	-	[116]
Zn,Al	2	$[H_2W_{12}O_{40}]^{6-}$	Anion exchange	Epoxidation	Cyclooctene	17	[101]
Zn,Al	2	$[H_2W_{12}O_{40}]^{6-}$	Co-precipitation	Peroxidation	Cyclohexene	22	[68]
Zn,Al	2	$[W_4Nb_2O_{19}]^{4-}$	Anion exchange	Epoxidation	Cyclooctene	5	[102]
Zn,Al	2	$PMo_{12}O_{40}^{3-}$	Anion exchange	Esterification	Acetic Acid	77.23	[86]
Zn,Al	2	$[SiW_{11}O_{39}]^{8-}$	Anion exchange	Photodegradation	Hexachlorocyclohexene	-	[117]
Zn,Al	2	$SiW_{11}O_{39}Mn(H_2O)^{6-}$	Anion exchange	Photodegradation	Hexachlorocyclohexene	-	[117]
Zn,Al	2	$[WCo_3(CoW_9O_{34})_2]^{12-}$	Anion exchange	Epoxidation	Prenol	68.8	[109]
Zn,Al	2	$[WZnMn_2(ZnW_9O_{34})_2]^{12-}$	Anion exchange	Epoxidation	Prenol	88.3	[109]
Zn,Al	3	$[SiW_{11}O_{39}]^{8-}$	Anion exchange	Epoxidation	Cyclohexene	<10	[114]
Zn,Al	3	$[SiW_{12}O_{40}]^{4-}$	Anion exchange	Epoxidation	Cyclohexene	<10	[114]

Additionally, the results of the epoxidation reaction indicated that $[H_2W_{12}O_{40}]^{6-}$-intercalated ZnAl-LDH gave the best results in terms of epoxide yield (17% at 24 h). Epoxidation was used also by Liu et al. to verify the catalytic activity of a series of self-assembled polyoxometalate (POM) catalysts that were directly immobilized into layered double hydroxides by a selective ion-exchange method [109]. The sandwich-type POM species are more favorable for the direct immobilization in the LDH structure when compared to Keggin-type POM, because there is no need for rigorous control of the pH [109]. Regarding the catalytic perfromance, the self-assembled polyoxometalate directly immobilized into LDH have achieved up to 99% selectivity of epoxide, 95% H_2O_2 efficiency, and 37,200 h^{-1} TOF [109].

Wei et al. [103] have obtained a new polyoxometalate anion-pillared LDH by the ionic exchange of an MgAl-LDH precursor in nitrate form with the Keggin-type tungstocobaltate anions $[CoW_{12}O_{40}]^{5-}$, as shown in Figure 13. Powder XRD, together with IR, TG, and cyclic voltammetry, strongly indicated that the guest anions were intercalated in the interlayer space of the resultant material [103]. The catalytic properties of the solid were tested in the oxidation of benzaldehyde with hydrogen peroxide, with high selectivities being obtained [103].

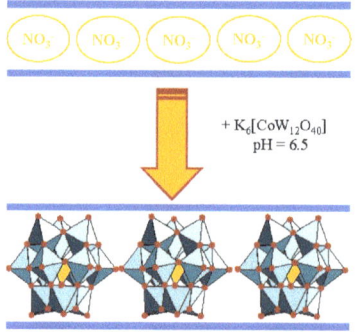

Figure 13. Synthesis route of polyoxotungstates-LDH (POW-LDH) Adapted from Ref. [103], Copyright 2008, Elsevier.

Maciuca et al. [35] performed the catalytic oxidation of different thioethers and thiophene derivatives with H_2O_2 in the presence of W-, V-, and Mo-containing LDH. The solids that were prepared by anion exchange showed good catalytic activity and selectivity but the nature of the intercalated anion influenced their performances [35]. Thus, the W-based LDH was more active and more stable than the V-LDH and Mo-LDH catalysts. Moreover, the conversion of the substrates also depended on their nucleophilicity and, consequently, the following order of reactivity resulted: benzothiophene < dibenzothiophene < diphenyl-sulfide < benzyl-phenyl-sulfide < methyl-phenyl-sulfide [35]. The same group [86] reported the intercalation of tungstate and paratungstate anions in the LDH structure by direct ionic exchange and their use as catalysts for DMSO oxidation in dimethyl sulfone with dilute H_2O_2. The catalysts performances were strongly influenced by the nature of the cations (Mg^{2+} or Zn^{2+}) in the brucite-like layer and by the type of W-anion species ($[WO_2]^{4-}$ or $[W_7O_{24}]^{6-}$) that are present in the interlayer gallery [107]. The MgAl-LDH-based catalysts exhibited better activity than the ZnAl-LDH type catalysts and the $[WO_2]^{4-}$-containing LDH was more active than the $[W_7O_{24}]^{6-}$-containing LDH [107].

Hulea et al. have also studied the oxidation of thiophenes and thioethers on tungstate-containing LDHs, while analyzing the effect of the relative content of tungstate $[WO_2]^{4-}$ or heptatungstate $[W_7O_{24}]^{6-}$ species on the catalytic performance [108]. A nitrate MgAl-LDH precursor (Mg/Al = 2) that was prepared at pH 9.5 was used to incorporate tungstate by anion exchange at pH 6.5 or 9.5 to obtain $W_7O_{24}^{6}$-LDH or WO_2^{4}-LDH, respectively. The UV–Vis spectroscopy confirmed the presence of both tungstate species in both samples. There were some bands at 215–230 nm (tetrahedral [WO$_4$] species in WO$_4{}^{2-}$) and at 250 nm (octahedral [WO$_6$] species in W$_7$O$_{24}{}^{6-}$) with relative intensities [WO$_4$]/[WO$_6$] of 5/1 and 3/1 for the W-LDH and PW-LDH samples, respectively [108]. Finally, Raman spectroscopy showed bands at 980 cm^{-1} ($\nu_{W=O}$) and 530, 405, and 250 cm^{-1}(δ_{W-O-W}) [108]. Overall, both of the spectroscopies indicated that at pH 9.5 the predominant species is tungstate and at pH 6.5 heptatungstate. Gardner et al. [68] studied the peroxidation of cyclohexene while using [W$_7$O$_{24}$]$^{6-}$-intercalated MgAl-LDH. The poor selectivity observed was corelated to the lack of intracrystalline microporosity under condensed phase reaction conditions [68].

Polyoxotungstates are a challenging and attractive part of the polyoxometalate possessing a variety of structures and topologies that can be used not only in catalysis, but also in photochemistry [104,110,115–117] and electrochemistry [105,112].

4.4. Layered Double Hydroxides Intercalated with Niobium Based Polyoxoanions

Niobium chemistry still remains into semi-obscurity, despite the growing interest of using niobium-based compounds in different applications, ranging from the production of magnets to optical devices and bone implants [118,119]. Niobium materials have also brought important contributions to heterogeneous catalysis, where they are used as catalyst supports or as the active phase [18]. There are not many studies concerning polyoxoniobiates, but those existing (Table 10) sustain that their synthesis occurs under basic pH conditions when compared to polyoxometalates containing V, Mo, or W [120].

The intercalation of niobium-based anions in the structure of layered double hydroxides is a laborious process and it mostly depends on the pH (Table 11) and niobium concentration. Although most of the niobium-based anions intercalated in the interlamellar space of the hydrotalcites are heteropolyoxometalates that are formed by the condensation of WO$_x$ polyhedra around a central niobium atom [102] or NbO$_x$ polyhedra around W [121], there is also reported, in the literature, the synthesis of a POM with the Lindqvist structure, i.e. [H$_3$Nb$_6$O$_{19}$]$^{5-}$, intercalated in a MgAl-LDH [121]. Evans et al. [121] studied the intercalation of some mixed Nb-W POMs (Nb$_x$W$_{6-x}$O$_{19}{}^{(x+2)-}$ where x = 2, 3, 4) within the interlayer region of MgAl-LDH. To avoid the hydrolysis of the precursor materials (LDH or POM) during the anionic exchange, good pH compatibility must exist between them. It was discovered that the niobotungstates [Nb$_3$W$_3$O$_{19}$]$^{5-}$ and [Nb$_4$W$_2$O$_{19}$]$^{6-}$ are hydrolytically stable when the pH is between 5.5 and 11.5 for the first anion and lower than 8.5 for the second one, which makes them perfectly compatible with MgAl-LDH, which present strong basic properties. [Nb$_2$W$_4$O$_{19}$]$^{4-}$ is more suitable for the acidic ZnAl-LDH, due to its increased stability at a pH, ranging from 4.5 to 7.5 [121].

Carriazo et al. also studied mixed Nb-W POM intercalated in LDH [102]. Their work was focused on the synthesis of layered double hydroxides containing different cations within the layers (Al and Mg or Zn) and [W$_4$Nb$_2$O$_{19}$]$^{4-}$ as anion in the interlayers. The PXRD patterns of the prepared materials are characteristic of hydrotalcite-type solids. The intercalation of the tungstoniobate anions has led to higher gallery heights (7.2 Å) when compared to regular hydrotalcites intercalated with nitrate (4 Å) and, as a consequence, the compounds presented microporosity and an increased specific surface area [102]. Moreover, the FTIR results showed that the tungstoniobate anion did not depolymerize and it is entirely located in the interlamellar space. The obtained materials were tested as catalysts in the epoxidation reaction of cyclooctene while using H$_2$O$_2$ as oxidant. It was observed that the compounds possess a rather low epoxidation activity (Table 10), but increased catalase activity (decomposition of H$_2$O$_2$ to H$_2$O and O$_2$) [102]. Carriazo has studied not only the intercalation of mixed Nb-W POMs between the layers of the LDH, but also the synthesis of a new MgAl-LDH intercalated this time with hexaniobate [H$_3$Nb$_6$O$_{19}$]$^{5-}$.

Table 10. Summary of LDH with intercalated niobium–based anions.

Layer Cations	Molar Ratio (M^{II}:M^{III})	Intercalated Species	Synthesis Method	Reaction	Substrate	Yield (%)	Reference
Mg,Al	1	$[Nb_2W_4O_{19}]^{4-}$	Anion exchange	-	-	-	[121]
Mg,Al	1	$[Nb_3W_3O_{19}]^{5-}$	Anion exchange	-	-	-	[121]
Mg,Al	1	$[Nb_4W_3O_{19}]^{6-}$	Anion exchange	-	-	-	[121]
Mg,Al	2	$[W_4Nb_2O_{19}]^{4-}$	Anion exchange	Epoxidation	Cyclooctene	7	[102]
Mg,Al	2.1	$[H_3Nb_6O_{19}]^{5-}$	Anion exchange	-	-	-	[120]
Zn,Al	2	$[W_4Nb_2O_{19}]^{4-}$	Anion exchange	Epoxidation	Cyclooctene	5	[102]

Table 11. The polymerization degree of oxoniobiate versus pH. Information taken from Ref. [119].

pH	Species
>14.5	$NbO_2(OH)_4{}^{3-}$
14	$Nb_6O_{19}{}^{8-}$
11.5	$H_xNb_6O_{19}{}^{(8-x)-}$
6.5	$Nb_{12}O_{36}{}^{12-}$, $Nb_2O_5 \cdot nH_2O$
3.65	$Nb_{12}O_{36}{}^{12-}$, $Nb_2O_5 \cdot nH_2O$
0.55	$Nb_2O_5 \cdot nH_2O$

The solid was prepared by anion exchange method from the corresponding nitrate LDH precursor and an aqueous solution of $K_8Nb_6O_{19}$ [120]. The PXRD results have indicated the existence of a microporous hydrotalcite-type material, having a gallery height of 7.2 Å, which has the hexaniobate anions oriented with their C_3 axes perpendicular to the layers. The obtained material has a rather small surface area (24 m^2g^{-1}), which considerably increases by calcination at 500 °C (157 m^2g^{-1}). However, calcination above 700 °C leads to a decrease of surface area (45 m^2g^{-1} for the sample calcined at 800 °C) due to crystallization of magnesium niobiate $Mg_4Nb_2O_9$ [120]. Furthermore, the FT-IR studies on the acid-basic properties that were carried out by pyridine and 2-propanol adsorption, showed that the sample obtained through calcination present Lewis-type acid and Brønsted-type basic sites [120]. The niobium chemistry is still a scarcely researched field, which is expected to produce findings that are able to expand the catalysis domain in the near future.

5. Conclusions

Herein, a wide screening of the scientific literature related to the expanded family of POM-LDH composite materials and their use as catalysts for fine organic synthesis is presented. For this investigation, the following three systems hosting polyoxometalates with different nuclearity were analyzed:

- Systems hosting low-nuclearity polyoxometalates, being mainly represented by LDH intercalated with chromate, dichromate, and other different chromium-based anions, such as $Cr^{III}(EDTA)$, $[Cr^{III}(SO_3\text{-salen})]^{2-}$. The prepared materials seemed to offer promising catalytic results, especially in the oxidation of benzyl alcohol and alkoxylation of n-butanol.
- Systems hosting medium-nuclearity polyoxometalates containing LDH intercalated either with molybdenum or vanadium-based anions. Not only Mo-LDH, but also V-LDH, proved to be efficient catalysts for the oxidation and epoxidation reactions. V-LDH was a better catalyst for the epoxidation of alcohols than Mo-LDH. The hydrotalcite containing molybdenum-based anions was also successfully used as a catalyst in reactions, such as condensation, esterification, and halodecarboxylation.
- Systems hosting high-nuclearity polyoxometalates whose primarily representative is LDH intercalated with tungsten-based anions. The catalytic results suggested that the solid is suitable for the epoxidation of alcohols and oxidation of S-containing compounds, such as benzothiophene, dibenzothiophene, benzyl-phenyl sulfide, etc.

The engineering of the POM-LDH materials has been realized through different methods: anionic exchange, co-precipitation, reconstruction, and adsorption mechanisms. All of the studies cited, based on the intercalation of different polyoxometalates in the interlayer space of the hydrotalcites, revealed that the synthesis method can provide POM-LDH materials that are tailored with impressionable redox and acid-base properties. These characteristics can be further improved through the modification of both components (the cations in the brucite-like layers and the polyoxometalate in the interlayer), which makes them suitable materials in many different catalytic processes. POM-LDH materials have an incredibly bright future ahead, being able to exceed other typical catalysts (when it comes to thermal stability, the size, and distribution of the pores). Therefore, it can be strongly sustained that their applications in the catalytic field are theoretically unlimited.

Author Contributions: Conceptualization, A.-E.S., I.-C.M. and R.Z.; writing-original draft preparation, A.-E.S.; writing-review and editing, R.Z., O.D.P. and I.-C.M.; visualization, O.D.P.; supervision, I.-C.M. All authors have read and agreed to the published version of the manuscript.

Funding: This research was funded by the Romanian Ministry of Research and Innovation, CCCDI – UEFISCDI, project number PN-III-P1-1.2-PCCDI-2017-0387 / 80PCCDI/2018.

Conflicts of Interest: The authors declare no conflict of interest.

References

1. Li, S.; Zhu, M.; Whitcombe, M.J.; Piletsky, S.A.; Turner, A.P.F. Molecularly Imprinted Polymers for Enzyme-like Catalysis: Principle,Design, and Applications. In *Molecularly Imprinted Catalysts: Principles, Syntheses, and Applications*, 1st ed.; Li, S., Cao, S., Piletsky, S.A., Turner, A.P.F., Eds.; Elsevier: Kidlington/Oxford, UK, 2016; pp. 1–3.
2. Tongamp, W.; Zhang, Q.; Saito, F. Mechanochemical route for synthesizing nitrate form of layered double hydroxide. *Powder Technol.* **2008**, *185*, 43–48. [CrossRef]
3. Baskaran, T.; Christopher, J.; Sakthivel, A. Progress on layered hydrotalcite (HT) materials as potential support and catalytic materials. *RSC Adv.* **2015**, *5*, 98853–98875. [CrossRef]
4. Li, T.; Miras, H.N.; Song, Y.F. Polyoxometalate (POM)-Layered Double Hydroxides (LDH) Composite Materials: Design and Catalytic Applications. *Catalysts* **2017**, *7*, 260. [CrossRef]
5. Cavani, F.; Trifiro, F.; Vaccari, A. Hydrotalcite-type anionic clays: Preparation, properties and applications. *Catal. Today* **1991**, *11*, 173–301. [CrossRef]
6. Mobley, J.K.; Crocker, M. Catalytic oxidation of alcohols to carbonyl compounds over hydrotalcite and hydrotalcite supported catalysts. *RSC Adv.* **2015**, *5*, 65780–65797. [CrossRef]
7. Sheldon, R.A.; Arends, I.; Hnefeld, U. *Green Chemistry and Catalysis*, 1st ed.; Wiley-VCH Verlag GmbH & Co. KGaA: Weinheim, Germany, 2007; pp. 1–14.
8. Yu, H.; Xu, B.; Bian, L.; Gao, H. Influence on Structure of Layered Double Hydroxides with Different Methods Synthesis. *Adv. Mater. Res.* **2011**, *160*, 656–660.
9. Kajdas, C. Importance of anionic reactive intermediates for lubricant component reactions with friction surfaces. *Lubr. Sci.* **1994**, *6*, 203–228. [CrossRef]
10. Knözinger, H. Infrared Spectroscopy for the characterization of Surface Acidity and Basicity. In *Handbook of Heterogeneous Catalysis*; Ertl, G., Knözinger, H., Schüth, F., Waitkamp, J., Eds.; Wiley-VCH Verlag GmbH & Co. KGaA: Weinheim, Germany, 2008; pp. 1155–1156.
11. Zhang, X.; Li, X. Mechanochemical approach for synthesis of layered double hydroxides. *Appl. Surf. Sci.* **2013**, *274*, 158–163. [CrossRef]
12. Marcu, I.C.; Popescu, I.; Urda, A.; Hulea, V. Layered Double Hydroxides-Based Materials as Oxidation Catalysts. In *Sustainable Nanosystems Development, Properties, and Applications*, 1st ed.; Putz, M.V., Mirica, M.C., Eds.; IGI Global: Hershey, PA, USA, 2017; pp. 59–121.
13. Qu, J.; Zhang, Q.; Li, X.; He, X.; Song, S. Mechanochemical approaches to synthesize layered double hydroxides: A review. *Appl. Clay Sci.* **2016**, *119*, 185–192. [CrossRef]
14. Mahapatra, D.K.; Bharti, S.K.; Asati, V. Anti-cancer Chalcones: Structural and Molecular Target Perspectives. *Eur. J. Med. Chem.* **2015**, *101*, 496–524. [CrossRef]
15. Omwoma, S.; Chen, W.; Tsunashima, R.; Song, Y.F. Recent Advances on Polyoxometalates Intercalated Layered Double Hydroxides: From Synthetic Approaches to Functional Material Applications. *Coord. Chem. Rev.* **2014**, *258–259*, 58–71. [CrossRef]
16. Izarova, N.V.; Pope, M.T.; Kortz, U. Noble Metals in Polyoxometalates. *Angewandte Chemie International Edition* **2012**, *51*, 9492–9510. [CrossRef] [PubMed]
17. Schoonheydt, R.A.; Pinnavaia, T.J.; Lagaly, G.; Gangas, N. Pillared clays and pillared layered solids. *Pure Appl. Chem.* **1999**, *71*, 2367–2371. [CrossRef]
18. Rives, V.; Carriazo, D.; Martín, C. Heterogeneous Catalysis by Polyoxometalate-Intercalated Layered Double Hydroxides. In *Pillared Clays and Related Catalysts*, 1st ed.; Gil, A., Korili, S.A., Trujillano, R., Vicente, M.A., Eds.; Springer Science and Business Media: New York, NY, USA, 2010; pp. 319–393.
19. Depège, C.; Forano, C.; De Roy, A.; Besse, J.P. [Cu-Cr] Layered Double Hydroxides Pillared by CrO_4^{2-} and $Cr_2O_7^{2-}$ Oxometalates. *Mol. Cryst. Liq. Cryst.* **1994**, *244*, 161–166. [CrossRef]
20. Malherbe, F.; Depège, C.; Forano, C.; Besse, J.P.; Atkins, M.P.; Sharma, B.; Wade, S.R. Alkoxylation reaction catalyzed by layered double hydroxides. *Appl. Clay Sci.* **1998**, *13*, 451–466. [CrossRef]
21. Tsyganok, A.; Green, R.G.; Giorgi, J.B.; Sayari, A. Non-oxidative dehydrogenation of ethane to ethylene over chromium catalysts prepared from layered double hydroxide precursors. *Catal. Commun.* **2007**, *8*, 2186–2193. [CrossRef]

22. Del Arco, M.; Carriazo, D.; Martin, C.; Perez-Grueso, A.M.; Rives, V. Acid and redox properties of mixed oxides prepared by calcination of chromate-containing layered double hydroxides. *J. Solid State Chem.* **2005**, *178*, 3571–3580. [CrossRef]

23. Prasanna, S.V.; Rao, R.A.P.; Kamath, P.V. Layered double hydroxides as potential chromate scavengers. *J Colloid Interef. Sci.* **2006**, *304*, 292–299. [CrossRef]

24. Prasanna, S.V.; Kamath, P.V.; Shivakumara, C. Synthesis and characterization of layered double hydroxides (LDHs) with intercalated chromate ions. *Mater. Res. Bull.* **2007**, *42*, 1028–1039. [CrossRef]

25. Das, N.; Das, R. Insertion of chromium (III) ascorbate complex into layered double hydroxide through reduction of intercalated chromate by ascorbic acid. *Appl. Clay Sci.* **2008**, *42*, 90–94. [CrossRef]

26. Randarevich, L.S.; Zhuravlev, I.Z.; Strelko, V.V.; Patrilyak, N.M.; Shaposhnikova, T.A. Synthesis, Anion Exchange Properties, and Hydrolytic Stability of Mg–Fe(III) Layered Double Hydroxides. *J. Water Chem. Technol.* **2009**, *31*, 110–114. [CrossRef]

27. Malherbe, F.; Bigey, L.; Forano, C.; De Roy, A.; Besse, J.P. Structural aspects and thermal properties of takovite-like layered double hydroxides pillared with chromium oxo-anions. *J. Chem. Soc. Dalton Trans.* **1999**, 3831–3839. [CrossRef]

28. Wu, G.; Wang, X.; Li, J.; Zhao, N.; Wei, W.; Sun, Y. A new route to synthesis of sulphonatosalen-chromium (III) hydrotalcites: Highly selective catalysts for oxidation of benzyl alcohol to benzaldehyde. *Catal. Today* **2008**, *131*, 402–407. [CrossRef]

29. Manayil, J.C.; Sankaranarayanan, S.; Bhadoria, D.S.; Srinivasan, K. CoAl-CrO$_4$ Layered Double Hydroxides as Selective Oxidation Catalysts at Room Temperature. *Ind. Eng. Chem. Res.* **2011**, *50*, 13380–13386. [CrossRef]

30. Wang, X.; Wu, G.; Wang, F.; Ding, K.; Zhang, F.; Liu, X.; Xue, Y. Base-free selective oxidation of glycerol with 3% H$_2$O$_2$ catalyzed by sulphonato-salen-chromium (III) intercalated LDH. *Catal. Commun.* **2012**, *28*, 73–76. [CrossRef]

31. Rives, V.; Ulibarri, M.A. Layered double hydroxides (LDH) intercalated with metal coordination compounds and oxometalates. *Coord. Chem. Rev.* **1999**, *181*, 61–120. [CrossRef]

32. Villa, A.L.; de Vos, D.E.; Verpoort, F.; Sels, B.F.; Jacobs, P.A. A Study of V-Pillared Layered Double Hydroxides as Catalysts for the Epoxidation of Terpenic Unsaturated Alcohols. *J. Catal.* **2001**, *198*, 223–231. [CrossRef]

33. Menetrier, M.; Han, K.S.; Guerlou-Demourgues, L.; Delmas, C. Vanadate-Inserted Layered Double Hydroxides: A 51V NMR Investigation of the Grafting Process. *Inorg. Chem.* **1997**, *36*, 2441–2445. [CrossRef]

34. Twu, J.; Dutta, P.K. Decavanadate Ion-Pillared Hydrotalcite: Spectroscopic Studies of the Thermal Decomposition Process. *J. Catal.* **1990**, *124*, 503–510. [CrossRef]

35. Maciuca, A.L.; Dumitriu, E.; Fajula, F.; Hulea, V. Mild oxidation of tetrahydrothiophene to sulfolane over V-, Mo- and W-containing layered double hydroxides. *Appl. Catal. A* **2008**, *338*, 1–8. [CrossRef]

36. Dobrea, I.D.; Ciocan, C.E.; Dumitriu, E.; Popa, M.I.; Petit, E.; Hulea, V. Raman spectroscopy—Useful tool for studying the catalysts derived from Mo and V-oxyanion-intercalated layered double hydroxides. *Appl. Clay Sci.* **2015**, *104*, 205–210. [CrossRef]

37. Malherbe, F.; Besse, J.P.; Wade, S.R.; Smith, W.J. Highly selective synthesis of 2-butoxy ethanol over Mg/Al/V mixed oxides catalysts derived from hydrotalcites. *Catal. Lett.* **2000**, *67*, 197–202. [CrossRef]

38. Bahranowskia, K.; Bueno, G.; Corberan, V.C.; Kooli, F.; Serwicka, E.M.; Valenzuela, R.X.; Wcislo, K. Oxidative dehydrogenation of propane over calcined vanadate-exchanged Mg, Al-layered double hydroxides. *Appl. Catal. A* **1999**, *185*, 65–73. [CrossRef]

39. Ulibarri, M.A.; Labajos, F.M.; Rives, V.; Trujillano, R.; Kagunya, W.; Jones, W. Comparative Study of the Synthesis and Properties of Vanadate-Exchanged Layered Double Hydroxides. *Inorg. Chem.* **1994**, *25*, 2592–2599. [CrossRef]

40. Bahranowski, K.; Kooli, F.; Poltowicz, J.; Serwicka, E.M. Polyvanadate-exchanged hydrotalcites as catalysts for selective oxidation of cyclohexene with molecular oxygen. *React. Kinet. Catal. Lett.* **1998**, *64*, 3–8. [CrossRef]

41. Dula, R.; Wcisło, K.; Stoch, J.; Grzybowska, B.; Serwicka, E.M.; Kooli, F.; Bahranowski, K.; Gawełc, A. Layered double hydroxide-derived vanadium catalysts for oxidative dehydrogenation of propane. Influence of interlayer-doping versus layer-doping. *Appl. Catal. A* **2002**, *230*, 282–291. [CrossRef]

42. Wegrzyn, A.; Rafalska-Łasocha, A.; Dudek, B.; Dziembaj, R. Nanostructured V-containing hydrotalcite-like materials obtained by non-stoichiometric anion exchange as precursors of catalysts for oxidative dehydrogenation of n-butane. *Catal. Today* **2006**, *116*, 74–81. [CrossRef]

43. Palmer, S.J.; Soisonard, A.; Frost, R.L. Determination of the mechanism(s) for the inclusion of arsenate, vanadate, or molybdate anions into hydrotalcites with variable cationic ratio. *J. Colloid Interface Sci.* **2009**, *329*, 404–409. [CrossRef]

44. Kooli, F.; Crespo, I.; Barriga, C.; Ulibarri, M.A.; Rives, V. Precursor dependence of the nature and structure of non-stoichiometric magnesium aluminium vanadates. *J. Mater. Chem.* **1996**, *6*, 1199–1206. [CrossRef]

45. Valverde, J.A.; Echavarria, A.; Eon, J.G.; Faro, A.C., Jr.; Palacio, L.A. V–Mg–Al catalyst from hydrotalcite for the oxidative dehydrogenation of propane. *React. Kinet. Mech. Catal.* **2014**, *111*, 679–696. [CrossRef]

46. Palmer, S.J.; Nguyen, T.; Frost, R.L. Synthesis and Raman spectroscopic characterization of hydrotalcite with CO32– and VO3– anions in the interlayer. *J. Raman Spectrosc.* **2007**, *38*, 1602–1608. [CrossRef]

47. Guo, L.; Zhang, F.; Lu, J.C.; Zeng, R.C.; Li, S.Q.; Song, L.; Zeng, J.M. A comparison of corrosion inhibition of magnesium aluminum and zinc aluminum vanadate intercalated layered double hydroxides on magnesium alloys. *Front. Mater. Sci.* **2018**, *12*, 198–206. [CrossRef]

48. De Castro, F.R.; Lam, Y.L.; Herbst, M.H.; Pereira, M.M.; Da Silva, T.C.; Homs, N.; De la Piscina, P.R. VO2+ Reaction with Hydrotalcite and Hydrotalcite-Derived Oxide: The Effect of the Vanadium Loading on the Structure of Catalyst Precursors and on the Vanadium Species. *Eur. J. Inorg. Chem.* **2013**, *2*, 241–247. [CrossRef]

49. Kooli, F.; Rives, V.; Ulibarri, M.A. Preparation and Study of Decavanadate-Pillared Hydrotalcite-like Anionic Clays Containing Transition Metal Cations in the Layers. 1. Samples Containing Nickel- Aluminum Prepared by Anionic Exchange and Reconstruction. *Inorg. Chem.* **1995**, *34*, 5114–5121. [CrossRef]

50. Valverde, J.A.; Echavarría, A.; Ribeiro, M.F.; Palacioa, L.A.; Eon, J.G. Decavanadate-intercalated Ni–Al hydrotalcites as precursors of mixed oxides for the oxidative dehydrogenation of propane. *Catal. Today* **2012**, *192*, 36–43. [CrossRef]

51. Vega, J.M.; Granizo, N.; de la Fuente, D.; Simancas, J.; Morcillo, M. Corrosion inhibition of aluminum by coatings formulated with Al–Zn–vanadate hydrotalcite. *Prog. Org. Coat.* **2011**, *70*, 213–219. [CrossRef]

52. Alvarez, M.G.; Urda, A.; Rives, V.; Carrazan, S.R.G.; Martin, C.; Tichit, D.; Marcu, I.C. Propane oxidative dehydrogenation over V-containing mixed oxides derived from decavanadate-exchanged ZnAl-layered double hydroxides prepared by a sol-gel method. *Comptes Rendus Chimie* **2017**, *21*, 210–220. [CrossRef]

53. Barriga, C.; Jones, W.; Malet, P.; Rives, V.; Ulibarri, M.A. Synthesis and Characterization of Polyoxovanadate-Pillared Zn-Al Layered Double Hydroxides: An X-ray Absorption and Diffraction Study. *Inorg. Chem.* **1998**, *37*, 1812–1820. [CrossRef]

54. Tedim, J.; Zheludkevich, M.L.; Salak, A.N.; Lisenkov, A.; Ferreira, M.G.S. Nanostructured LDH-container layer with active protection functionality. *J. Mater. Chem.* **2011**, *21*, 15464–15470. [CrossRef]

55. Salak, A.N.; Tedim, J.; Kuznetsova, A.I.; Ribeiro, J.L.; Vieira, L.G.; Zheludkevich, M.L.; Ferreira, M.G.S. Comparative X-ray diffraction and infrared spectroscopy study of Zn–Al layered double hydroxides: Vanadate vs nitrate. *Chem. Phys.* **2012**, *397*, 102–108. [CrossRef]

56. Salak, A.N.; Tedim, J.; Kuznetsova, A.I.; Vieira, L.G.; Ribeiro, J.L.; Zheludkevich, M.L.; Ferreira, M.G.S. Thermal Behavior of Layered Double Hydroxide Zn–Al–Pyrovanadate: Composition, Structure Transformations, and Recovering Ability. *J. Phys. Chem. C* **2013**, *117*, 4152–4157. [CrossRef]

57. Zhang, Y.; Li, Y.; Ren, Y.; Wang, H.; Chen, F. Double-doped LDH films on aluminum alloys for active protection. *Mater. Lett.* **2017**, *192*, 33–35. [CrossRef]

58. Del Arco, M.; Rives, V.; Trujillano, R.; Malet, P. Thermal characterization of Zn-Cr layered double hydroxides with hydrotalcite-like structures containing carbonate or decavanadate. *J. Mater. Chem.* **1996**, *6*, 1419–1428. [CrossRef]

59. Nejati, K.; Akbari, A.R.; Davari, S.; Asadpour-Zeynali, K.; Rezvanic, Z. Zn–Fe-layered double hydroxide intercalated with vanadate and molybdate anions for electrocatalytic water oxidation. *New J. Chem.* **2018**, *42*, 2889–2895. [CrossRef]

60. Davantès, A.; Lefèvre, G. In Situ Real Time Infrared Spectroscopy of Sorption of (Poly)molybdate Ions into Layered Double Hydroxides. *J. Phys. Chem. A* **2013**, *117*, 12922–12929. [CrossRef]

61. Smith, H.D.; Parkinson, G.M.; Hart, R.D. In situ absorption of molybdate and vanadate during precipitation of hydrotalcite from sodium aluminate solutions. *J. Cryst. Growth* **2005**, *275*, 1665–1671. [CrossRef]

62. Van Laar, F.M.P.R.; de Vos, D.E.; Pierard, F.; Kirsch-de Mesmaeker, A.; Fiermans, L.; Jacobs, P.A. Generation of Singlet Molecular Oxygen from H₂O₂ with Molybdate-Exchanged Layered Double Hydroxides: Effects of Catalyst Composition and Reaction Conditions. *J. Catal.* **2001**, *197*, 139–150. [CrossRef]

63. Krishnan, C.V.; Garnett, M.; Hsiao, B.; Chu, B. Electrochemical Measurements of Isopolyoxomolybdates: 1. pH Dependent Behavior of Sodium Molybdate. *Int. J. Electrochem. Sci.* **2007**, *2*, 29–51.

64. Simon Ng, K.Y.; Gulari, E. Spectroscopic and scattering investigation of isopoly-molybdate and tungstate solutions. *Polyhedron* **1984**, *3*, 1001–1011.

65. Del Arco, M.; Carriazo, D.; Gutierrez, S.; Martin, C.; Rives, V. Synthesis and Characterization of New Mg₂Al-Paratungstate Layered Double Hydroxides. *Inorg. Chem.* **2004**, *43*, 375–384. [CrossRef]

66. Soled, S.; Levin, D.; Miseo, S.; Ying, J. Soft chemical synthesis of mixed metal molybdate oxidation catalysts and their structural relationship to hydrotalcite. In *Studies in Surface Science and Catalysis*; Delmon, B., Jacobs, P.A., Maggi, R., Martens, J.A., Grange, P., Poncelet, G., Eds.; Elsevier Science B.V.: Louvain-la-Neuve, Belgium, 1998; Volume 118, pp. 359–367.

67. Palmer, S.J.; Frost, R.L.; Nguyen, T. Thermal decomposition of hydrotalcite with molybdate and vanadate anions in the interlayer. *J. Therm. Anal. Calorim.* **2008**, *92*, 879–886. [CrossRef]

68. Gardner, E.; Pinnavaia, T.J. On the nature of selective olefin oxidation catalysts derived from molybdate- and tungstate-intercalated layered double hydroxides. *Appl. Catal. A Gen.* **1998**, *167*, 65–74. [CrossRef]

69. Carriazo, D.; Martin, C.; Rives, V. An FT-IR study of the adsorption of isopropanol on calcined layered double hydroxides containing isopolymolybdate. *Catal. Today* **2007**, *126*, 153–161. [CrossRef]

70. Sels, B.F.; de Vos, D.E.; Jacobs, P.A. Tungstate and Molybdate exchanged Layered Double Hydroxides (LDHs) as catalysts for selective oxidation of organics and for bleaching. In *Studies in Surface Science and Catalysis*; Sayari, A., Jaroniec, M., Pinnavaia, T.J., Eds.; Elsevier Science B.V.: Banff, AB, Canada, 2000; Volume 129, pp. 845–850.

71. Ciocan, C.E.; Dumitriu, E.; Cacciaguerra, T.; Fajula, F.; Hulea, V. New approach for synthesis of Mo-containing LDH based catalysts. *Catal. Today* **2012**, *198*, 239–245. [CrossRef]

72. Carriazo, D.; Martin, C.; Rives, V.; Popescu, A.; Cojocaru, B.; Mandache, I.; Parvulescu, V.I. Hydrotalcites composition as catalysts: Preparation and their behavior on epoxidation of two bicycloalkenes. *Microporous Mesoporous Mater.* **2006**, *95*, 39–47. [CrossRef]

73. Wahlen, J.; de Vos, D.E.; Sels, B.F.; Nardello, V.; Aubry, J.M.; Alsters, P.L.; Jacobs, P.A. Molybdate-exchanged layered double hydroxides for the catalytic disproportionation of hydrogen peroxide into singlet oxygen: Evaluation and improvements of ¹O₂ generation by combined chemiluminescence and trapping experiments. *Appl. Catal. A* **2005**, *293*, 120–128. [CrossRef]

74. Zavoianu, R.; Cruceanu, A.; Pavel, O.D.; Angelescu, E.; Soares Dias, A.P.; Birjega, R. Oxidation of tert-butanethiol with air using Mo containing hydrotalcite-like compounds and their derived mixed oxides as catalysts. *React. Kinet. Mech. Catal.* **2012**, *105*, 145–162. [CrossRef]

75. Mitchell, P.C.H.; Wass, S.A. Propane dehydrogenation over molybdenum hydrotalcite catalysts. *Appl. Catal. A* **2002**, *225*, 153–165. [CrossRef]

76. Wahlen, J.; de Vos, D.; Jary, W.; Alsters, P.; Jacobs, P. Glycol-modified molybdate catalysts for efficient singlet oxygen generation from hydrogen peroxide. *Chem. Commun.* **2007**, *23*, 2333–2335. [CrossRef]

77. Klemkaite-Ramanauske, K.; Zilinskas, A.; Taraskevicius, R.; Khinsky, A.; Kareiva, A. Preparation of Mg/Al layered double hydroxide (LDH) with structurally embedded molybdate ions and application as a catalyst for the synthesis of 2-adamantylidene(phenyl)amine Schiff base. *Polyhedron* **2014**, *68*, 340–345. [CrossRef]

78. Zeng, R.G.; Liu, Z.G.; Zhang, F.; Li, S.Q.; Cui, H.Z.; Han, E.H. Corrosion of molybdate intercalated hydrotalcite coating on AZ31 Mg alloy. *J. Mater. Chem. A* **2014**, *2*, 13049–13057. [CrossRef]

79. Zavoianu, R.; Birjega, R.; Pavel, O.D.; Cruceanu, A.; Alifanti, M. Hydrotalcite like compounds with low Mo-loading active catalysts for selective oxidation of cyclohexene with hydrogen peroxide. *Appl. Catal. A* **2005**, *286*, 211–220. [CrossRef]

80. Choudary, B.M.; Someshwar, T.; Lakshmi Kantam, M.; Venkat Reddy, C. Molybdate-exchanged Mg–Al–LDH catalyst: An eco-compatible route for the synthesis of β-bromostyrenes in aqueous medium. *Catal. Commun.* **2004**, *5*, 215–219. [CrossRef]

81. Frost, R.L.; Musumeci, A.W.; Bostrom, T.; Adebajo, M.O.; Weier, M.L.; Martens, W. Thermal decomposition of hydrotalcite with chromate, molybdate or sulphate in the interlayer. *Thermochim. Acta* **2005**, *429*, 179–187. [CrossRef]

82. Palmer, S.J.; Frost, R.L.; Ayoko, G.; Nguyen, T. Synthesis and Raman spectroscopic characterisation of hydrotalcite with CO_3^{2-} and $(MoO_4)^{2-}$ anions in the interlayer. *J. Raman Spectrosc.* **2008**, *39*, 395–401. [CrossRef]

83. Bonifacio-Martinez, J.; Serano-Gomez, J.; del Carmen Lopez-Reyes, M.; Grandos-Correa, F. Mechano-chemical effects on surface properties and molybdate exchange on hydrotalcite. *Clay Miner.* **2009**, *44*, 311–317. [CrossRef]

84. Han, N.; Zhao, F.; Li, L. Ultrathin Nickel-Iron Layered Double Hydroxide Nanosheets Intercalated with Molybdate Anions for Electrocatalytic Water Oxidation. *J. Mater. Chem. A* **2015**, *3*, 16348–16353. [CrossRef]

85. Colombo, K.; Maruyama, S.A.; Yamamoto, C.I.; Wypych, F. Intercalation of Molybdate Ions into Ni/Zn Layered Double Hydroxide Salts: Synthesis, Characterization, and Preliminary Catalytic Activity in Methyl Transesterification of Soybean Oil. *J. Braz. Chem. Soc.* **2017**, *28*, 1315–1322. [CrossRef]

86. Das, J.; Parida, K.M. Heteropoly acid intercalated Zn/Al HTlc as efficient catalyst for esterification of acetic acid using n-butanol. *J. Mol. Catal.* **2007**, *264*, 248–254. [CrossRef]

87. Yu, X.; Wang, J.; Zhang, M.; Yang, L.; Li, J.; Yang, P.; Cao, D. Synthesis, characterization and anticorrosion performance of molybdate pillared hydrotalcite/in situ created ZnO composite as pigment for Mg–Li alloy protection. *Surf. Coat. Technol.* **2008**, *203*, 250–255. [CrossRef]

88. Yu, X.; Wang, J.; Zhang, M.; Yang, P.; Yang, L.; Cao, D.; Li, J. One-step synthesis of lamellar molybdate pillared hydrotalcite and its application for AZ31 Mg alloy protection. *Solid State Sci.* **2009**, *11*, 376–381. [CrossRef]

89. Levin, D.; Soled, S.L.; Ying, J.Y. Chimie Douce Synthesis of a Layered Ammonium Zinc Molybdate. *Chem. Mater.* **1996**, *8*, 836–843. [CrossRef]

90. Mostafa, M.S.; Bakr, A.A.; Eshaq, G.; Kamel, M.M. Novel Co/Mo layered double hydroxide: Synthesis and uptake of Fe (II) from aqueous solutions (Part 1). *Desalin. Water Treat.* **2014**, *56*, 239–247. [CrossRef]

91. Mostafa, M.S.; Bakr, A.A.; Eshaq, G.; Kamel, M.M. Kinetics of uptake of Fe(II) from aqueous solutions by Co/Mo layered double hydroxide (Part 2). *Desalin. Water Treat.* **2014**, *56*, 248–255.

92. Mostafa, M.S.; Bakr, A.A.; El Naggar, A.M.A.; Sultan, E.A. Water decontamination via the removal of Pb (II) using a new generation of highly energetic surface nano-material: $Co^{+2}Mo^{+6}$LDH. *J. Colloid Interface Sci.* **2016**, *461*, 261–272. [CrossRef]

93. Muramatsu, K.; Saber, O.; Tagaya, H. Preparation of new layered double hydroxide, Zn-Mo LDH. *J. Porous Mater.* **2007**, *14*, 481–484. [CrossRef]

94. Narita, E.; Kaviratna, P.D.; Pinnavaia, T.J. Direct synthesis of a polyoxometallate-pillared layered double hydroxide by coprecipitation. *J. Chem. Soc. Chem. Commun.* **1993**, *1*, 60–62. [CrossRef]

95. Carriazo, D.; Domingo, C.; Martin, C.; Rives, V. Structural and texture evolution with temperature of layered double hydroxides intercalated with paramolybdate anions. *Inorg. Chem.* **2006**, *45*, 1243–1251. [CrossRef]

96. Liu, J.; Han, Q.; Chen, L.; Zhao, J. A brief review of the crucial progress on heterometallic polyoxotungstates in the past decade. *Cryst. Eng. Comm.* **2016**, *18*, 842–862. [CrossRef]

97. Yun, S.K.; Pinnavaia, T.J. Layered Double Hydroxides Intercalated by Polyoxometalate Anions with Keggin (α-$H_2W_{12}O_{40}^{6-}$), Dawson (α-$P_2W_{18}O_{62}^{6-}$), and Finke ($Co_4(H_2O)_2(PW_9O_{34})_2^{10-}$) Structures. *Inorg. Chem.* **1996**, *35*, 6853–6860. [CrossRef]

98. Choudary, B.M.; Someshwar, T.; Reddy, C.V.; Kantam, M.L.; Ratnam, K.J.; Sivaji, L.V. The first example of bromination of aromatic compounds with unprecedented atom economy using molecular bromine. *Appl. Catal. A Gen.* **2003**, *251*, 251–397. [CrossRef]

99. Sels, B.F.; de Vos, D.E.; Buntinx, M.; Pierard, F.; Kirsch-de Mesmaeker, A.; Jacobs, P. Layered double hydroxides exchanged with tungstate as biomimetic catalysts for mild oxidative bromination. *Nature* **1999**, *400*, 855–857. [CrossRef]

100. Choudary, B.M.; Bharathi, B.; Venkat Reddy, C.; Lakshmi Kantam, M. Tungstate exchanged Mg-Al-LDH catalyst: An eco-compatible route for the oxidation of sulfides in aqueous medium. *J. Chem. Soc. Perkin Trans.* **2002**, *1*, 2069–2074. [CrossRef]

101. Palomeque, J.; Figueras, F.; Gelbard, G. Epoxidation with hydrotalcite-intercalated organotungstic complexes. *Appl. Catal. A Gen.* **2006**, *300*, 100–108. [CrossRef]

102. Carriazo, D.; Lima, S.; Martín, C.; Pillinger, M.; Valente, A.A.; Rives, V. Metatungstate and tungstoniobate-containing LDHs: Preparation, characterization and activity in epoxidation of cyclooctene. *J. Phys. Chem. Solids* **2007**, *68*, 1872–1880. [CrossRef]

103. Wei, X.; Fu, Y.; Xu, L.; Li, F.; Bi, B.; Liu, X. Tungstocobaltate-pillared layered double hydroxides: Preparation, characterization, magnetic and catalytic properties. *J. Solid State Chem.* **2008**, *181*, 1292–1297. [CrossRef]

104. Guo, Y.; Li, D.; Hu, C.; Wang, Y.; Wang, E.; Zhou, Y.; Feng, S. Photocatalytic degradation of aqueous organocholorine pesticide on the layered double hydroxide pillared by Paratungstate A ion, $Mg_{12}Al_6(OH)_{36}(W_7O_{24})\cdot 4H_2O$. *Appl. Catal. B Environ.* **2001**, *30*, 337–349. [CrossRef]

105. Del Arco, M.; Carriazo, D.; Gutierrez, S.; Martin, C.; Rives, V. An FT-IR study of the adsorption and reactivity of ethanol on systems derived from Mg_2Al–$W_7O_{24}{}^{6-}$ layered double hydroxides. *Phys. Chem.* **2004**, *6*, 465–470. [CrossRef]

106. Sels, B.F.; Levecque, P.; Brosius, R.; de Vos, D.E.; Jacobs, P.; Gammon, D.W.; Kinfe, H.H. A new catalytic route for the oxidative halogenation of cyclic enol ethers using tungstate exchanged on takovite. *Adv. Synth. Catal.* **2005**, *347*, 93–104. [CrossRef]

107. Maciuca, A.L.; Dumitriu, E.; Fajula, F.; Hulea, V. Catalytic oxidation processes for removing dimethylsulfoxide from wastewater. *Chemosphere* **2007**, *68*, 227–233. [CrossRef]

108. Hulea, V.; Maciuca, A.L.; Fajula, F.; Dumitriu, E. Catalytic oxidation of thiophenes and thioethers with hydrogen peroxide in the presence of W-containing layered double hydroxides. *Appl. Catal. A Gen.* **2006**, *313*, 200–207. [CrossRef]

109. Liu, P.; Wang, C.; Li, C. Epoxidation of allylic alcohols on self-assembled polyoxometalates hosted in layered double hydroxides with aqueous H_2O_2 as oxidant. *J. Catal.* **2009**, *262*, 159–168. [CrossRef]

110. Zhang, Y.; Su, J.; Wang, X.; Pan, Q.; Qu, W. Photocatalytic performance of polyoxometallate intercalated layered double hydroxide. *Mater. Sci. Forum* **2010**, *663*, 187–190. [CrossRef]

111. Jana, S.K.; Kubota, Y.; Tatsumi, T. Cobalt-substituted polyoxometalate pillared hydrotalcite: Synthesis and catalysis in liquid-phase oxidation of cyclohexanol with molecular oxygen. *J. Catal.* **2008**, *255*, 40–47. [CrossRef]

112. Vaysse, C.; Guerlou-Demourgues, L.; Demourgues, A.; Delmas, C. Thermal Behavior of Oxometalate (Mo,W)-Intercalated Layered Double Hydroxides: Study of the Grafting Phenomenon. *J. Solid State Chem.* **2002**, *167*, 59–72. [CrossRef]

113. Sels, B.F.; de Vos, D.E.; Jacobs, P.A. Use of $WO_4{}^{2-}$ on layered double hydroxides for mild oxidative bromination and bromide-assisted epoxidation with H_2O_2. *J. Am. Chem. Soc.* **2001**, *123*, 8350–8359. [CrossRef] [PubMed]

114. Watanabe, Y.; Yamamoto, K.; Tatsumi, T. Epoxidation of alkenes catalyzed by heteropolyoxometalate as pillars in layered double hydroxides. *J. Mol. Catal. A Chem.* **1999**, *145*, 281–289. [CrossRef]

115. Kwon, T.; Pinnavaia, T.J. Synthesis and properties of anionic clays pillared by $[XM_{12}O_{40}]^{n-}$ Keggin ions. *J. Mol. Catal.* **1992**, *74*, 23–33. [CrossRef]

116. Ivanova, A.S.; Korneeva, E.V.; Bondareva, V.M.; Glazneva, T.S. Gas-phase dehydration of glycerol over calcined tungsten-modified Zn-Al-O hydrotalcite-type catalysts. *J. Mol. Catal. A Chem.* **2015**, *408*, 98–106. [CrossRef]

117. Guo, Y.; Li, D.; Hu, C.; Wang, E.; Zou, Y.; Ding, H.; Feng, S. Preparation and photocatalytic behavior of Zn/Al/W(Mn) mixed oxides via polyoxometalates intercalated layered double hydroxides. *Microporous Mesoporous Mater.* **2002**, *56*, 153–162. [CrossRef]

118. Dey, K.C.; Kumari, P. Historical development, synthetic variables, classification and characterization techniques of polyoxometalates. *Int. J. Res. Sci. Technol.* **2017**, *7*, 154–177.

119. Nowak, I.; Ziolek, M. Niobium Compounds: Preparation, Characterization, and Application in Heterogeneous Catalysis. *Chem. Rev.* **1999**, *99*, 3603–3624. [CrossRef] [PubMed]

120. Carriazo, D.; Martin, C.; Rives, V. Thermal Evolution of a MgAl Hydrotalcite-Like Material Intercalated with Hexaniobate. *Eur. J. Inorg. Chem.* **2006**, *22*, 4608–4615. [CrossRef]

121. Evans, J.; Pillinger, M.; Zhang, J. Structural studies of polyoxometalate-anion-pillared layered double hydroxides. *J. Chem. Soc. Dalton Trans.* **1996**, *27*, 2963–2974. [CrossRef]

Article

Ultrasonic-Assisted Michael Addition of Arylhalide to Activated Olefins Utilizing Nanosized CoMgAl-Layered Double Hydroxide Catalysts

Nada S. Althabaiti [1], Fawzia M. Al-Nwaiser [1], Tamer S. Saleh [2] and Mohamed Mokhtar [1,*]

[1] Chemistry Department, Faculty of Science, King Abdulaziz University, Jeddah 21589, Saudi Arabia; nadoosh-72@hotmail.com (N.S.A.); falnowaiser@kau.edu.sa (F.M.A.-N.)

[2] Chemistry Department, Faculty of Science, University of Jeddah, P.O. Box 80329, Jeddah 21589, Saudi Arabia; tamsaid@yahoo.com

* Correspondence: mmoustafa@kau.edu.sa; Tel.: +966-500558045

Received: 27 December 2019; Accepted: 8 February 2020; Published: 11 February 2020

Abstract: An efficient cobalt-based layered double hydroxide (LDH)-catalyzed Michael addition of an aryl halide compound onto activated olefin as a Michael acceptor is described. The synthesized catalytic materials were characterized using different techniques to investigate their physicochemical, morphological, and textural properties. The partial isomorphic substitution of magnesium by cobalt ions in the cationic sheets of the layered double hydroxide (CoMgAl-LDH) appears to be an appropriate catalyst to cause this reaction. This technique enables compound synthesis resulting from the 1,4-addition in good to excellent yields. Moreover, ultrasound was found to have beneficial effect on this reaction due to the cavitation phenomenon.

Keywords: Michael addition; cobalt-based LDHs; ultrasonic irradiation; synergistic effect

---

1. Introduction

The new formation of carbon–carbon bonds using the conjugate addition of functionalized aryl compounds to Michael acceptors is considered a key point for synthesizing useful organic models for further transformation [1]. Researchers around the world made great efforts to develop new methods of synthesizing this critical reaction for more stable and adaptable problems [2]. Different organometallic catalysts were applied for the regioselective 1,4-addition of aryl halides [3–5]. For these types of reactions, several classical methods were applied, including the use of homogeneous catalysis with palladium salts [6,7], rhodium complexes [8,9], and copper salts [10,11]. Additional transition-metal conjugate additions of major groups of organometallic reagents such as aryl-aluminum, -tin, -silicon, -titanium, -indium, and -boron compounds were also documented [12]. However, these compounds are moisture-sensitive under classical working conditions in air/moisture atmosphere. A direct activation of the aryl halides using chemical and electrochemical processes was developed to avoid such difficulties. On the other hand, homogeneous catalysis remains of highest interest utilizing in situ reduction of transition-metal complexes [13]. The direct electrochemical arylation of electron-deficient olefins carried out utilizing cobalt catalysts in the presence of a sacrificial anode was successful for the synthesis of various aromatic halides [14]. In addition to the difficulties of handling the electrochemical approaches in assessment of the classical ones, the cobalt-based catalyst, $CoBr_2$, is only effective for electrochemical addition of aryl bromides attached with electron-withdrawing groups. Muriel et al. developed a direct chemical procedure for the conjugate addition of different substituted aryl bromides, chlorides, and triflates, bearing an electron-withdrawing or -donating group, on various activated olefins [15]. Although this one-step chemical process was novel and solved the problems of pre-preparation of organometallic reagents, using $CoBr_2(2,2'\text{-bipyridine})$ as a catalyst, it suffered from separation and regeneration difficulties of the selected designed catalysts. A "green" alternative to such reactions that provides high yields and good selectivity with waste reduction can be achieved by using cobalt-based solid heterogeneous catalysts [16].

Layered double hydroxide-based materials (LDHs) exhibit great potential as heterogeneous catalysts due to their versatility, defined by the combination of different catalytically active transition-metal species (e.g., Cu, Co, Fe, Ni, V, Rh), robust alkaline nature, and relaxed product departure [17–19]. Our group of research succeeded in synthesizing many organic synthons via carbon–carbon coupling reactions using LDH-based catalysts [20–22].

In addition, sonochemical synthesis is an important synthesis technique. Unlike traditional methods, this procedure has certain advantages, such as simplicity, short reaction time, controllability, convenience, and good product yields [23–27].

To the best of our knowledge, few reports dealt with a heterogeneous catalyst system for this 1,4-conjugate addition [28–30]. Bearing in mind all of the abovementioned factors, we continued our interest in introducing a benign protocol for organic reactions utilizing heterogeneous catalysts [20,31–34]. We introduce here a facile sonochemical synthesis for the conjugate addition of aryl halides onto electron-deficient olefins utilizing a CoMgAl LDH.

2. Results

2.1. Fourier-Transform Infrared Spectroscopy (FTIR)

FTIR spectra of all the solid materials investigated (Figure 1) showed a 3484 cm^{-1} wide band assigned to the O–H stretching vibration, credited to interlayer water and hydroxyl groups in the layered double hydroxide. The OH stretching modes of weak hydrogen bonds occurred in the area between 3600 and 3500cm^{-1} for water adsorbed on clay minerals. The OH stretching vibrations and a stretching vibration of interlayer water were due to a broad band about 3300–3000 cm^{-1} with a shoulder, often apparent, containing two or three overlapping bands. A low peak in the infrared spectrum of around 1630 cm^{-1} was due to the interlayer water mode of υ H_2O. A low peak in the infrared spectrum of around 1630 cm^{-1} was due to the interlayer water mode of υ H_2O. Due to the interaction of carbonate with interlayer water molecules and/or hydroxyl groups from the Mg–Al LDH base, the integration of the carbonate species into the layered structure showed a change toward lower wave numbers. Layered double carbonate hydroxides typically show infrared bands at about 1360–1400, 875, and 670 cm^{-1}, which are generally absent from the given spectrum in the interlayer. The obtained results indicate that a pure layered double hydroxide with hydroxyl groups and water molecules in the interlayer gallery was synthesized using the hydrothermal post-treated coprecipitation method.

The spectra of the CoMgAl-LDH and CoAl-LDH samples revealed bands below 1000 cm^{-1} as a result of stretching/deformation modes of the M–O, M–O–M, and O–M–O layers [35]. The peak at 620 cm^{-1} could be attributed to the existence of Co^{II} and Co–OH conversion, while the signal at 565 cm^{-1} remained due to the Al–OH translational mode [36].

2.2. Thermal Gravimetric Analysis (TGA)

Thermal gravimetric analysis was conducted for all of the catalysts investigated (Figure 2). For the MgAl-LDH study, four significant mass loss measures could be identified: (a) a poor mass loss at 50–140 °C with a shoulder at 132 °C owing to the loss of physisorbed water; (b) a strong mass loss signal at 135–300 °C at a height of 222 °C, primarily due to the loss of understructure losses [37–39]; (c) a large mass loss between 250 and 600 °C (maximum at 330, 352, and 410 °C) due to the evolution of H_2O vapor, which confirmed the process of dehydroxylation of the brucite structure by dihydroxylation; (d) one last mass loss at 499 °C for the decomposition of metal hydroxides, leading to the formation of metal oxides.

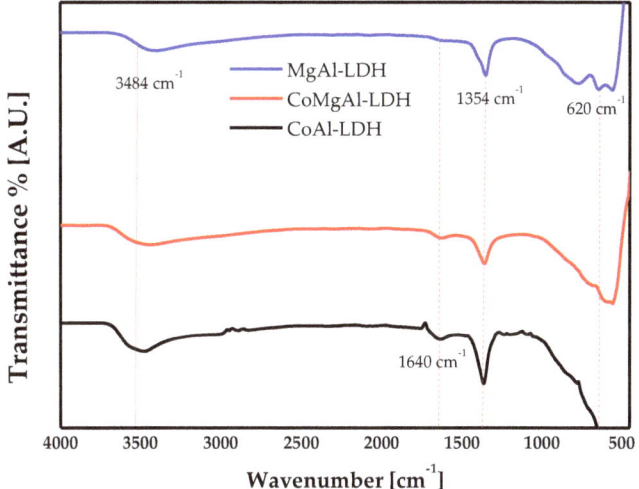

Figure 1. Fourier-transform infrared (FTIR) spectra of all the investigated samples.

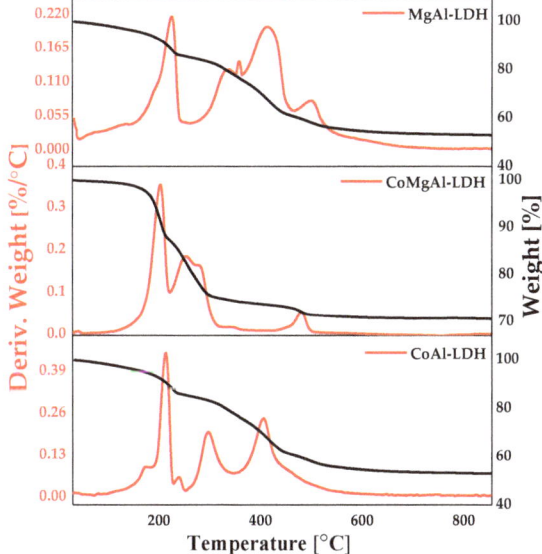

Figure 2. Thermal gravimetric and differential thermal gravimetric analyses of all catalysts.

The thermal behavior of the MgAl-LDH was significantly changed upon the inclusion of cobalt ions in the cationic sheet. The thermal decomposition at lower temperatures was sharply steep, and the Co-Mg-Al LDH sample was recognized as having three main mass loss steps. In the CoAl-LDH sample, the isomorphic substitution of cobalt cations to the magnesium ions in the cationic layer resulted in a similar thermal activity to that of the material MgAl-LDH. The obtained results revealed that the presence of both cobalt and magnesium ions in the cationic sheet favored the thermal decomposition of the catalyst.

2.3. X-ray Diffraction (XRD)

The X-ray powder diffraction pattern of synthetic Mg-Al LDH is shown in Figure 3. A phase analysis showed that only the double layered hydroxide meixnerite phase structure was obtained [40] with chemical formula $Mg_6Al_2(OH)_{18}\cdot4H_2O$, as indexed by Ref. Pattern 38-0478, JCPDS. Rhombohedral with R-3 m space group was the crystal structure of the investigated MgAl-LDH. The lattice parameters were measured, and the average crystallite sizes from the Scherrer equation were estimated using the basal reflection plane (003) and non-basal line (110) full width at half maximum (FWHM), as used by Mokhtar et al. [17]. The calculated lattice parameter a, which is mainly related to the composition of the cation, was equal to 3.0463 nm. The measured cell parameter c, which is directly linked to the distance between the interlayers, was equivalent to 22.93 nm. The size of crystallite attained was 22 nm. The extended dimension in the c directions was primarily due to the relatively large distance between meixnerite cationic sheets.

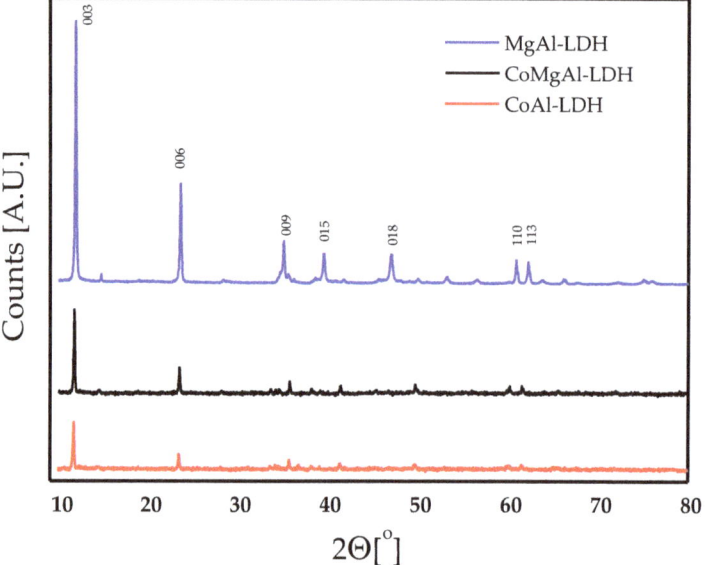

Figure 3. X-ray diffraction (XRD) patterns of all the investigated solid samples.

The CoMgAl-LDH system showed reflections typical of an LDH structure without the appearance of any phase impurities. The stacking of the double hydroxide layers in the BC–CA–AB–BC stacking caused a rhombohedral hydrotalcite-like (meixnerite) structure to form. The relatively sharp XRD reflections for the CoMgAl-LDH catalyst indicate the bigger crystal domain size compared to the MgAl-LDH catalyst. Scherrer analysis of the LDH basal reflection at around $2\theta = 12°$ indicates the apparent crystallite sizes in the c-direction (stacking direction of the LDH sheets) of around 45 nm for the CoMgAl-LDH (Figure 3). CoAl-LDH's XRD pattern indicates the development of a layered double hydroxide phase in three sections: (a) low-angle area (<30° 2θ) covering basal reflections (00l); (b,c) far-angle areas (>55° 2θ) comprising the (110) and (113) reflections representative of metal hydroxide layers. The high crystallinity of CoAl-LDH indicates the increase in the crystallite size upon isomorphic substitution of Mg by Co in the cationic sheet. Scherrer analysis of the LDH basal reflection at around $2\theta = 12°$ indicates apparent crystal sizes in the c-direction (stacking direction of the LDH sheets) of around 55 nm.

2.4. Scanning Electron Microscopy and Energy-Dispersive X-ray Spectroscopy (SEM–EDS)

Figure 4 shows SEM images of the as-synthesized (A) MgAl-LDH, (B) CoMgAl-LDH, and (C) CoAl-LDH samples. The morphology of the synthesized layered double hydroxides shows hexagonal platelets with uniform thickness of about 18 nm. The obtained results complement the crystallite size obtained from XRD analysis for pure Mg-Al LDH. The addition of cobalt to the cationic sheet with an isomorphic substitution of magnesium stabilized the hexagonal platelets with a pronounced increase in the platelet thickness (~25 nm) (Figure 4B,C). The hydrothermal preparation method [40] was successful in synthesizing non-sized layered double hydroxide materials of expected high catalytic activity.

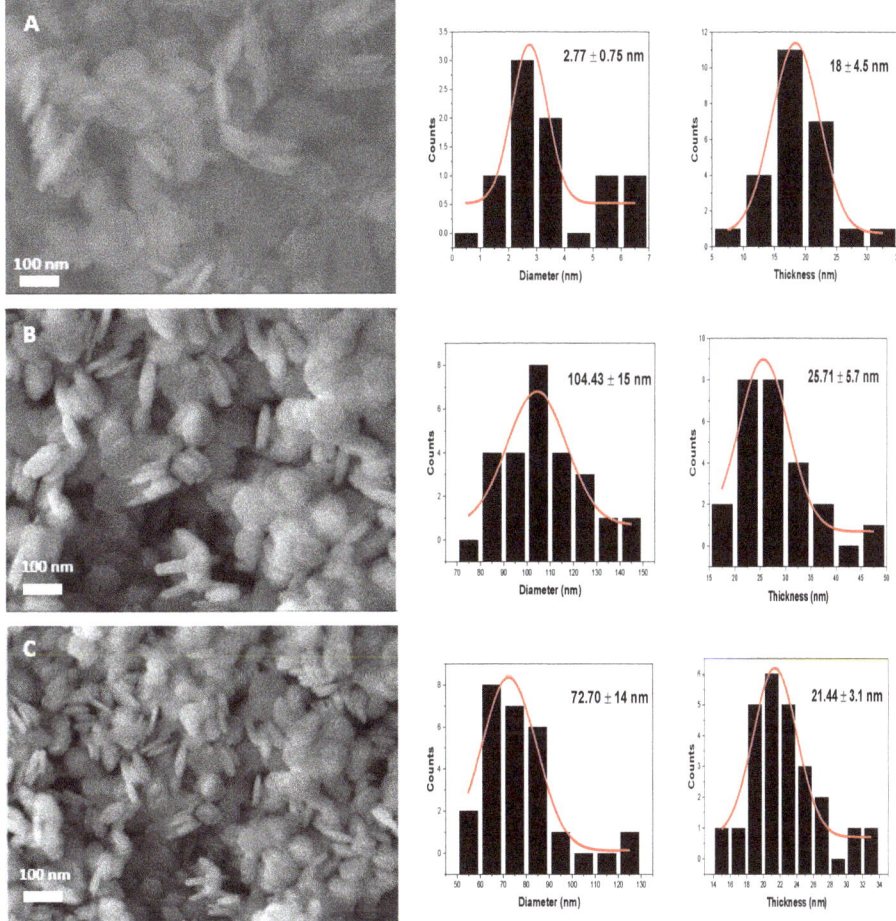

Figure 4. SEM images and particle size distribution analysis for all the investigated samples: (**A**) MgAl layer double hydroxide (LDH), (**B**) CoMgAl-LDH, and (**C**) CoAl-LDH samples. Energy-dispersive spectroscopy (EDS) analyses show pure materials free from any sodium ions with a stoichiometric ratio matching the calculated molar ratios of different cations. EDS analyses confirm the presence of cobalt cations in the chemical structure of the meixnerite phase, detected by XRD analysis. The absence of any diffraction peaks for cobalt compounds in the XRD analysis confirm the uniform isomorphic substitution of magnesium ions by cobalt ions in the cationic sheets (Table 1).

Table 1. Data obtained from EDS analysis for all the investigated samples.

Sample	OK		MgK		AlK		CoK	
	wt.%	at.%	wt.%	at.%	wt.%	at.%	wt.%	at.%
MgAl-LDH	67.47	74.85	25.15	19.44	07.38	05.71	-	-
CoMgAl-LDH	52.95	72.09	10.45	09.36	11.37	09.18	25.23	09.36
CoAl-LDH	50.33	74.64	-	-	11.07	09.74	38.06	15.61

2.5. N₂ Physisorption

Experiments on N_2 adsorption/desorption were performed to analyze the texture characteristics of various investigated LDH products. N_2 isotherms for the adsorption and desorption of examined samples are shown in Figure 5. All the materials displayed a type II isotherm according to International Union for Pure and Applied Chemistry(IUPAC) classification [41] with an H3 hysteresis loop for the entire studied sample except for CoAl-LDH, which showed an H4 hysteresis type. The samples also showed type B behavior, which is typical for clay materials [42]. The BET surface areas, derived from the isotherms for MgAl-LDH, CoMgAl-LDH, and CoAl-LDH samples were 73, 54, and 64 m^2/g, respectively. The LDHs with a high anion population in the water layer between sheets, as shown from FTIR spectra for the cobalt-containing LDH samples, showed particle sizes of 24 and 21 nm, for CoMgAl-LDH and CoAl-LDH, respectively. The relatively small surface area values could be attributed to the relatively large particle size and higher anion population, which partially prevented the penetration of N_2 into the pores [43]. The distribution of pores in LDHs is known to be regulated by crystallite size and crystallite packing arrangement [44]. A relatively narrow distribution of pores was observed for MgAl-LDH and CoMgAl-LDH samples, whereas a broader distribution of pores could be observed for the CoAl-LDH sample. This was indeed reflected in the pronounced change in the surface area and the pore nature upon the isomorphic substitution of Mg ions by Co ions in the cationic sheet.

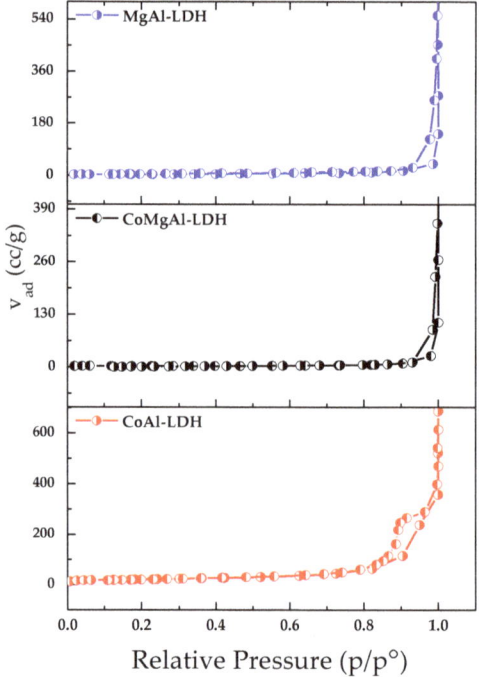

Figure 5. N_2 adsorption/desorption isotherms for different investigated materials.

2.6. Temperature-Programmed Desorption (TPD-CO_2)

Michael addition reactions are usually catalyzed using basic catalysts [15]. In the present study, we selected layered double hydroxide-based catalysts as they are well-known basic solid catalysts [39]. The combination of cobalt active species and basicity of the hydrotalcite-like structure could improve the catalytic efficiency of the catalysts toward Michael addition reactions. Therefore, CO_2-TPD was applied to study the basic nature of the synthesized samples. Figure 6 shows the CO_2-TPD profile signals for MgAl-LDH, CoMgAl-LDH, and CoAl-LDH solids. As these catalysts are carbonate-free in the interlayer gallery, the desorption CO_2 peaks are directly related to the adsorbed CO_2 species to the solid surface. It is shown from this figure that MgAl-LDH adsorbed more CO_2 species than the other two samples. The meixnerite phase of MgAl-LDH had terminal hydroxyl groups and more Brønsted basic sites. Therefore, the desorption temperature started at a relatively low temperature (~300 °C). Moreover, the two overlapping peaks with maxima at 330 and 400 °C were assigned as the Brønsted basic sites. The isomorphic partial substitution of cobalt species to the magnesium cations in the cationic sheet did not affect the basic nature of the solid catalyst (CoMgAl-LDH). There was one single sharp peak at 330 °C for CoAl-LDH, as a result of complete substitution of magnesium by cobalt ions in the cationic sheet.

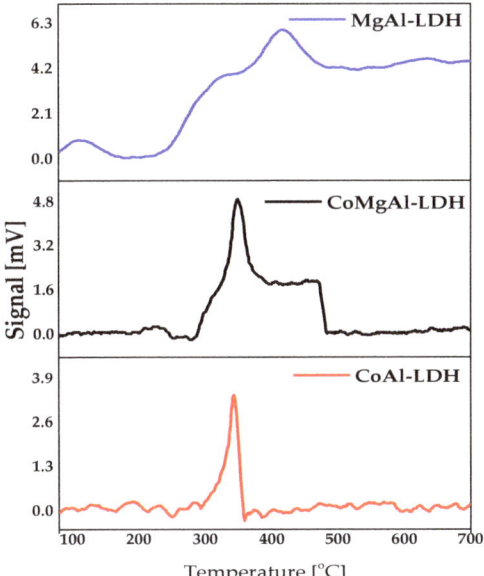

Figure 6. CO_2-TPD profile for all the investigated solid materials.

2.7. Catalytic Test Reaction

The reaction between ethyl *p*-bromobenzoate (1) and ethyl acrylate (2) was investigated through modification of different parameters. The respective conjugate adduct was synthesized in the presence of various synthesized catalysts (Scheme 1).

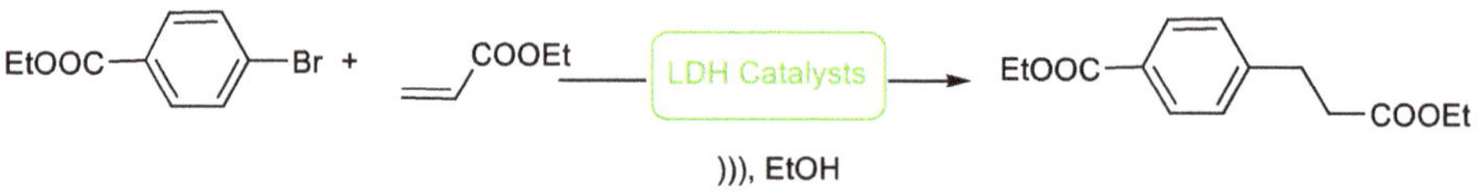

Scheme 1. Optimizing the reaction conditions.

To improve the reaction conditions, three separate LDH catalysts (0.35 g) were tested using ultrasonic irradiation to react with an equimolar mixture of ethyl *p*-bromobenzoate (1) and ethyl acrylate (2) (Scheme 1). Such solid catalysts, MgAl-LDH, CoMgAl-LDH, and CoAl-LDH, were chosen because they have low environmental impact and are easy to recycle (Table 2).

Table 2. Synthesis of compound 3 using different catalysts under ultrasonic irradiation.

Entry	Catalyst	Conventional Heating		Ultrasonic Irradiation	
		Time (h)	Yield (%)	Time (min)	Yield (%)
1	Catalyst-free	12	-	120	-
2	MgAl-LDH	12	-	120	-
3	CoMgAl-LDH	6	71	45	85
4	CoAl-LDH	8	58	60	72

In order to find the beneficial effect of ultrasound irradiation on the above reaction, we performed the above reaction on the same scale under conventional heating.

From the findings cited in Table 2, it is clear that no substance was formed in the absence of a catalyst under ultrasonic irradiation or traditional conditions, even after 2 h or 12 h, respectively, (entry 1). In addition, the best yield 85% of the desired product 3 was reached using the CoMgAl-LDH catalyst for 45 min. Furthermore, no product was formed using MgAl-LDH under conventional heating or ultrasonic irradiations (entry 2). Introducing cobalt into the cationic sheet of the layered double hydroxide improved the catalytic efficacy of those catalytic materials under classical or ultrasonic irradiation. Utilizing CoMgAl-LDH catalyst resulted in the formation of high yield of the desired product in a short reaction time (entry 3). In contrast, the complete isomorphic substitution of Mg ions by Co ions in the cationic sheet, i.e., in CoAl-LDH, reduced the desired product yield (72%) and increased the reaction time (60 min). Obviously, in the presence of CoMgAl-LDH catalysts, ultrasonic irradiation had a beneficial effect on the formation of the Michael adduct 3 compared to conventional heating, which showed a longer reaction time and lower product yield. To improve the reaction conditions using the best CoMgAl-LDH catalyst, a series of experiments were performed in variable catalyst amounts, and the progress of the reaction was monitored using thin-layer chromatography (TLC) for a typical reaction (Scheme 1) under ultrasonic irradiation. The results obtained from the catalytic test reaction are cited in Table 3.

Table 3. Optimization of reaction condition for synthesis of 3 utilizing CoMgAl-LDH catalyst under ultrasonic irradiation.

Entry	Catalyst Weight (g)	Solvent	Time (min)	Yield (%)
1	0.3	EtOH	45	82
2	0.35	EtOH	45	85
3	0.4	EtOH	45	85
4	0.35	DMF	30	76
5	0.35	Butanol	15	81
6	0.35	Methanol	15	63
7	0.35	Propanol	25	76

Initially, different quantities of the CoMgAl-LDH catalyst were evaluated under the same reaction situations for optimization of the catalyst mass, where the greatest yield was found using 0.35 g of the catalyst with the process running for 45 min (Table 3, entry 2). Moving forward to the investigation of the effect of solvent on this catalytic system, four reactions were performed with DMF, butanol, methanol, and propanol, in addition to ethanol, with the best catalyst amount (entries 2, 4–7). The best solvent for this reaction protocol proved to be EtOH (entry 2), in which the desired product **3** was isolated in 85% yield.

The structure of product **3** was confirmed according to elemental analysis and spectral data. The ^{1}H-NMR spectrum exhibited two doublets signals at 7.91 and 7.32 due to four aromatic protons, two quartet signals due to two methylene groups of esters at 4.32 and 4.12, two triplet signals at 3.11 and 2.65 due to two adjacent methylene groups, and two triplets at 1.39 and 1.29 due to two methyl groups of esters. This spectroscopic data of the reaction product and the satisfactory elemental analysis supported the structure ethyl 4-(3-ethoxy-3-oxopropyl)benzoate **3**.

Under optimized reaction conditions, the reusability of the CoMgAl-LDH catalyst was tested for several reaction cycles for synthesis of compound **3** (Figure 7), where the catalyst was washed with hot ethanol and dried under vacuum after completion of the reaction. Using the same reaction conditions, the recuperated catalyst was reused five times. Figure 6 shows that, even after being used up to five times, the regenerated catalyst conducted the reactions effectively under the same conditions. The slight decay observed at the fourth and fifth repeats in the catalytic activity of the CoMgAl-LDH catalyst could be attributed to the catalyst's weight loss.

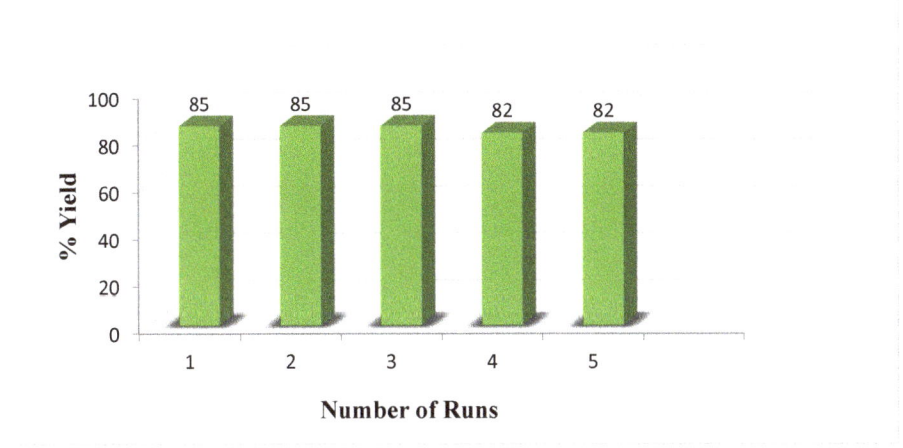

Figure 7. Re-usability of CoMgAl-LDH catalyst for formation of compound **3**.

It is obvious, in the light of the above findings, that the addition of cobalt to MgAl-LDH had a key rule for this reaction. The mechanism for this reaction was, therefore, postulated, in which low-valent cobalt species are most likely to develop with an aryl halide via oxidative addition to form an aryl-cobalt intermediate that could react via nucleophilic attack in a six-member transition state [A] with electron-deficient olefin, leading to an intermediate cobalt enolate [B] [45] (Scheme 2). The latter species reacts with the basic catalytic site to restore the expected 1,4-adduct release of the Co^{II} species to the catalyst.

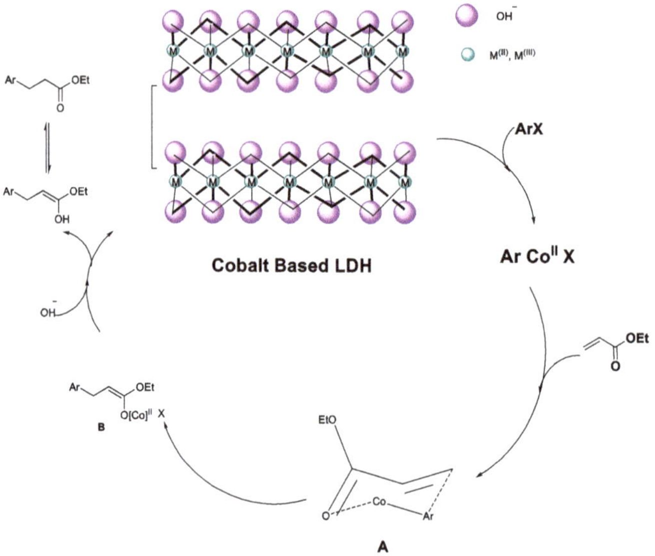

Scheme 2. Suggested mechanism for the conjugate addition reaction.

Noteworthy, the higher catalytic efficacy and the accountable yield of the anticipated products were due to different parameters in a short reaction time: (i) the physicochemical characteristics of the catalyst, i.e., their chemical, structural, permeable, and basic characteristics; (ii) reaction conditions, i.e., ultrasonic irradiation. Introducing cobalt to the cationic sheet improved the catalytic efficiency of the cobalt0containing catalysts. However, this superior activity was reduced upon complete substitution of magnesium cations by cobalt cations in the cationic sheet, i.e., CoAl-LDH catalyst. These results revealed the synergistic effect between cobalt and the basic nature of meixnerite phase in tuning the catalytic efficiency of the Co-containing layered double hydroxide catalyst in conjugate addition of functionalized aryl compounds into Michael acceptors. The results of physicochemical analyses and crystallography clearly display that the Co-treatment samples maintained a crystalline structure as the isomorphic substitution of Mg^{2+} ions by Co^{2+} ions occurred. The partial substitution of the basic Mg^{2+} cations in the cationic sheet by other cations such as Co^{2+} cations affected the thermal behavior of the CoMgAl-LDH catalyst, as seen from TGA data. The synergistic effect between the Co^{2+} ions and the basic component played a role in the catalytic activity. Moreover, ultrasound irradiation had a beneficial effect on the reaction in the presence of the CoMgAl-LDH catalyst, lowering the reaction time and increasing the product yield. This obvious effect can be reasonably interpreted in terms of the physical phenomenon known as acoustic cavitation (at the solid/liquid interface) [46,47]. It was suggested that the micro-jet impact and shockwave damage on the surface of the solid (catalyst), along with the shock wave associated with the cavitation collapse, would cause localized deformation and surface erosion, which would increase the size of the possible reaction area [47].

3. Materials and Methods

3.1. Materials

Chemical reagents used in this work included sodium hydroxide solution (Fluka AG, Chemische Fabrik CH-9470 Buchs, Switzerland), aluminum nitrate nonahydrate (PRS, Panreac Química SLU, Barcelona, Spain), magnesium nitrate hexahydrate, cobalt(II) nitrate hexahydrate, and sodium carbonate (BDH Chemical Ltd., Dubai, UAE). All organic solvents were purchased from (Fluka AG, Chemische Fabrik CH-9470 Buchs, Switzerland) and used as received unless otherwise stated.

3.2. Preparation of Catalysts

All catalytic samples were prepared using a coprecipitation method followed by post-hydrothermal treatment under autogenous pressure. Three different solutions containing nitrate salts Mg/Al ratio of 3:1, Co/Mg/Al ratio of 1.5:1.5:1, and Co/Al ratio of 3:1 were dissolved in deionized water (250 mL) to give a solution concentration of 1 M, and this solution was denoted as solution A. Another solution denoted solution B was prepared using NaOH and Na_2CO_3 dissolved in deionized water (250 mL) to give a solution concentration of 1 M. Solutions A and B were added simultaneously to form the mixed base solution (as precipitating agent) at 30 °C and pH = 11 under N_2 atmosphere. All precipitates were aged overnight prior to transferring them to an autoclave and hydrothermally treating them at 170 °C for 24 h. The cake was centrifuged and then dried at 100 °C for 16 h. The synthesized solids were named MgAl-LDH, CoMgAl-LDH, and CoAl-LDH.

3.3. Measurements and Characterization

FTIR spectrometer (PerkinElmer) was used to perform Fourier-infrared transfer (FTIR) spectroscopy in transmission mode. The measurements were carried out in the wavenumber range 4000–400 cm^{-1} with 4-cm^{-1} resolution.

Thermogravimetric analysis (TGA) runs on a TA TGA-Q500 model were carried out. In each cycle, 6–10 mg was put in the TGA pan and heated to 800 °C at 10 °C·min^{-1} in N_2 atmosphere (40 mL·min^{-1} flow rate).

Measurements for powder X-ray diffraction (PXRD) were carried out at room temperature using a Bruker diffractometer (Bruker D8 advance target, Karlsruhe, Germany). The patterns were collected using Cu Kα1 radiation and a 40 kV and 40 mA monochromator (λ = 1.5405 Å), with 2θ varying between 2° and 80°. X'Pert HighScore Plus software was used to analyze the spectra. The crystallite size of Ni-LDH was calculated using the Scherrer equation.

$$d\ (nm) = \frac{0.9 \times \lambda}{B \times Cos(\theta)},$$

where *d* is the mean Ni-LDH crystallite size under investigation, λ is wavelength of the X-ray beam used, 0.9 is the Scherrer constant, B is the full width at half maximum (FWHM) of the diffraction peak, and θ is the diffraction angle.

High-resolution field-emission (FEG quanta 250) scanning electron microscopy (SEM) allowed the chemistry of materials to be investigated on the sub-micron scale.

The texture properties of the prepared samples were calculated from nitrogen adsorption/desorption isothermal measurements at 77 K using the automated gas sorption device NOVA3200e (Quantachrome, Boynton Beach, FL, USA) model. Every sample was degassed at 100 °C for 12 h prior to the measurement. The actual surface area, S_{BET}, was determined using the Brunauer–Emmett–Teller (BET) equation [48]. The distribution of the pore size over the mesopore range was generated by Barrett–Joyner–Halenda (BJH) [49] analysis of the desorption branches, and mean pore size values were estimated.

Measurements of carbon dioxide temperature-programmed desorption (TPD-CO2) were carried out using CHEMBET 3000 (Quantachrome, FL, USA). A sample exhaust at 100 °C (1 h) was performed when helium was transferred to physical water detachment. The sample's CO_2 saturation then occurred at 120 °C. The temperature-programmed desorption was easily achieved by ramping the adsorbent temperature to 800 °C at 10 °C/min.

Thin-layer chromatography (TLC) was performed on precoated Merck 60 GF254 silica gel plates with a fluorescent indicator and UV light detection at 254 and 360 nm. The melting points of the organic materials were determined on a Stuart melting point apparatus and were not corrected.

IR spectra were recorded on the Smart iTR, an ultra-high-performance, flexible attenuated total reflectance (ATR) sampling system on the Nicolet iS10 FT-IR spectrometer.

The NMR spectra were registered on a Bruker Avance III 400 spectrometer (9.4 T, 400.13 MHz for ^{1}H and 100.62 MHz for ^{13}C) with a 5-mm BBFO probe at 298 K. Chemical changes (δ in ppm) are given in relation to the internal solvent, DMSO; 2.50 was used as an external norm for ^{1}H and 39.50 was used for [13]C. Mass spectra were recorded on a single-quadrupole GC–MS Thermo ISQ. Elemental analyses were performed for C, H, N, and S on an EA3000 Series EuroVector instrument. Ultrasonic irradiation experiments were carried out using an ultrasonic cleaning bath Elma P30H.

3.4. Typical Procedure for Conjugate Addition Reaction

3.4.1. Conventional Condition

To a stirred solution of 25 mL of ethanol, an equimolar amount of ethyl *p*-bromobenzoate (**1**) and ethyl acrylate (**2**) was added, followed by 0.35 g of LDH catalyst 80 °C. The reaction progress was monitored by thin-layer chromatography until the aryl halide was consumed. The mixture was filtrated to separate the catalyst, and the filtrate was evaporated to get a yellow oil which was purified by column chromatography on a silica gel (pentane/diethylether) to afford the conjugate adduct **3** in pure form.

3.4.2. Ultrasonic Condition

The reaction here was done on the same scale as described in above conventional condition, whereby a round-bottom flask was placed in the cleaning bath.

The spectroscopic data of compound **3** are as follows:

IR (ν_{max}, cm^{-1}) 2980, 1713, 1611, 1272; ^{1}H-NMR (400 MHz, DMSO) δ 7.98 (d, J = 8.3 Hz, 2H), 7.28 (d, J = 8.3 Hz, 2H), 4.37 (q, J = 7.1 Hz, 2H), 4.13 (q, J = 7.1 Hz, 2H), 3.01 (t, J = 7.7 Hz, 2H), 2.65 (t, J = 7.7 Hz, 2H), 1.39 (t, J = 7.1 Hz, 3H), 1.24 (t, J = 7.1 Hz, 3H); ^{13}C-NMR (100 MHz, DMSO) δ 172.6, 166.6, 145.9, 129.8 (2C), 128.6, 128.3 (2C), 60.9, 60.6, 35.4, 30.9, 14.3, 14.2; ESI-MS *m/z* (relative intensity) 250 (M$^+$).

4. Conclusions

Coprecipitation under inert atmosphere followed by hydrothermal treatment under autogenous pressure was a successful tool for the preparation of hydrotalcite-like materials (meixnerite). Partial isomorphic substitution of magnesium by cobalt in the cationic sheet of a layered double hydroxide resulted in the formation of a catalyst for the coupling of ethyl *p*-bromobenzoate with ethyl acrylate. This catalytic method uses a simple CoMgAl-LDH, which tends to be an extremely suitable catalyst for a wide variety of substrates ranging from aryl halides to many reactive groups. The reaction affords excellent yields in a rapid reaction period under ultrasound irradiation. The superior catalytic efficiency was allocated to the synergistic effect between Co ions in the cationic sheet and the basicity of the layered materials.

Author Contributions: M.M. conceptualized and designed the experiments; N.S.A. performed the experiments; F.M.A.-N. analyzed the data; T.S.S. contributed reagents/materials/analysis tools; M.M. wrote the paper. All authors edited and revised the manuscript before submission. All authors have read and agreed to the published version of the manuscript.

Funding: This research received no external funding.

Acknowledgments: The authors acknowledge Huda Sherbini and Ghalia Alzahrani for their contribution to TPD measurements. The group of surface chemistry and catalytic studies in the Chemistry Department, Faculty of Science, KAU, is acknowledged for the facilities provided to carry out this research.

Conflicts of Interest: The authors declare no conflicts of interest.

References

1. Perlmutter, P. *Conjugate Addition Reactions in Organic Synthesis*, 1st ed.; Pergamon Press: Oxford, UK, 1992.

2. Muriel, A.; Gosmini, C.; Périchon, J. CoBr$_2$(Bpy): An Efficient Catalyst for the Direct Conjugate Addition of Aryl Halides or Triflates onto Activated Olefins. *J. Org. Chem.* **2006**, *71*, 6130–6134.

3. Tucker, C.E.; Majid, T.N.; Knochel, P. Preparation of highly functionalized magnesium, zinc, and copper aryl and alkenyl organometallics via the corresponding organolithiums. *J. Am. Chem. Soc.* **1992**, *114*, 3983–3985. [CrossRef]

4. Varchi, G.; Ricci, A.; Cahiez, G.; Knochel, P. Copper Catalyzed Conjugate Addition of Highly Functionalized Arylmagnesium Compounds to Enones. *Tetrahedron* **2000**, *56*, 2727–2731. [CrossRef]

5. Klement, I.; Stadtmüller, H.; Knochel, P.; Cahiez, G. Preparation and reactivity of functionalized aryl and alkenylmanganese halides. *Tetrahedron Lett.* **1997**, *38*, 1927–1930. [CrossRef]

6. Lu, X.; Lin, S. Pd(II)-Bipyridine Catalyzed Conjugate Addition of Arylboronic Acid to α,β-Unsaturated Carbonyl Compounds. *J. Org. Chem.* **2005**, *70*, 9651–9653. [CrossRef] [PubMed]

7. Liu, R.; Yang, Z.; Ni, Y.; Song, K.; Shen, K.; Lin, S.; Pan, Q. Pd(II)/Bipyridine-Catalyzed Conjugate Addition of Arylboronic Acids to α,β-Unsaturated Carboxylic Acids. Synthesis of β-Quaternary Carbons Substituted Carboxylic Acids. *J. Org. Chem.* **2017**, *82*, 8023–8030. [CrossRef] [PubMed]

8. Vautravers, N.R.; Breit, B. Rhodium(I)-Catalyzed 1,4-Addition of Arylboronic Acids to Acrylic Acid in Water: One-Step Preparation of 3-Arylpropionic Acids. *Synlett* **2011**, *17*, 2517–2520. [CrossRef]

9. Kamikawa, K.; Tseng, Y.-Y.; Jian, J.-H.; Takahashi, T.; Ogasawara, M. Planar-Chiral Phosphine-Olefin Ligands Exploiting a (Cyclopentadienyl)manganese(I) Scaffold to Achieve High Robustness and High Enantioselectivity. *J. Am. Chem. Soc.* **2017**, *139*, 1545–1553. [CrossRef] [PubMed]

10. Wu, C.; Yue, G.; Nielsen, C.D.-T.; Xu, K.; Hirao, H.; Zhou, J. Asymmetric Conjugate Addition of Organoboron Reagents to Common Enones Using Copper Catalysts. *J. Am. Chem. Soc.* **2016**, *138*, 742–745. [CrossRef] [PubMed]

11. Lerebours, R.; Wolf, C. Palladium(II)-Catalyzed Conjugate Addition of Arylsiloxanes in Water. *Org. Lett.* **2007**, *9*, 2737–2740. [CrossRef] [PubMed]

12. Westermann, J.; Imberg, U.; Nguyen, A.T.; Nickisch, K. Nickel-Catalysed 1,4-Addition of Aryl Groups to Enones Using Aryldialkylaluminum Compounds. *Eur. J. Inorg. Chem.* **1998**, *295*, 295–298. [CrossRef]

13. Subburaj, K.; Montgomry, J. A New Catalytic Conjugate Addition/Aldol Strategy That Avoids Preformed Metalated Nucleophiles. *J. Am. Chem. Soc.* **2003**, *125*, 11210–11211. [CrossRef] [PubMed]

14. Gomes, P.; Gosmini, C.; Nédélec, J.Y.; Périchon, J. Electrochemical vinylation of aryl and vinyl halides with acrylate esters catalyzed by cobalt bromide. *Tetrahedron Lett.* **2002**, *43*, 5901–5903. [CrossRef]

15. Amatore, M.; Gosmini, C.; Périchon, J. Process for Forming Carbon-Carbon Bonds by a Coupling Reaction between Unsaturated Compounds in the Presence of a Cobalt-Based Catalyst. Centre National de la Recherche Scientifique CNRS Rhodia Chimie SAS. Patent FR2865203, 22 July 2005.

16. Ahmed, N.S.; Menzel, R.; Wang, Y.; Garcia-Gallastegui, A.; Bawaked, S.M.; Obaid, A.Y.; Basahel, S.N.; Mokhtar, M. Graphene-oxide-supported CuAl and CoAl layered double hydroxides as enhanced catalysts for carbon-carbon coupling via Ullmann reaction. *J. Solid State Chem.* **2017**, *246*, 130–137. [CrossRef]

17. Mokhtar, M.; Inayat, A.; Ofili, J.; Schwieger, W. Thermal decomposition, gas phase hydration and liquid phase reconstruction in the system Mg/Al hydrotalcite/mixed oxide: A comparative study. *Appl. Clay Sci.* **2010**, *50*, 176–181. [CrossRef]

18. Mokhtar, M.; Saleh, T.S.; Ahmed, N.S.; Al-Thabaiti, S.A.; Al-Shareef, R.A. An eco-friendly N-sulfonylation of amines using stable and reusable Zn–Al–hydrotalcite solid base catalyst under ultrasound irradiation. *Ultrason. Sonochem.* **2011**, *18*, 172–176. [CrossRef] [PubMed]

19. Mokhtar, M.; Saleh, T.S.; Basahel, S.N. Mg–Al hydrotalcites as efficient catalysts for aza-Michael addition reaction: A green protocol. *J. Mol. Catal. A Chem.* **2012**, *353–354*, 122–131. [CrossRef]

20. Narasimharao, K.; Al-Sabban, E.; Saleh, S.T.; Garcia-Gallastegui, A.; Sanfiz, A.C.; Basahel, S.; Al-Thabaiti, S.; Alyoubi, A.; Obaid, A.; Mokhtar, M. Microwave assisted efficient protocol for the classic Ullmann homocoupling reaction using Cu–Mg–Al hydrotalcite catalysts. *J. Mol. Catal. A Chem.* **2013**, *379*, 152–162. [CrossRef]

21. Abdellattif, M.; Mokhtar, H.M. MgAl-Layered Double Hydroxide Solid Base Catalysts for Henry Reaction: A Green Protocol. *Catalysts* **2018**, *8*, 133. [CrossRef]

22. Alzhrani, G.; Ahmed, N.S.; Aazam, E.S.; Saleh, T.S.; Mokhtar, M. Novel Efficient Pd-Free Ni-Layered Double Hydroxide Catalysts for a Suzuki C–C Coupling Reaction. *Chem. Select* **2019**, *4*, 7904–7911. [CrossRef]

23. Abbasi, A.R.; Akhbari, K.; Morsali, A. Dense coating of surface mounted CuBTC metal–organic framework nanostructures on silk fibers, prepared by layer-by-layer method under ultrasound irradiation with antibacterial activity. *Ultrason. Sonochem.* **2012**, *19*, 846–852. [CrossRef] [PubMed]

24. Safarifard, V.; Morsali, A. Applications of ultrasound to the synthesis of nanoscale metal–organic coordination polymers. *Coord. Chem. Rev.* **2015**, *292*, 1–14. [CrossRef]

25. Li, Z.-Q.; Qiu, L.-G.; Wang, W.; Xu, T.; Wu, Y.; Jiang, X. Fabrication of nanosheets of a fluorescent metal–organic framework [Zn(BDC)(H$_2$O)]$_n$ (BDC= 1,4-benzenedicarboxylate): Ultrasonic synthesis and sensing of ethylamine. *Inorg. Chem. Commun.* **2008**, *11*, 1375–1377. [CrossRef]

26. Abdollahi, N.; Masoomi, M.Y.; Morsali, A.; Junk, P.C.; Wang, J. Sonochemical synthesis and structural characterization of a new Zn(II) nanoplate metal–organic framework with removal efficiency of Sudan red and Congo red. *Ultrason. Sonochem.* **2018**, *45*, 50–56. [CrossRef] [PubMed]

27. Eghbali-Arani, M.; Sobhani-Nasab, A.; Rahimi-Nasrabadi, M.; Ahmadi, F.; Pourmasoud, S. Ultrasound-assisted synthesis of YbVO$_4$ nanostructure and YbVO$_4$/CuWO$_4$ nanocomposites for 20 enhanced photocatalytic degradation of organic dyes under visible light. *Ultrason. Sonochem.* **2018**, *43*, 120–135. [CrossRef]

28. Choudhary, B.M.; Lakshmi, K.M.; Neeraja, V.; Koteswara, R.K.; Figueras, F.; Delmotte, L. Layered double hydroxide fluoride: A novel solid base catalyst for C-C bond formation. *Green Chem.* **2001**, *3*, 257–260. [CrossRef]

29. Kantam, M.L.; Ravindra, A.; Reddy, C.V.; Sreedhar, B.; Choudary, B.M. Layered Double Hydroxides-Supported Diisopropylamide: Synthesis, Characterization and Application in Organic Reactions. *Adv. Synth. Catal.* **2006**, *348*, 569–578. [CrossRef]

30. Varga, G.; Kozma, V.; Kolcsár, V.J.; Kukovecz, Á.; Kónya, Z.; Sipos, P.; Szőllősi, G. β-Isocupreidinate-CaAl-layered double hydroxide composites—Heterogenized catalysts for asymmetric Michael addition. *Mol. Catal.* **2019**, *482*, 110675. [CrossRef]

31. El-bendary, M.M.; Saleh, T.S.; Al-Bogami, A.S. Ultrasound Assisted High-Throughput Synthesis of 1,2,3-Triazoles Libraries: A New Strategy for "Click" Copper-Catalyzed Azide-Alkyne Cycloaddition Using Copper(I/II) as a Catalyst. *Catal. Lett.* **2018**, *148*, 3797–3810. [CrossRef]

32. Al-Bogami, A.S.; Saleh, T.S.; Moussa, T.A.A. Green synthesis, antimicrobial activity and cytotoxicity of novel fused pyrimidine derivatives possessing a trifluoromethyl moiety. *Chemistryselect* **2018**, *3*, 8306–8311. [CrossRef]

33. Saleh, T.S.; Narasimharao, K.; Ahmed, N.S.; Basahel, S.N.; Al-Thabaiti, S.A.; Mokhtar, M. Mg–Al hydrotalcite as an efficient catalyst for microwave assisted regioselective 1,3-dipolar cycloaddition of nitrilimines with the enaminone derivatives: A green protocol. *J. Mol. Catal. A Chem.* **2013**, *367*, 12–22. [CrossRef]

34. Mokhtar, M.; Saleh, T.S.; Ahmed, N.S.; Al-Bogami, A.S. A Green Mechanochemical One-Pot Three-Component Domino Reaction Synthesis of Polysubstituted Azoloazines Containing Benzofuran Moiety: Cytotoxic Activity against HePG2 Cell Lines. *Polycycl. Aromat. Compd.* **2018**, 1–15.

35. Béres, A.; Palinko, I.; Kiricsi, I.; Nagy, J.B.; Kiyozumi, Y.; Mizukami, F. Layered double hydroxides and their pillared derivatives—Materials for solid base catalysis; synthesis and characterization. *Appl. Catal. A* **1999**, *182*, 237–247. [CrossRef]

36. Velu, S.; Shah, N.; Jyothi, T.M.; Sivasanker, S. Effect of manganese substitution on the physicochemical properties and catalytic toluene oxidation activities of Mg–Al layered double hydroxides. *Microporous Mesoporous Mater.* **1999**, *33*, 61–75.

37. Di Cosimo, J.I.; Díez, V.K.; Xu, M.; Iglesia, E.; Apesteguía, C.R. Structure and surface and catalytic properties of Mg–Al basic oxides. *J. Catal.* **1998**, *178*, 499–510. [CrossRef]

38. Tichit, D.; Naciri Bennani, M.; Figueras, F.; Ruiz, J.R. Decomposition processes and characterization of the surface basicity of Cl- and CO$_3^{2-}$ hydrotalcites. *Langmuir* **1998**, *14*, 2086–2091. [CrossRef]

39. Rocha, J.; del Arco, M.; Rives, V.; Ulibarri, M.A. Reconstruction of layered double hydroxides from calcined precursors: A powder XRD and ^{27}Al MAS NMR study. *J. Mater. Chem.* **1999**, *9*, 2499–2503. [CrossRef]

40. Liao, H.; Jia, Y.; Wang, L.; Yin, Q.; Han, J.; Sun, X.; Wei, M. Size Effect of Layered Double Hydroxide Platelets on the Crystallization Behavior of Isotactic Polypropylene. *ACS Omega* **2017**, *2*, 4253–4260.

41. Sing, K.S.W.; Everett, D.H.; Haul, R.A.W.; Moscou, L.; Pierotti, R.A.; Rouquérol, J.; Siemieniewska, T. Reporting physisorption data for gas/solid systems with special reference to the determination of surface area and porosity. *Pure Appl. Chem.* **1985**, *57*, 603–619. [CrossRef]

42. Bergada, O.; Vicente, I.; Salagre, P.; Cesteros, Y.; Medina, F.; Sueiras, J.E. Microwave effect during aging on the porosity and basic properties of hydrotalcites. *Microporous Mesoporous Mater.* **2007**, *101*, 363–373. [CrossRef]
43. Lowell, S.; Shields, J.E.; Thomas, M.A.; Thommes, M. *Characterization of Porous Solids and Powders: Surface Area, Pore Size and Density*; Kluwer Academic Publishers: Dordrecht, The Netherlands, 2004.
44. Prinetto, F.; Ghiotti, G.; Graffin, P.; Tichit, D. Synthesis and characterization of sol-gel Mg/Al and Ni/Al layered double hydroxides and comparison with co-precipitated samples. *Microporous Mesoporous Mater.* **2000**, *39*, 229–247. [CrossRef]
45. Amatore, M.; Gosmini, C.; Pe´richon, J. Cobalt-Catalyzed Vinylation of Functionalized Aryl Halides with Vinyl Acetates. *J. Eur. J. Org. Chem.* **2005**, *2005*, 989–992. [CrossRef]
46. Mason, T.J.; Tiehm, A. *Advances in Sonochemistry: Ultrasound in Environmental Protection*, 1st ed.; Elsevier: Amsterdam, The Netherlands, 2001.
47. Shah, Y.T.; Pandit, A.B.; Moholkar, V.S. *Cavitation Reaction Engineering*, 1st ed.; Springer Science & Business Media, LLC: New York, NY, USA, 1999.
48. Brunauer, S.; Emmett, P.H.; Teller, E. Adsorption of Gases in Multimolecular Layers. *J. Am. Chem. Soc.* **1938**, *60*, 309–319. [CrossRef]
49. Barrett, E.P.; Joyner, L.G.; Halenda, P.P. The Determination of Pore Volume and Area Distributions in Porous Substances. I. Computations from Nitrogen Isotherms. *J. Am. Chem. Soc.* **1951**, *73*, 373–380. [CrossRef]

Article

Activity in the Photodegradation of 4-Nitrophenol of a Zn,Al Hydrotalcite-Like Solid and the Derived Alumina-Supported ZnO

Raquel Trujillano *, César Nájera and Vicente Rives *

GIR-QUESCAT, Departamento de Química Inorgánica, Universidad de Salamanca, 37008 Salamanca, Spain; cesteban@usal.es
* Correspondence: rakel@usal.es (R.T.); vrives@usal.es (V.R.)

Received: 7 May 2020; Accepted: 19 June 2020; Published: 22 June 2020

Abstract: A Zn,Al layered double hydroxide (LDH), with the hydrotalcite structure and the mixed oxide obtained upon its calcination at 650 °C, was tested in the adsorption and photocatalytic degradation of 4-Nitrophenol in aqueous solution. The Zn,Al LDH was fast and easily obtained by the coprecipitation method. Hydrothermal treatment under microwave irradiation was applied to compare the effect of the ageing treatment on the photocatalytic behavior. The efficiency of the synthetized solids was compared to that of a commercial ZnO. The ageing treatment did not improve the performance of the original samples in the degradation of 4-nitrophenol. The activity of the synthetized solids tested exceeded that observed for the reaction with commercial ZnO. The photocatalytic performance of the original non-calcined hydrotalcite is similar to that of commercial ZnO. The calcined hydrotalcite showed a better performance in the adsorption-degradation of the contaminant than ZnO, and its reusability would be possible as it recovered the hydrotalcite-like structure during the reaction.

Keywords: photocatalysis; nitrophenol degradation; Zn,Al-hydrotalcite; ZnO dispersed on alumina; reusability; layered double hydroxide; LDH

---

1. Introduction

Although the use as photocatalysts of layered systems, and the mixed oxides derived from them by thermal decomposition, has been described in the literature, they are not as widely used as TiO_2. Discoloration of methylene blue using hydrotalcite-based catalysts was studied by Abderrazek et al. [1], while Barhoum et al. [2], among others, reported the photocatalytic activity of ZnO particles in the oxidation of methanol. More recently, Zhang et al. [3] synthetized nanostructured Zn-Al layered double hydroxides and have tested their photocatalytic activity, and that of the mixed oxides formed upon their calcination under different conditions, on the photodegradation of rhodamine dye in aqueous solutions.

Hydrotalcite is a layered hydroxycarbonate belonging to the so-called layered double hydroxides (LDHs) family, sometimes identified as anionic clays because of the similarity between their structure and that of clays, but with negative layers, electrically balanced by anions intercalated in the interlayers. These solids are transformed into well dispersed mixed oxides after being calcined at temperatures at or above 500 °C. Both the original materials and the derived oxides can act as catalysts or catalyst precursors in different industrially interesting reactions [3–7]. There are multiple synthesis routes to produce these solids [8–10]; among them, the co-precipitation method is the less expensive and the simplest one, and it is easily applied to manufacture this material at a large scale [11].

On the other hand, ZnO and modified ZnO have been widely studied in photocatalysis since the band gap energy of ZnO is similar to that of TiO_2 and it is cheaper than this one; deep and broad studies about ZnO-based photocatalysts, their synthesis, reaction mechanisms and applications have been made available in recent years [12–14]. Lee et al. conclude that, in addition to its high photocatalytic activity, modification of ZnO photocatalysts is essential to improve its performance, especially in the field of organic dye degradation and, more specifically, in phenolic compounds removal [12].

There are also many papers about the use of LDHs as adsorbents for removal of phenolic contaminants [15]; the studies have pointed out the increase of the adsorption capacity of the mixed oxides when they are calcined at mild (below 700 °C) temperatures and recover the layered structure when they are suspended in a water solution containing anions (the potential contaminants in many cases), which are initially adsorbed on the solid nanoparticles to finally "rebuild" the original layered structure. This ability of hydrotalcites is called "memory effect" [16,17]. Chen et al. studied the kinetics of adsorption of phenol and 4-nitrophenol on Mg,Al-mixed oxides formed by thermal decomposition of a Mg,Al-LDH with the hydrotalcite structure [18]. These authors concluded that the adsorption mechanism involves the reconstruction by incorporating the nitrophenolate or phenolate ions in the interlayer and by adsorption on the surface.

4-nitrophenol (4-NP) is one of the most common contaminants in industrial waters, since it is used for the synthesis of drugs, fungicides, dyes, explosives, coloring agents for leather, etc., and it is also generated during formulation or degradation of some pesticides [19–22]. The aim of this work is to synthetize and characterize a Zn,Al-hydrotalcite with carbonate in the interlayer space, and to calcine it in order to obtain a Zn,Al mixed oxide to test both compounds in the photocatalytic degradation of 4-NP. Co-precipitation at constant pH is used to prepare the original layered sample [8] which, upon calcination, leads to a solid with a ZnO phase well dispersed in an amorphous Al_2O_3 phase; the synthesis procedure is the same as reported elsewhere [10]. The main interest of this work is in the study of the two possible ways of retention/degradation of these compounds, namely, adsorption and photocatalytic oxidation. The effect of microwave (MW) post-synthesis treatment of the original LDH is also studied. The performance of the materials prepared is compared to that of a commercial, unsupported ZnO.

2. Results and Discussion

2.1. Characterization of the Samples

Results from element chemical analysis for Zn and Al for both initial samples show the same content of zinc and aluminum (i.e., 50.28 and 6.45% (*w/w*), respectively). From these values, it is concluded that the composition was not affected by the microwave treatment applied to prepare sample ZnAlMW31, and taking into account the atomic masses of these cations, the real Zn/Al molar ratio is 3.17 (expected 3.00). FT-IR spectroscopy (see below) show that the only interlayer anion is carbonate, while thermogravimetric (TG) analysis data permitted to calculate the content of interlayer water, leading to the formula $[Zn_{0.76}Al_{0.24}(OH)_2](CO_3)_{0.12} \cdot 0.71H_2O$ for both samples.

The Powder X-Ray Diffraction (PXRD) diagrams (Figure 1) for both the original, MW-treated and calcined samples show intense peaks in the characteristic positions of the hydrotalcite-like structure [10], corresponding to the rhombohedral hydrotalcite phase. The Miller indexes for the most significant signals are shown in the figure. Concerning the uncalcined samples, the peaks are better defined and are sharper and more symmetric for sample ZnAl31MW, evidencing its better crystallinity and layer stacking, probably as a result of the microwave post-synthesis treatment. Very weak peaks (not labelled in the figure) recorded in both diagrams are probably due to a very small portion of ZnO or $Zn(OH)_2$, which are preferentially segregated after the MW treatment.

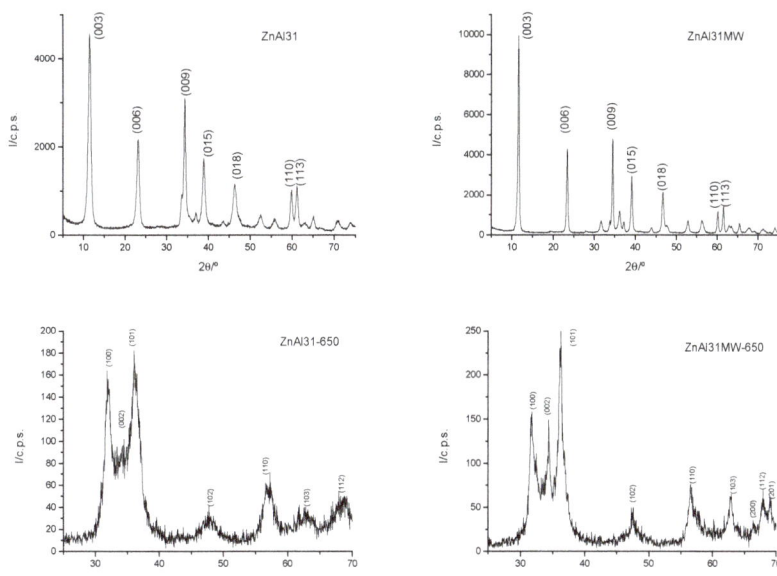

Figure 1. PXRD diagrams of: (**top**) Original samples (**left**) ZnAl31 and (**right**) ZnAl31MW, and (**bottom**) calcined samples (**left**) ZnAl31-650 and (**right**) ZnAl31MW-650. Ascription of the diffraction maxima to the corresponding planes, indicated by their Miller indexes, is shown.

Crystallographic data for these samples are included in Table 1, together with other data. Crystallographic parameters c and a were calculated from the positions of the diffraction peaks due to planes (003) and (110), respectively, assuming a rhombohedral phase; then, $c = 3$ d(003) and $a = 2$ d(110). The values are rather similar for both the original and the MW-treated samples, differences being only 1% for parameter c and ca. 0.3% for parameter a. The crystallite size values, also included in Table 1, were calculated by the Scherrer equation, from the full half-maximum width (FHMW, β) estimated by using commercial software Diffract Plus. Despite the crystallographic parameters are coincident within experimental error, the crystal size, both in the (003) and (110) directions (two directions were taken, due to the anisotropic nature of the material), is much greater for sample ZnAl31MW. The microwave hydrothermal treatment entails a much better crystallization, as it was observed by Benito et al. [10] for samples comparable to those here synthesized.

Table 1. Crystallographic parameters and crystallite sizes of the uncalcined samples.

Sample	c	a	β		D	
			(003)	(110)	(003)	(110)
ZnAl31	23.09	3.086	0.36	0.26	230	365
ZnAl31MW	22.85	3.074	0.17	0.19	480	510

β = full half-maximum width (FHMW) in °(2θ); all other values in Å; D rounded to ±5 Å.

In order to check if the only anion in the interlayer balancing the positive layer charge in the uncalcined samples is carbonate, the FT-IR spectra of the samples were recorded; that for sample ZnAl31 is included in Figure 2. Several intense bands typical of hydrotalcite-like compounds [23] are recorded. The first one, centered around 3470 cm^{-1}, is due to the O-H stretching mode of the layer hydroxyl groups and interlayer water molecules. Broadening of this band toward the low wavenumber side indicates the existence of hydrogen bonds between the hydroxyl groups and carbonate anions [23,24]. The vibrational bending mode band of the interlayer water molecules is observed at 1634 cm^{-1}, while the band at 1364 cm^{-1} is due to the v_3 vibrational mode of interlayer carbonate species. The bands recorded below 800 cm^{-1} correspond to the stretching vibrations involving metal-hydroxyl, and metal-oxygen-metal bonds of the brucite layer cations [10,23,25,26]. No band due to intercalated nitrate (the counteranion of the Zn^{2+} and Al^{3+} salts used in the synthesis) is recorded.

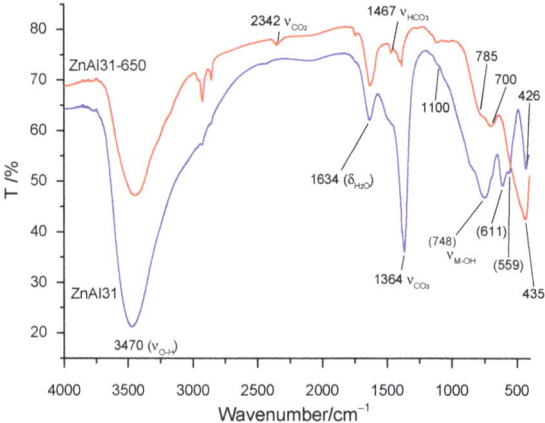

Figure 2. FT-IR spectra of samples ZnAl31 (blue line) and ZnAl31-650 (red line).

The results of the thermal analyses are included in Figure 3. The thermogravimetric (TG) curve shows a first mass loss amounting ca. 25% of the initial sample mass. The corresponding DTG (derivative thermogravimetric analysis) and DTA (differential thermal analysis) curves are also included in the same figure. The DTG curve is useful to identify the inflection points of the TG curve (where the mass loss rate reaches a maximum); in our case, these maxima are recorded at 160, 220 and 555 °C. The first two correspond also to the positions of the minima appearing in the DTA curve (the third minimum in the DTA curve is very poorly defined because of it is extremely width), indicating that all these mass losses are due to endothermic processes. From mass spectrometric analysis of gases evolved during thermal decomposition of this type of solids [27], it was concluded that releasing of interlayer water molecules is responsible for the first mass loss, while further mass losses correspond to evolution of CO$_2$ (from interlayer carbonate species) and of H$_2$O, formed by condensation of layer hydroxyl groups. Another mass loss recorded between 300 and 650 °C, less than 10% of the initial sample mass, is related to a very broad and smooth endothermic effect in the DTA curve. This mass loss corresponds to the slow elimination of residual water and mainly CO$_2$ molecules [28], which migrate to the outside of the crystallites through the pores and channels (chimneys) generated during the previous intense dehydration–dehydroxylation–decarbonation processes. Elimination of these components results in the collapse of the structure, destroying the layered framework and forming an essentially amorphous phase that, when the temperature is further raised, leads to the formation of the spinel-type oxides, together with the oxide of the divalent cation [29].

Overall, the thermal decomposition of this sample follows the same trends as previously observed for other LDHs with easily decomposable interlayer species (e.g., carbonate or nitrate) and without oxidizable cations in the brucite-like layers [9,30]. From these results, we chose calcination at 650 °C as the calcination temperature to obtain well-dispersed ZnO, without segregation of new phases, to prepare new photocatalysts.

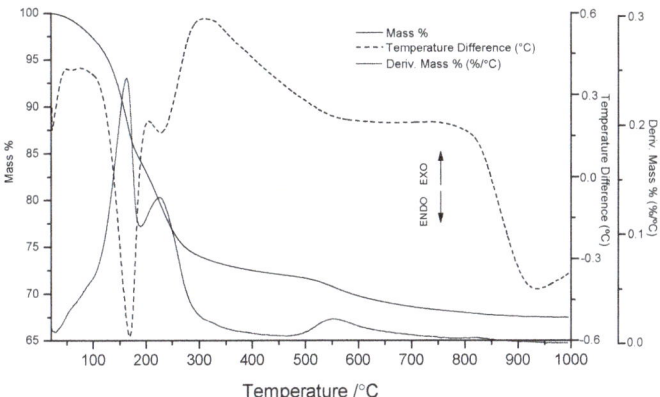

Figure 3. TG (solid), DTG (dotted) and DTA (dashed) curves of sample ZnAl31.

The PXRD diagrams for the calcined samples are included in Figure 1. The peaks of the LDH structure are no longer recorded, and the positions of the peaks are very close to those of ZnO, zincite (JCPDS standard n° 36-1451) [31].

The peaks are again better defined for sample ZnAl31MW-650, probably because of the higher crystallinity of the LDH precursor. The reflection planes (JCPDS 75-0576) are identified in the figure and correspond to the wurtzite-type structure [32–34]. No peaks corresponding to crystalline phases of Al oxides are recorded, meaning that, after calcination, aluminum remains forming amorphous phases, which might correspond to some sort of alumina, or even singly dispersed within the ZnO phase.

The FT-IR spectrum of sample ZnAl31-650 (Figure 2) confirms removal of interlayer anions upon calcination, as expected from the thermal analysis results, and confirming the collapse of the structure, as concluded from the PXRD results. On comparing the relative intensities of the bands due to water molecules bending and hydroxyl groups stretching modes, it is clear that in the calcined samples such bands are much weaker than in the uncalcined sample; at the same time, the bands due to the carbonate anion are completely (or almost completely) removed. Despite that carbonate has evolved as CO_2 during calcination, it is obvious that, due to the strong basic nature [35] of the oxides formed upon calcination, adsorption of atmospheric CO_2 and formation of surface carbonates species can be hardly avoided, thus it is not completely unexpected to record weak bands due to carbonate species in the spectra of calcined samples.

The band around 1365 cm^{-1}, the shoulder at 1467 cm^{-1} and the weak feature at 1100 cm^{-1} are due to vibrational modes of residual carbonate and probably hydrogencarbonate; the weak absorption close to 2342 cm^{-1} is due to confined CO_2 molecules. The stability of these species is due to the strong basicity of this solid. The bands recorded around 785 and 700 cm^{-1} are similar to the typical doublet due to the vibrations of "isolated" MO_4 tetrahedra at 785, 700 and 485 cm^{-1}, confirming the tetrahedral environment of the Zn^{2+} cations in the ZnO wurtzite-like phase [35].

Table 2 summarizes the surface texture data of these solids (two original samples with the LDH structure, and two calcined samples) concerning their surface areas: The Brunauer-Emmett-Teller (BET) specific surface data (S_{BET}), the external surface area (S_t) and the surface area equivalent to adsorption in micropores. First of all, the surface area equivalent to adsorption in micropores, S_{mp}, is rather low in all cases (even zero for sample ZnAl31, and around 10% in other cases), suggesting an insignificant presence of micropores. Secondly, the specific surface areas of the calcined samples are larger, in both cases, than for the corresponding uncalcined ones, in agreement with development of pores and channels as the escaping routes of internal species, water and CO_2 [36–39].

Table 2. Surface area data (m^2g^{-1}) for the samples studied *.

Sample	S_{BET}	S_t	S_{mp}
ZnAl31	51	51	0
ZnAl31MW	26	23	3
ZnAl31-650	73	66	7
ZnAl31MW-650	34	30	3

* S_{BET}: BET specific surface area; S_t = external surface area; S_{mp} = microporous-equivalent surface area.

A significant decrease in the specific surface area of the sample submitted to MW treatment, in comparison to that which had not been submitted to any post-synthesis treatment, is also observed. Actually, the specific surface area of sample ZnAl31MW is only one half of that for sample ZnAl31. This difference should be ascribed to the increase in the crystallinity of the sample after the MW treatment, as concluded from the larger sharpness and intensity of the PXRD peaks of sample ZnAl31MW compared to those of sample ZnAl31, as well as in the crystallite size, Table 1. Such a decrease in the specific surface area is "dragged" to the calcined samples: Although these have a larger specific surface area than the samples before calcination, the surface is still larger for the calcined sample prepared from the sample not submitted to the microwave hydrothermal treatment.

2.2. Degradation Studies

The catalytic activity of the samples in the photocatalytic degradation of 4-NP was measured; for this purpose, 750 mL of an aqueous solution with 25 ppm of 4-NP were introduced into the photoreactor and the catalyst was added to this solution. The amount of solid used in the degradation reactions was calculated in order to refer the measurements in all experiments to the activity of 200 ppm of Zn in different forms. The original solids, ZnAl31 and ZnAl31MW, were tested as catalysts and, in view of the results obtained therewith, one of the calcined sample, ZnAl31-650, was also tested. The activity of a commercial ZnO sample as a reference catalyst was also analyzed for comparison. The photolysis of the 4-nitrophenol solution, that is, the degradation of the contaminant under the action of ultraviolet light in the same experimental conditions, but in the absence of any catalyst, was studied as well. In all cases the solution (with or without dispersed catalyst) was continuously stirred and small volumes (aliquots) were taken periodically in order to follow the reaction by UV–Vis spectroscopy, by monitoring the decrease in the intensity of the typical band of 4-NP at 400 nm.

To study the adsorption, the suspension is stirred in the reaction chamber and kept in the darkness until a constant concentration of the contaminant was reached; at that time the suspension is irradiated with ultraviolet light to study the photocatalytic reaction.

When the catalyst is added to the solution, 4-nitrophenol is deprotonated due to the basicity of these solids and the band typical of the phenolate group at 400 nm is recorded [40,41]. For this reason, the degradation of the contaminant was studied taking, as a reference, the maximum absorbance of this band and following its decrease as the reaction time elapses.

The results obtained for the adsorption of 4-NP in the absence of light using ZnAl31 and ZnAl31MW are shown in Figure 4; those obtained in the same conditions with a commercial ZnO are included for comparison.

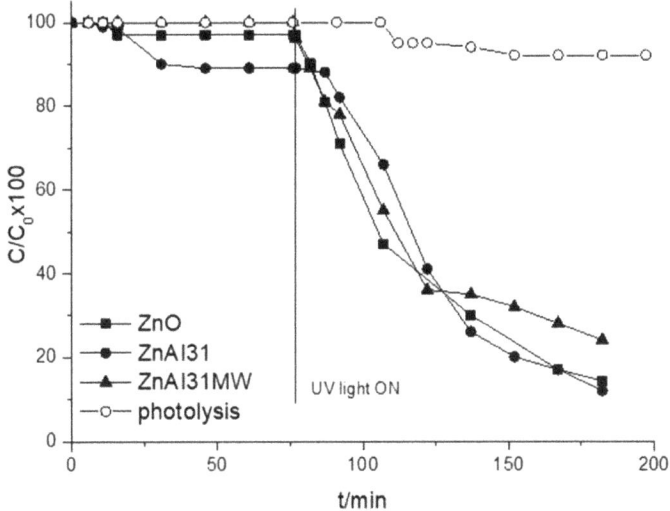

Figure 4. 4-NP degradation using ZnAl31, ZnAl31MW and commercial ZnO. Photolysis (open circles) is included for comparison.

The decrease in the contaminant concentration in the darkness is probably due to its adsorption or absorption on the catalyst used. It can be concluded that in the darkness there is no significant adsorption on sample ZnAl31MW. However, the relative concentration of the contaminant decreases by about 3 and 10%, respectively, on samples ZnO and ZnAl31 after 30 min and remains constant for longer times. This decrease is probably due to the adsorption of the contaminant on the surface of the solids, and it should be noticed that it is greater on sample ZnAl31, perhaps because of its larger specific surface area and also because, in this case, part or the contaminant can enter the interlayer space. On the contrary, the curve for sample ZnAl31MW shows that the adsorption of the contaminant is constant and almost negligible during all time tested.

After 75 min in the darkness the reaction was continued under UV light, in order to check the photocatalytic activity of the solids. After 40 min of illumination, between 40 and 50% of the initial 4-NP was degraded and after 50 min the amount of 4-NP degraded is the same for all samples. When the reaction time is further extended, the degradation practically stabilizes for sample ZnAl31MW and increases slightly for ZnO and ZnAl31, both curves becoming almost coincident when the reaction time is further prolonged. In other words, the experimental results indicate that the MW ageing treatment to which the ZnAl31MW sample was submitted does not improve its photocatalytic activity if compared to that of ZnAl31, which matched the degradation capacity of ZnO. For this reason, the study was continued only with the calcined sample, ZnAl31-650, prepared from sample ZnAl31.

The degradation curve obtained in the darkness using sample ZnAl31-650 is shown in Figure 5, together with that for ZnO. It is worth noting that for sample ZnAl31-650 the adsorption reaches 54% after 45 min and 67% after 75 min and then stabilizes. When the reaction continues under UV light, removal of 4-NP after 30 min reaches 75% for sample ZnAl31-650, while it is 53% for pure ZnO, and after 60 min it is 90% for sample ZnAl31-650 and only 70% for sample ZnAl31 or ZnO.

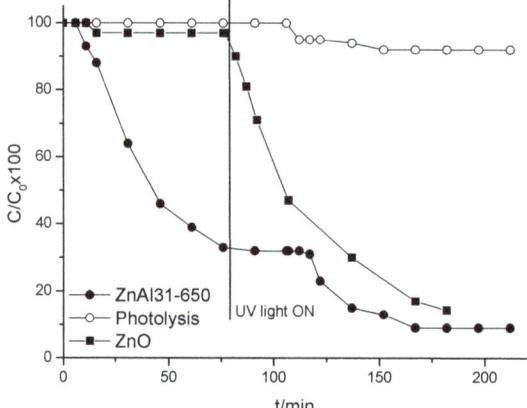

Figure 5. 4-NP degradation using ZnAl31-650 and ZnO. Photolysis (open circles) results are included for comparison.

In order to confirm the photocatalytic capacity of sample ZnAl31-650, some aliquots of solution were withdrawn from the reaction medium at different reaction times (in the darkness and under UV light) and analyzed by mass spectrometry. Aliquot D5 was withdrawn after 5 min in the darkness, while aliquots L10, L30 and L165 were withdrawn after 10, 30 and 165 min under UV light, respectively.

The mass spectrum of the solution collected after 5 min of reaction in the darkness (Supplementary material, Figure S1), D5, shows a weak signal at about m/z 139 corresponding to a small portion of the nitrophenol molecule together with an intense signal recorded at m/z close to 138, with intensity 1.4×10^7 (in arbitrary units) which would correspond to deprotonated 4-nitrophenol. This deprotonation is not unexpected, since the catalyst provides a basic character to the solution of the contaminant and in this medium 4-NP (molecular mass 139.1) loses the proton of the alcohol function being transformed into the nitrophenolate anion (mass 138.1), as already confirmed by the analysis of the absorption spectra previously discussed. A very weak signal recorded at m/z 168 may be due to the combination of fragments formed in the mass spectrometer chamber. These results confirm that in the darkness there is no degradation of 4-nitrophenol and only adsorption/absorption can occur on the catalyst surface. The signal at m/z 108 (less than 0.1×10^7 of relative intensity) indicates that diquinone begins to form in the reactor, the first by-product that suggests the incipient oxidation of the substrate that will accelerate when the ultraviolet light strikes the reaction medium.

In the mass spectrum of aliquot L10 (Supplementary material, Figure S2), it is noteworthy that the decrease in the intensity in the signal due to the molecular fragment of the contaminant, which decreases from ca. 1.5×10^7 for sample D5, to 6×10^6 for sample L10, that is, after 10 min of reaction under UV light. In this spectrum, very weak signals are recorded at m/z 124.9, 174.7 and 201.1, which may correspond to the hydroxybenzoquinone molecule, and to fragments formed by aggregates of by-products during the reaction or in the mass spectrometer chamber. These data indicate that after 10 min of UV irradiation, by-products of the oxidation of 4-nitrophenol are already formed [42,43].

In the mass spectrum of aliquot L30 (Supplementary material, Figure S3), the previously described signals are again recorded, in addition to another one at m/z 153.9. According to results reported in the literature [42], this signal could be due to deprotonated 4-nitrocatechol, which is formed during degradation by hydroxylation prior to the oxidation of the contaminant.

The mass spectrum of the last aliquot collected (Supplementary material, Figure S4) confirms that the concentration of the pollutant molecule is practically null and the spectrum shows signals of reaction by-products or aggregates thereof. These signals indicate that after this reaction time the oxidation of the contaminant was completed.

Figure 6 shows the evolution of the intensity of the molecular signal of the contaminant (m/z = 137.9) with the reaction time in the aliquots analyzed by mass spectrometry. As it can be seen, at the end of the reaction this signal has completely disappeared, that is, after 165 min of photocatalytic reaction 4-NP was completely degraded.

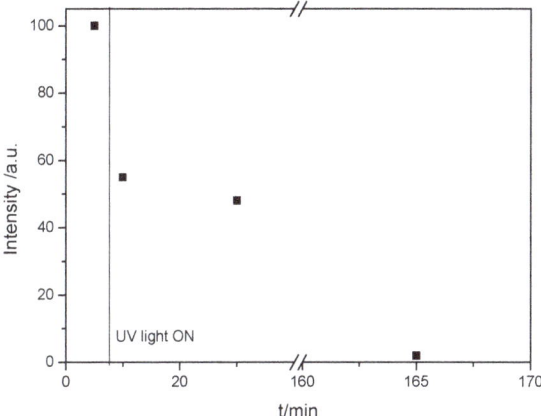

Figure 6. Evolution of the relative intensity of the m/z 137.9 signal for the aliquots analyzed by mass spectrometry. Dots (from left to right) correspond to aliquots D5, L10, L30 and L165, respectively.

In order to study the possible reusability of the catalyst, sample ZnAl31-650 was collected and dried after the whole experiment (reaction in the darkness and under UV light) to study the structural changes undergone by the catalyst during the reaction (the sample was named as ZnAl31-650Rec). The PXRD diagram of this sample reveals that the layered structure of the original hydrotalcite is, in a large extent, recovered during the reaction and reveals the possibility of reusing it after a new calcination cycle. Figure 7 shows the PXRD diagrams of the calcined (before photocatalytic tests) and recovered samples, and the inset includes a comparison of the diagram of the original hydrotalcite (ZnAl31) with that of the recovered catalyst, ZnAl31-650Rec. These results undoubtedly demonstrate that the catalysts could be reusable.

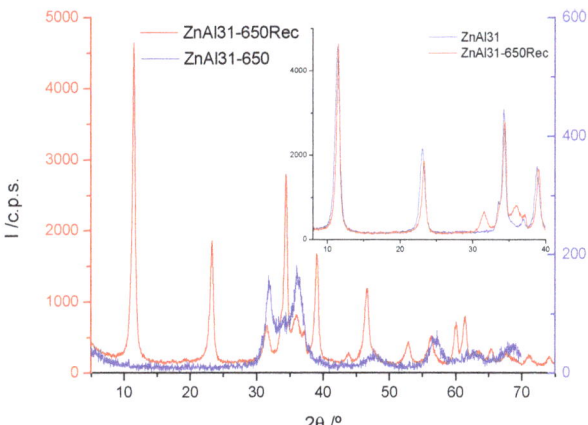

Figure 7. PXRD diagrams of samples ZnAl31-650Rec and ZnAl31-650. Inset: ZnAl31-650Rec and ZnAl31 samples.

The FT-IR spectrum of sample ZnAl31-650Rec (Figure 8) was recorded in order to confirm the reconstruction of the layered structure after the reaction and also to determine if bands belonging to the contaminant molecule are present. The spectra of samples ZnAl31 and ZnAl31-650 were included in the figure for comparison. It is remarkable that after the reaction the sample shows an FT-IR spectrum similar to that of the original, uncalcined sample. The increase in intensity of the carbonate band of the ZnAl31-650Rec compared to that for ZnAl31-650 indicates that the sample has absorbed this anion (probably formed upon complete degradation of 4-NP) to reconstruct the layered structure.

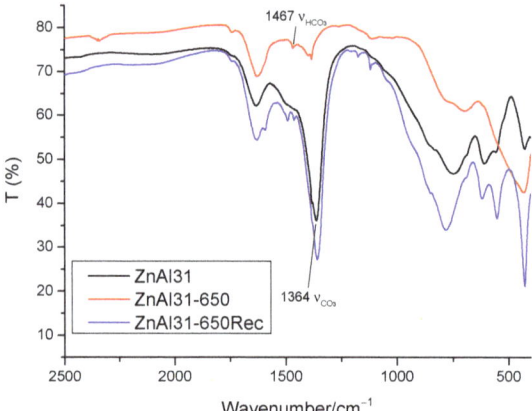

Figure 8. FT-IR spectra of samples ZnAl31-650Rec, ZnAl31 and ZnAl31-650.

3. Materials and Methods

All inorganic reagents, zinc nitrate hexahydrate (98%), aluminum nitrate nonahydrate (98%), zinc oxide (99.9%), sodium carbonate, sodium hydroxide and potassium bromide (analytical purity) were supplied by PanReac AppliChem (Castellar del Vallés, Barcelona, Spain); 4-nitrophenol was purchased to Merck KGaA (Darmstadt, Germany) and liquid nitrogen and gases used for analysis (N_2, O_2, and He) were supplied by L'Air Liquide, S. A. (Madrid, Spain).

The Zn,Al hydrotalcite with carbonate in the interlayer was synthesized by slowly adding an aqueous solution of the cation salts (0.5 M Zn^{2+} and 0.165 M Al^{3+}) with a Zn:Al molar ratio of 3:1 to a 0.11 M solution of sodium carbonate (CO_3^{2-}/Al^{3+} molar ratio = 1); the pH of the mixture was kept constant at 10 by adding 1 M NaOH using a pH-burette 24 from Crison. Once the addition was complete, a portion of the suspension was kept under stirring in air for 24 h and the other one was submitted to hydrothermal treatment in a microwave oven (Milestone Ethos Plus) at 100 °C for 2 h. Both suspensions obtained were centrifuged and the solids were washed with distilled water to remove the counterions (nitrate and sodium). Then, they were dried at room temperature in open air and hand ground in an agate mortar. The samples thus obtained were named as ZnAl31 and ZnAl31MW, respectively, where "MW" stands for microwave (MW) treatment. Both samples were calcined at 650 °C (leading to samples named ZnAl31-650 and ZnAl31MW-650, respectively) in order to obtain mixtures of well dispersed oxides. This calcination temperature was selected after recording the thermogravimetric analysis of the original samples.

The samples were characterized by the techniques described below. Element chemical analyses were performed by atomic absorption in a Perkin Elmer Elan 6000 ICP Mass Spectra apparatus at Servicio General de Análisis Químico Aplicado (Universidad de Salamanca, Spain). The powder X-ray diffraction (PXRD) analysis was carried out using a Siemens D-5000 diffractometer, with an electric power of 1200 W (30 mA and 40 kV), 0.05 °/step and 1.5 s/step (scanning rate 2 °/min). The FT-IR spectra were recorded using the KBr pressed pellet method in a Perkin-Elmer Spectrum-One spectrometer, with a nominal resolution of 4 cm^{-1} and averaging 50 scans to improve the signal-to-noise ratio. Specific surface area measurement and porosity analysis were carried out from the nitrogen adsorption-desorption isotherms at −196 °C. The isotherms were recorded in a Micromeritics Gemini VII 2390t equipment. Before analysis the samples were treated under dry nitrogen flow for 2 h at 110 °C in a Micromeritics FlowPrep 060 Sample Degass System apparatus to remove weakly adsorbed species. Thermogravimetric and differential thermal analyses (TGA and DTA) were recorded from room temperature to 1000 °C at a heating rate of 10 °C/min under a continuous oxygen flow in an SDT Q600 apparatus from TA INSTRUMENTS.

A MPDS-Basic system from Peschl Ultraviolet, with a PhotoLAB Batch-L reactor and a TQ150-Z0 lamp (power 150 W), integrated in a photonCABINET was used to check the photoactivity. Its spectrum is continuous, with the main peaks at 366 nm (radiation flux, ϕ 6.4 W) and 313 nm (ϕ 4.3 W). The solutions were analyzed by ultraviolet–visible spectroscopy in order to control the progress of the reaction in a Perkin-Elmer LAMBDA 35 spectrophotometer coupled to a computer with UV WINLAB 2.85 software.

Aiming to determine the by-products generated during UV degradation, some selected solutions were analyzed after various reaction times by mass spectrometry. The equipment used was an Agilent 1100 HPLC mass spectrometer coupled to an ultraviolet detector and an Agilent Trap XCT mass spectrometer. These analyses were carried out at Servicio Central de Análisis Elemental, Cromatografía y Masas (Universidad de Salamanca, Spain).

4. Conclusions

The synthesis method used without further treatment facilitates a rapid and effective production of the original solid, ZnAl31, whose calcination at 650 °C allows obtaining a solid with higher specific surface area and consisting of a ZnO phase dispersed on an Al_2O_3 amorphous phase.

The activity of the synthetized solids tested for the degradation reaction of 4-nitrophenol in aqueous solution exceeded that observed for the reaction with a commercial ZnO. The photocatalytic performance of the original non-calcined hydrotalcite is similar to that of commercial ZnO. The ZnAl31-650 sample shows the best performance in the adsorption-degradation of the contaminant, perhaps because of its high specific surface area and the high dispersion of the active phase. These two characteristics are a consequence of the use of a synthetic hydrotalcite as the ZnO active phase precursor, thus the final solid acts by adding its capacity as a photocatalyst to its capacity as an adsorbent.

Supplementary Materials: The following are available online at http://www.mdpi.com/2073-4344/10/6/702/s1, Figure S1. Mass spectrum of aliquot D5, after 5 min in the darkness, Figure S2. Mass spectrum of aliquot L10, after 10 min under UV light, Figure S3: Mass spectrum of aliquot L30, after 30 min under UV light, Figure S4: Mass spectrum of aliquot L165, after 165 min under UV light.

Author Contributions: Conceptualization, methodology and validation V.R. and R.T.; investigation C.N., resources V.R. and R.T.; original draft, C.N.; supervision, R.T. and V.R. All authors have read and agreed to the published version of the manuscript reported.

Funding: The project leading to this submission received funding from MAT2016-78863-C2-2-R.

Conflicts of Interest: The authors declare no conflicts of interest.

References

1. Abderrazek, K.; Najoua, F.S.; Srasra, S. Synthesis and characterization of [Zn–Al] LDH: Study of the effect of calcination on the photocatalytic activity. *Appl. Clay Sci.* **2016**, *119*, 229–235. [CrossRef]
2. Barhoum, A.; Melcher, J.; Van Assche, G.; Rahier, H.; Bechelany, M.; Fleisch, M.; Bahnemann, D. Synthesis, growth mechanism, and photocatalytic activity of Zinc oxide nanostructures: Porous microparticles versus nonporous nanoparticles. *J. Mater. Sci.* **2017**, *52*, 2746–2762. [CrossRef]
3. Zhang, Z.; Hua, Z.; Lang, J.; Song, Y.; Zhang, Q.; Han, Q.; Fan, H.; Gao, M.; Li, X.N.; Yang, J. Eco-friendly nanostructured Zn–Al layered doublehydroxide photocatalysts with enhancedphotocatalytic activity. *Cryst. Eng. Commun.* **2019**, *21*, 4607–4619. [CrossRef]

4. Basile, F.; Benito, P.; Fornasari, G.; Rosetti, V.; Scavetta, E.; Tonelli, D.; Vaccari, A. Electrochemical synthesis of novel structured catalysts for H2 production. *Appl. Catal. B Environ.* **2009**, *91*, 563–572. [CrossRef]

5. Chai, R.; Li, Y.; Zhang, Q.; Fan, S.; Zhang, Z.; Chen, P.; Zhao, G.; Liu, Y.; Lu, Y. Foam-Structured NiO-MgO-Al$_2$O$_3$ Nanocomposites Derived from NiMgAl Layered Double Hydroxides In Situ Grown onto Nickel Foam: A Promising Catalyst for High-Throughput Catalytic Oxymethane Reforming. *Chem. Catal. Chem.* **2017**, *9*, 268–272.

6. Chai, R.; Fan, S.; Zhang, Z.; Chen, P.; Zhao, G.; Liu, Y.; Lu, Y. Free-Standing NiO-MgO-Al$_2$O$_3$ Nanosheets Derived from Layered Double Hydroxides Grown onto FeCrAl-Fiber as Structured Catalysts for Dry Reforming of Methane. *ACS Sustain. Chem. Eng.* **2017**, *5*, 4517–4522. [CrossRef]

7. Chai, R.; Zhang, Z.; Chen, P.; Zhao, G.; Liu, Y.; Lu, Y. Ni-foam-structured NiO-MOx-Al$_2$O$_3$ (M = Ce or Mg) nanocomposite catalyst for high throughput catalytic partial oxidation of methane to syngas. *Microporous Mesoporous Mater.* **2017**, *253*, 123–128. [CrossRef]

8. Rives, V. (Ed.) *Layered Double Hydroxides: Present and Future*; Nova Science Publishers: New York, NY, USA, 2001.

9. Trujillano, R.; González-García, I.; Morato, A.; Rives, V. Controlling the Synthesis Conditions for Tuning the Properties of Hydrotalcite-Like Materials at the Nano Scale. *ChemEngineering* **2018**, *2*, 31. [CrossRef]

10. Benito, P.; Guinea, I.; Labajos, F.M.; Rocha, J.; Rives, V. Microwave-hydrothermally aged Zn,Al hydrotalcite-like compounds: Influence of the composition and the irradiation conditions. *Micropor. Mesopor. Mat.* **2008**, *110*, 292–302. [CrossRef]

11. Mishraa, G.; Dasha, B.; Pandey, S. Layered double hydroxides: A brief review from fundamentals to application as evolving biomaterials. *Appl. Clay Sci.* **2018**, *153*, 172–186. [CrossRef]

12. Lee, K.M.; Lai, C.W.; Ngai, K.S.; Juan, J.C. Recent developments of zinc oxide based photocatalyst in water treatment technology: A review. *Water Res.* **2016**, *88*, 428–448. [CrossRef] [PubMed]

13. Ong, C.B.; Ng, L.Y.; Abdul, W.M. A review of ZnO nanoparticles as solar photocatalysts: Synthesis, mechanisms and applications. *Renew. Sustain. Energy Rev.* **2018**, *81*, 536–551. [CrossRef]

14. Serrà, A.; Zhang, Y.; Sepúlveda, B.; Gómez, E.; Nogués, J.; Michler, J.; Philippe, L. Highly active ZnO-based biomimetic fern-like microleaves for photocatalytic water decontamination using sunlight. *Appl. Catal. B Environ.* **2019**, *248*, 129–146. [CrossRef]

15. He, X.; Wang, B.; Zhang, Q. Phenols removal from water by precursor preparation for MgAl layered double hydroxide: Isotherm, kinetic and mechanism. *Mater. Chem. Phys.* **2019**, *221*, 108–117. [CrossRef]

16. Kwon, T.; Pinnavaia, T.J. Pillaring of a layered double hydroxide by polyoxometalate with Keggin-ion structures. *Chem. Mater.* **1989**, *14*, 381–383. [CrossRef]

17. Chibwe, K.; Jones, W. Intercalation of organic and inorganic anions into layered hydroxides. *J. Chem. Soc. Chem. Commun.* **1989**, *14*, 926–927. [CrossRef]

18. Chen, S.; Xu, Z.P.; Zhang, Q.; Lu, G.Q.M.; Hao, Z.P.; Liu, S. Studies on adsorption of phenol and 4-nitrophenol on MgAl-mixed oxide, derived from MgAl layered double hydroxides. *Sep. Pur. Technol.* **2009**, *67*, 194–200. [CrossRef]

19. Rajamanickam, D.; Shanthi, M. Photocatalytic degradation of an organic pollutant by zinc oxide—Solar process. *Arab. J. Chem.* **2016**, *9*, S1858–S1868. [CrossRef]

20. Zheng, Y.; Shu, J.; Wang, Z. AgCl@Ag composites with rough surfaces as bifunctional catalyst for the photooxidation and catalytic reduction of 4-nitrophenol. *Mater. Lett.* **2015**, *158*, 339–342. [CrossRef]

21. Uberoi, V.; Bhattacharya, S.K. Toxicity and degradability of nitrophenols in anaerobic systems. *Water Environ. Res.* **1997**, *69*, 146–156. [CrossRef]

22. Kidak, R.; Ince, N.H. Ultrasonic destruction of phenol and substituted phenols: A review of current research. *Ultrason. Sonochem.* **2006**, *13*, 195–199. [CrossRef]

23. Kloprogge, J.T.; Frost, R.L. Infrared and Raman Spectroscopic Studies of Layered Double Hydroxides (LDHs). In *Layered Double Hydroxides: Present and Future*; Rives, V., Ed.; Nova Science Publisher: New York, NY, USA, 2001; Chapter 5; pp. 139–192.

24. Kloprogge, T.; Hickey, L.; Frost, R. FT-Raman and FT-IR spectroscopic study of sinthetic Mg/Zn/Al hydrotalcites. *J. Raman Spectrosc.* **2004**, *35*, 967–974. [CrossRef]

25. Alzamora, L.E.; Ross, J.R.H.; Kruissink, E.C.; Van Reijden, L.L. Coprecipitated nickel–alumina catalysts for methanation at high temperature. Part 2—Variation of total and metallic areas as a function of sample composition and method of pretreatment. *J. Chem. Soc. Faraday Trans. I* **1981**, *77*, 665. [CrossRef]

26. Kloprogge, J.T. Infrared and Raman spectroscopy of naturally occurring hydrotalcites and their synthetic equivalents. In *The Application of Vibrational Spectroscopy to Clay Minerals and Layered Double Hydroxides*; Kloprogge, J.T., Ed.; CMS Workshop Lectures; The Clay Minerals Society: Aurora, CO, USA, 2005; Volume 13, pp. 203–238.

27. Rives, V. Comment on "Direct Observation of a Metastable Solid Phase of Mg/Al/CO$_3$-Layered Double Hydroxide by Means of High-Temperature in Situ Powder XRD and DTA/TG". *Inorg. Chem.* **1999**, *38*, 406–407. [CrossRef]

28. Miyata, S. Anion-exchange properties of hydrotalcite-like compounds. *Clays Clay Min.* **1983**, *31*, 305. [CrossRef]

29. Labajos, F.M.; Rives, V.; Ulibarri, M.A. Effect of hydrothermal and thermal treatments on the physicochemical properties of Mg-Al hydrotalcite-like materials. *J. Mater. Sci.* **1992**, *27*, 1546–1552. [CrossRef]

30. Rives, V. Study of Layered Double Hydroxides by Thermal Methods. In *Layered Double Hydroxides: Present and Future*; Rives, V., Ed.; Nova Science Publishers: New York, NY, USA, 2001; Chapter 4; pp. 115–137, ISBN 1-59033-060-9.

31. Basnet, P.; Samanta, D.; Inakhunbi Chanu, T.; Mukherjee, J.; Chatterjee, S. Assessment of synthesis approaches for tuning the photocatalytic property of ZnO nanoparticles. *SN Appl. Sci.* **2019**, *1*, 633. [CrossRef]

32. Amari, R.; Mahroug, A.; Boukhari, A.; Deghfel, B.; Selmi, N. Structural, Optical and Luminescence Properties of ZnO Thin Films Prepared by Sol-Gel Spin-Coating Method: Effect of Precursor Concentration. *Chin. Phys. Lett.* **2018**, *35*, 016801. [CrossRef]

33. Prasada Rao, T.; Santhoshkumar, M.C. Effect of thickness on structural, optical and electrical properties of nanostructured ZnO thin films by spray pyrolysis. *Appl. Surf. Sci.* **2009**, *255*, 4579–4584. [CrossRef]

34. Wang, L.; Lou, Z.; Fei, T.; Zhang, T. Templating synthesis of ZnO hollow nanospheres loaded with Au nanoparticles and their enhanced gas sensing properties. *J. Mater. Chem.* **2012**, *22*, 4767–4771. [CrossRef]

35. Montanari, T.; Sisani, M.; Nocchetti, M.; Vivani, R.; Herrera Delgado, M.C.; Ramis, G.; Busca, G.; Costantino, U. Zinc–aluminum hydrotalcites as precursors of basic catalysts: Preparation, characterization and study of the activation of methanol. *Catal. Today* **2010**, *152*, 104–109. [CrossRef]

36. Carriazo, D.; del Arco, M.; García-López, E.; Marcì, G.; Martín, C.; Palmisano, L.; Rives, V. Zn,Al hydrotalcites calcined at different temperatures: Preparation, characterization and photocatalytic activity in gas-solid regime. *J. Mol. Catal. A Chem.* **2011**, *342–343*, 83–90. [CrossRef]

37. Rives, V. Characterization of layered double hydroxides and their decomposition products. *Mater. Chem. Phys.* **2002**, *75*, 19–25. [CrossRef]

38. Rives, V. Surface Texture and Electron Microscopy Studies of Layered Double Hydroxides. In *Layered Double Hydroxides: Present and Future*; Nova Science Publishers: New York, NY, USA, 2001; Chapter 8; pp. 229–250.

39. Reichle, W.T.; Kang, S.Y.; Everhardt, D.S. The nature of the thermal decomposition of a catalytically active anionic clay mineral. *J. Catal.* **1986**, *101*, 352. [CrossRef]

40. Mancipe, S.; Tzompantzi, F.; Rojas, H. Photocatalytic degradation of phenol using MgAlSn. *Appl. Clay Sci.* **2016**, *129*, 71–78. [CrossRef]

41. Mancipe, S.; Tzompantzi, F.; Gómez, R. Photocatalytic reduction of 4-nitrophenol to 4-aminophenol over CdS/MgAl layered double hydroxide catalysts under UV irradiation. *React. Kinet. Mech. Catal.* **2017**, *122*, 625–634. [CrossRef]

42. Anuradha, G.; Meenukhurana Archana, C.; Masahiro, T.; Asit, K.C.; Rakesh, K.J. Degradation of 4-Nitrophenol, 2-Chloro-4-nitrophenol, and 2,4-initrophenol. *Environ. Sci. Technol.* **2010**, *44*, 1069–1077.

43. Wei, L.; Zhu, H.; Mao, X.; Gan, F. Electrochemical oxidation process combined with UV photolysis for the mineralization of nitrophenol in saline wastewater. *Sep. Purif. Technol.* **2011**, *77*, 18–25. [CrossRef]

MDPI

Article

Green Epoxidation of Olefins with Zn$_x$Al/Mg$_x$Al-LDH Compounds: Influence of the Chemical Composition

Rodica Zăvoianu [1,2], Anca Cruceanu [1,2], Octavian Dumitru Pavel [1,2,*], Corina Bradu [3,4], Mihaela Florea [5] and Ruxandra Bîrjega [6]

1   Faculty of Chemistry, University of Bucharest, 4-12 Regina Elisabeta Blv., 030018 Bucharest, Romania; rodica.zavoianu@chimie.unibuc.ro (R.Z.); anca.cruceanu@chimie.unibuc.ro (A.C.)
2   Research Center for Catalysts & Catalytic Processes, Faculty of Chemistry, University of Bucharest, 4-12 Regina Elisabeta Blv., 030018 Bucharest, Romania
3   Faculty of Biology, University of Bucharest, 91-95 Splaiul Independenţei, 050095 Bucharest, Romania; corina.bradu@g.unibuc.ro
4   Research Center for Environmental Protection and Waste Management (PROTMED), Research Platform in Biology and Systems Ecology, Faculty of Biology, University of Bucharest, 91-95 Splaiul Independenţei, 050095 Bucharest, Romania
5   National Institute of Materials Physics, 405A Atomistilor Street, 077125 Magurele, Romania; mihaela.florea@infim.ro
6   National Institute for Lasers, Plasma and Radiation Physics, 077125 Magurele, Romania; ruxandra.birjega@inflpr.ro
*   Correspondence: octavian.pavel@chimie.unibuc.ro; Tel.: +40-21-305-14-64

Citation: Zăvoianu, R.; Cruceanu, A.; Pavel, O.D.; Bradu, C.; Florea, M.; Bîrjega, R. Green Epoxidation of Olefins with Zn$_x$Al/Mg$_x$Al-LDH Compounds: Influence of the Chemical Composition. *Catalysts* 2022, 12, 145. https://doi.org/10.3390/catal12020145

Academic Editor: Anabela A. Valente

Received: 23 December 2021
Accepted: 21 January 2022
Published: 24 January 2022

Publisher's Note: MDPI stays neutral with regard to jurisdictional claims in published maps and institutional affiliations.

Abstract: This contribution concerns the effect of the chemical composition of the brucite-type layer of bi-cationic LDH materials Zn$_x$Al and Mg$_x$Al (x = 2–5) and tri-cationic LDH Mg$_y$Zn$_z$Al (y + z = 4, y = 1, 2, 3) on their catalytic activity for olefin epoxidation with H$_2$O$_2$ in the presence of acetonitrile. LDH materials were prepared by the standard method of co-precipitation at constant pH 10, using an aqueous solution of the corresponding metal nitrates and a basic solution containing NaOH and Na$_2$CO$_3$. The fresh LDHs were calcined to yield the corresponding mixed oxides and then the recovery of the LDH structure by hydration of the mixed oxides was performed. The resulting samples were characterized by AAS, XRD, DRIFT, DR-UV–Vis, BET and determination of basic sites. The results of the catalytic tests for olefin epoxidation were well correlated with the basicity of the samples, which was in turn related to the M^{2+}/Al^{3+} ratio and the electronegativity of different bivalent metals in the brucite-type layer.

Keywords: layered double hydroxides; base catalysts; oxidation; epoxide

1. Introduction

In recent decades, the fine chemical industry started to show a real interest in using base-type heterogeneous catalysts due to the fact that this class of materials presents numerous advantages compared with homogeneous catalysts [1–3], namely: the separation of the catalyst from the reaction products is achieved by a simple filtration, the temperature range in which it can work is often wide, the thermal regeneration of the catalyst is often possible, etc. The layered double hydroxides (LDHs) (discovered in the 19th century [4]), can substitute with good results the corrosive and not environmentally friendly base homogeneous catalysts [4–7]. The LDH general formula is $[M^{2+}_{1-x}M^{3+}_x(OH)_2]^{x+}[A^{n-}_{x/n}]\cdot mH_2O$ where M^{2+} and M^{3+} represent divalent and trivalent cations in the brucite-type layers, A is the interlayer anion with charge n which compensates the excedentary positive charge brought by the isomorphic substitution of M^{2+} with M^{3+}, x is the fraction of the trivalent cation (usually 0.20–0.33) and m is the number of crystallization water molecules [4]. In these structures, the cations adopt octahedral geometry. The mild calcination of LDHs converts them to mixed oxides having an important property named the *"memory effect"*, which consists in the spontaneous structural reconstruction of the original layered structure after hydration with water or aqueous solutions containing different anions [4,8,9]. Through this method, it is possible to include several anions in the interlayer space or to isomorphically exchange the existent cations with others having an adequate radius range. The mixed oxides obtained by calcinations at temperatures higher than 600 °C lose this property because the cations are placed in tetrahedral positions, which have a lower energy state compared to the octahedral ones [10]. Over time, these materials have found applications as catalysts in various fine chemical syntheses [4,11–15]. One class of important reactions are the oxidation ones, which can be classified into two types: one implying only a dehydrogenation and the second one involving both dehydrogenation and oxygen insertion into the hydrocarbon molecule [16]. The difficulties of controlling oxidation reactions in order to avoid total oxidation are due to the fact that usually any oxidation is accompanied by the release of heat, which can make undesirable side reactions thermodynamically possible. Until now, commonly used oxidation agents were NaOCl, alkylperoxides or peroxyacids [17]. However, due to environmental regulations, the actual trend is to replace these oxidants by others which are environmentally friendly, such as H_2O_2 or molecular oxygen, which are considered to be green oxidants [18]. Some oxidation reactions of alkenes give cyclic ethers in which both carbons of a double bond become bonded to the same oxygen atom, products called epoxides or oxiranes with a highly strained cycle which makes those epoxides more reactive than other ethers. Selective epoxidation of olefinic compounds is one of the important steps in organic synthesis of fine chemicals since, by ring opening reactions, epoxides are directly transformed into a wide variety of compounds with excellent yields [19–21]. Among epoxides, epoxycyclohexane is a valuable organic intermediate, used in the synthesis of pharmaceuticals, pesticides, epoxy paints, rubber promoters and stabilizers for chlorinated hydrocarbons [22]. Many publications present the oxidation of cyclohexene to cyclohexene oxide using different solids: magnetic core–shell type Fe_3O_4@chitosan-Schiff base Co(II), Cu(II) and Mn(II) complexes [23], vanadia-based catalysts [24], Ti(III)APO-5 materials [25], Fe nanocatalyst [26], nanostructured Au/SiO_2 [27], Fe(Salen) intercalated-zirconium phosphate [28], niobium oxyhydroxide [29], etc. The green epoxidation of olefins with H_2O_2 was first studied by Payne who showed that a peroxycarboximidic acid serves as a terminal oxygen donor when a nitrile is present as a co-reactant in the presence of a homogeneous base catalyst [30]. The new trend, from the environmental point of view, is using H_2O_2 as an oxidizing agent and different types of LDH solid catalysts, such as: hydrotalcite (HT) [1,31], Cu^{II}(Sal-Ala)/MgAl-LDH and Cu^{II}(Sal-Phen)/MgAl-LDH [32], hydrotalcite-like compounds with low Mo-loading [33], Mg/Al; Mg,Zn/Al; Mg/Al,Ga hydrotalcite-like compounds [34,35], etc., or molecular oxygen and M^{II}Mg/Al hydrotalcites and hydrotalcite-supported M(II) acetylacetonate (M(II) = Co, Cu or Ni) catalysts [36], cobalt-modified hydrotalcites [17], LDH hosted Fe and Mn sulfonato-salen complexes [37], etc.

The studies published until now concerning the Payne epoxidation with H_2O_2 using hydrotalcite catalysts were limited to the investigation of several factors affecting the selectivity to epoxide such as: the influence of the solvent and the activation agent [1], the influence of the Mg/Al ratio, the presence of pure hydrotalcite phase in the samples, the reconstruction rate of the HT-like phase during the reaction and the addition of water in the reaction mixture [33]. The Mg/Al ratio is a factor particularly influencing the basicity of the catalysts which increases with the Mg content. In a previous work, we investigated the effect of tuning the basicity of the HT structure by inclusion of very small amounts of Zn or Ga in the brucite-type layer on the epoxidation of cyclohexene in the presence of benzonitrile as a reductant agent, showing that Zn- and Ga-containing hydrotalcites presented higher activity than the Mg/Al hydrotalcite for the oxidation of cyclohexene to cyclohexene oxide in the initial stages of the process [34]. One of the issues occurring when Mg/Al LDHs are used as catalysts is that these structures are sensitive to the action of acids such as the peroxycarboximidic acid intermediate formed during Payne epoxidation with H_2O_2. An opportunity to overcome this issue would be the utilization of an LDH structure such as Zn/Al, which possesses the active basic sites but is also more resistant to acid action. To our knowledge, neither pure Zn/Al LDH nor ternary compositions of MgZn/Al with more than 5 mol% Zn in the brucite-type layer have been tested until now as catalysts for cyclohexene epoxidation. There are only two references concerning Zn/Al LDH used as a host for catalytic active guest species such as metallophtalocyanines [38], metatungstate and tungstoniobate [39] in the epoxidation of olefins with O_2.

Based on the above, the aim of this contribution was to bring new insights into the influence of the chemical composition of a brucite-type layer on the catalytic activity for selective cyclohexene epoxidation with H_2O_2 in the presence of acetonitrile by extending the studies from Mg_xAl to Zn_xAl (x = 1–5) and Mg_yZn_zAl (y + z = 4, y = 1, 2, 3)-type hydrotalcites. It was presumed that the utilization of LDH structures containing larger amounts of Zn in the brucite-type layer could bring the benefit of a better stability towards the action of the peroxycarboximidic acid intermediate. The catalytic performances were determined for all fresh, calcined and reconstructed LDH samples and were correlated with the physico-chemical characteristics of the solids in order to foresee a selection criterion of the optimal composition.

2. Results and Discussion

2.1. Characterization of Catalysts

The chemical compositions of the freshly synthesized samples are presented in Table 1. The results show that the amount of water in the interlayer ranges within 6–18% of the total weight, in concordance with the literature [4].

Table 1. The chemical composition of the prepared solid samples.

Samples	The Composition Percent (% Mass)						Mg/Al Molar Ratio	Zn/Al Molar Ratio	CO_3^{2-}/Al Molar Ratio	H_2O/Al Molar Ratio
	Mg	Zn	Al	CO_3^{2-}	NO_3^-	H_2O				
HT Mg_2Al	19.44	n.d.	11.19	14.17	n.d.	17.98	1.93	-	0.57	2.41
HT $Mg_{2.5}Al$	21.07	n.d.	9.55	11.68	n.d.	15.60	2.45	-	0.55	2.45
HT Mg_3Al	22.64	n.d.	8.98	10.98	n.d.	14.97	2.80	-	0.55	2.50
HT Mg_4Al	24.86	n.d.	7.46	8.79	n.d.	14.98	3.70	-	0.53	3.01
HT Mg_5Al	27.64	n.d.	6.40	7.39	n.d.	11.99	4.80	-	0.52	2.81
HT Zn_2Al	n.d.	37.83	8.87	10.25	0.87	12.42	-	1.76	0.52	2.10
HT $Zn_{2.5}Al$	n.d.	40.44	7.98	9.01	0.91	11.66	-	2.10	0.51	2.20
HT Zn_3Al	n.d.	43.20	7.08	8.34	2.12	10.29	-	2.52	0.53	2.18
HT Zn_4Al	n.d.	46.65	6.13	7.02	3.25	8.34	-	3.14	0.52	2.04
HT Zn_5Al	n.d.	50.75	4.81	5.87	4.48	6.44	-	4.36	0.55	2.01
HT $MgZn_3Al$	4.92	40.12	5.56	6.55	n.d.	10.24	0.98	2.98	0.53	2.76
HT Mg_2Zn_2Al	11.10	30.21	6.01	6.92	n.d.	10.55	2.05	2.07	0.52	2.63
HT Mg_3ZnAl	18.10	16.02	6.67	7.61	n.d.	11.03	3.01	0.99	0.51	2.48

HT Mg_xAl samples retain more water than **HT Zn_xAl** ones. Looking at the columns where the molar ratios between each component and Al are presented, it is noticed that the preparation by co-precipitation at pH 10 leads to a lower incorporation of Zn than the one expected considering the molar ratios in the starting solutions, which is in agreement with literature data [40,41]. This effect is less pronounced for Mg, where the molar ratios of Mg/Al are closer to the theoretical ones. An explanation for this effect is that at pH 10, Zn species are more soluble than Mg ones [42,43]. The ratio of CO_3^{2-}/Al always exceeds the theoretical value of 0.5, a fact that may be related to the excess carbonate used in the preparation as well as the high affinity of the cations to this anion. The presence of nitrate ions was evidenced in small concentrations only in the **HT Zn_xAl** samples. This may be a consequence of the formation of a hydroxynitrate impurity during the co-precipitation and aging of Zn-containing samples [44].

The XRD patterns of the synthesized samples are presented in Figures 1–3 and the calculated structural parameters are presented in the Supplementary Information (Tables S1–S3). The XRD patterns of **HT Mg_xAl** samples present the typical lines of layered materials, as shown in Figure 1 [4,8]. Their intensity decreases with the increase in the Mg/Al ratio because part of Mg forms an amorphous hydroxyde phase [45]. Meanwhile, in the diffraction patterns of **HT Zn$_x$Al**, besides the typical diffraction lines of layered materials, some of the characteristic lines for a zinc hydroxy nitrate Zn(OH)$_4$(NO$_3$)$_2$ impurity phase marked with * in Figure 1 are noticed. This additional phase increases its concentration along with x value from 2 to 5, but it is missing in the XRD patterns of **HT Mg$_y$Zn$_z$Al** samples. The partial substitution of Mg with Zn leads to the decrease in the crystallinity compared to Mg$_4$Al and Zn$_4$Al samples.

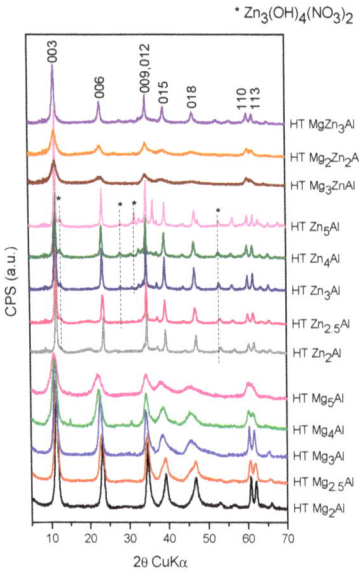

Figure 1. The XRD patterns of the freshly prepared HT materials.

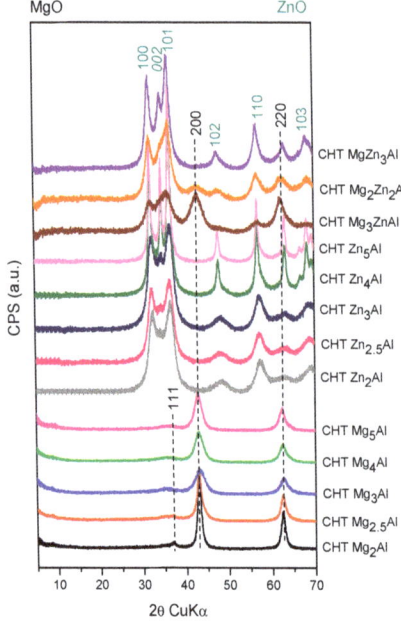

Figure 2. The XRD patterns of the calcined materials.

Catalysts **2022**, *12*, 145

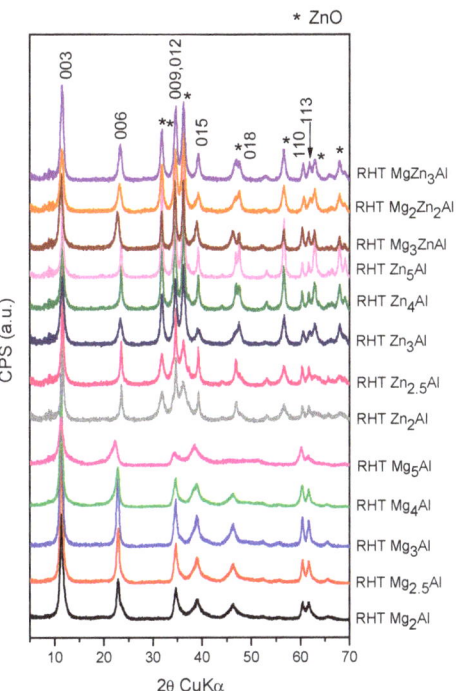

Figure 3. The XRD patterns of the reconstructed samples.

The diffraction patterns of the calcined samples, shown in Figure 2, present the characteristic reflections of the mixed oxides Mg(Al)O (for CHT Mg_xAl) and zincite Zn(Al)O (for **CHT Zn_xAl**) and a mixture of these oxides in the case of **CHT Mg_yZn_zAl** samples. The intensity of the diffraction lines for MgO-periclase-type (JCPDS-45-0946) in the patterns of **CHT Mg_xAl** decreases with the increase in x. Meanwhile, in the diffraction patterns of **CHT Mg_yZn_zAl** samples, the periclase-type phase intensity from the corresponding mixed oxides decreases with decreasing amounts of Mg cation to almost nothing for the CHT $MgZn_3$Al sample.

The diffraction patterns in Figure 3 show that a good reconstruction by the memory effect is possible for a Mg/Al ratio lower than 5, the optimum ratio for a good reconstruction being 3 which is in agreement with literature [4]. For **RHT Zn_xAl** samples, the reconstruction of the layered structure by the memory effect is less pronounced than that noticed for **RHT Mg_xAl** samples. The zincite phase is also always present as an independent phase in the XRD patterns of **RHT Zn_xAl** and **RHT Mg_yZn_zAl** samples, and the intensity of its corresponding maxima denoted with * in Figure 3 increases with Zn content. This aspect was also reported by other authors working with ZnAl LDHs [46].

The DR-UV–Vis spectra of the freshly prepared Zn-containing samples, **HT Zn_xAl** and **HT Mg_yZn_zAl**, shown in Figure 4A, show the presence of the band corresponding to Zn(Al)O at 208 nm, in agreement with the findings of XRD analyses. The intensity of this band increases with the Zn content. The spectra of the corresponding RHT samples, shown in Figure 4B, show multiple bands due to the preservation of the Zn(Al)O phase even after hydration. In the spectra of **RHT Mg_yZn_zAl**, it is noticed that the increasing Mg content in the brucite-type layer leads to a shifting of λ_{max} to higher wavelengths (274 < 292 < 300 < 304 nm), indicating a decrease in the M-O bond strength. As a partial conclusion, DR-UV–Vis spectra confirm the poor reconstruction of Zn-containing samples.

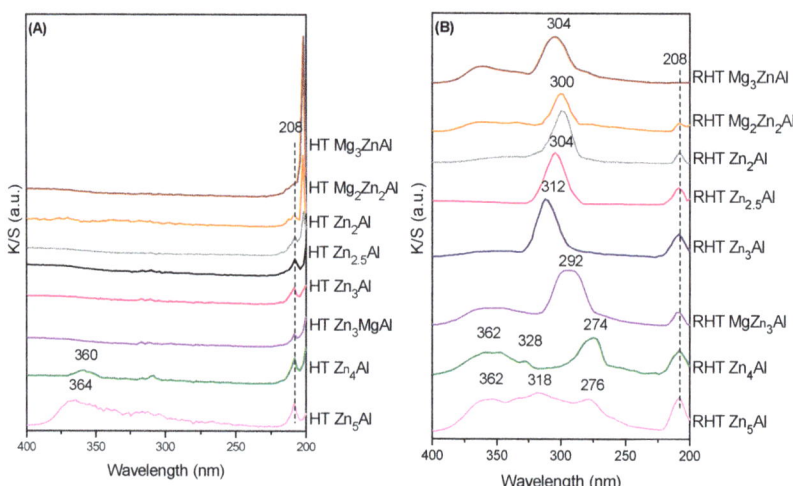

Figure 4. DR-UV–Vis spectra of the Zn-modified layered materials; (**A**) freshly prepared HT samples and (**B**) reconstructed samples.

For **HT Mg$_x$Al** samples, shown in Figure 5A, it was found that DRIFTS spectra present, in the region 3500–3700 cm^{-1}, a broad band due to the stretching vibration of the OH groups present in the lamellar structures. The position of the maximum is shifted to higher wavenumbers as the Mg/Al ratio increases. A shoulder of this broad band at approx. 3167 cm^{-1} is due to the hydrogen bonds formed with carbonate anions in the interlayer space. The characteristic deformation vibration of water at approx. 1600 cm^{-1} is slightly shifted to higher wavenumbers due to the bridged-type interaction of CO$_3$$^{2-}$-H$_2$O between water molecules and carbonate anions from the interlayer. The effects due to the interaction of carbonate counter ions with superficial hydroxyl groups observed in the spectral region beyond 3000 cm^{-1} are confirmed by specific bands characteristic of the asymmetric stretching vibration (ν_3) from free carbonate groups in D$_{3h}$ symmetry in the region from 1420 to 1360 cm^{-1}. The increasing intensity of this band characteristic for carbonate groups certifies the increasing concentration of this compensation anion in the system as the Mg/Al ratio decreases. This conclusion is also supported by the fact that the characteristic peak for carbonate ν_4 vibration (700 cm^{-1}) is observed only for samples with Mg/Al ratio < 3 and the characteristic maximum derived from the overlapping of carbonate ν_2 and ν_1 bands shifts its position from 935 to 1007 cm^{-1} when the Mg/Al ratio decreases (i.e., as the interlayer carbonate amount increases). An additional band at about 1500 cm^{-1} appears in the spectra of the samples that contain an excedentary brucite-like phase [47,48].

The DRIFT spectra profiles of **HT Zn$_x$Al**, shown in Figure 5B, are similar to those obtained for **Mg$_x$Al** compounds. In the region 3700–3000 cm^{-1}, the band corresponding to hydroxyl groups is shifted towards lower wavenumbers with decreasing Zn/Al ratio (3493; 3507; 3497; 3485 and 3489 cm^{-1} for Zn/Al ratios = 5; 4; 3; 2.5; 2) as in the case of **Mg$_x$Al** compounds. Compared to **Mg$_x$Al** solids, the position of the maximum of bands is located at smaller wavenumbers, suggesting a lower basicity of Zn-containing solids. The shift of the shoulder from 3167 cm^{-1} at 3064 cm^{-1} (which is more pronounced for **Zn$_3$Al**) indicates that the hydrogen bridges of OH ions with the counter anions and water molecules in the interlayer are less intense than in **HT Mg$_x$Al** series. The band at 1601 cm^{-1} (δ_{H2O}) is clearly visible only for the sample richest in Al, **Zn$_2$Al**, indicating a lower confinement of water in the interlayer of **HT Zn$_x$Al** compared to **HT Mg$_x$Al**. In the spectra of the **HT Zn$_x$Al** samples, the band characteristic of carbonate around 1423 cm^{-1} overlaps those characteristic of nitrate. A shift of the maximum to lower wavenumbers is clearly noticed for **HT Zn$_3$Al**, **HT Zn$_4$Al** and **HT Zn$_5$Al** samples which, according to the results of the chemical analyses, contain higher concentrations of nitrate (see Table 1). By comparing the DRIFT of **HT Mg$_x$Al** and **HT Zn$_x$Al**, shown in Figure 5A,B, it can be concluded that the first class of materials possess OH groups with a more pronounced basic character.

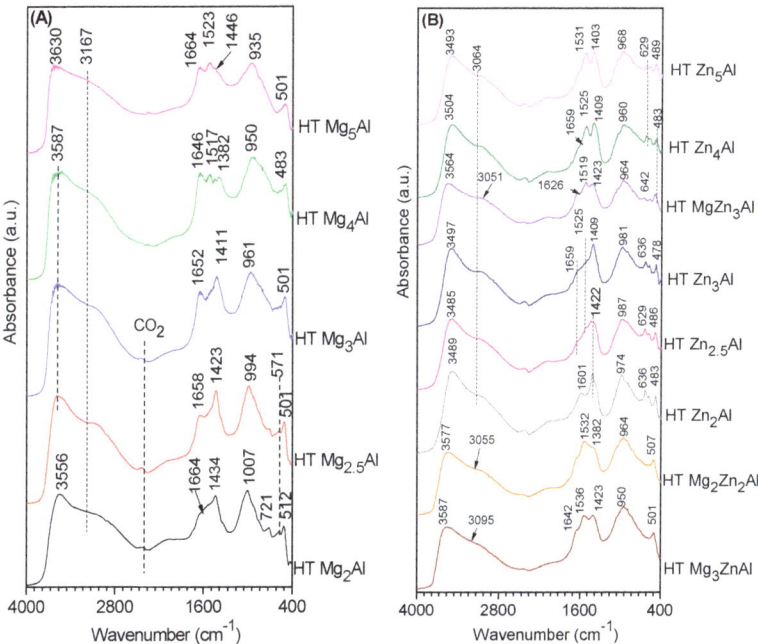

Figure 5. DRIFT spectra of freshly prepared HT samples: (**A**) Mg$_x$Al, (**B**) Zn$_x$Al and Mg$_y$Zn$_z$Al.

The DRIFT spectra of the corresponding mixed oxides, shown in Figure 6A,B, show that the relative intensity of the bands in the region 4000–3000 cm^{-1} and in the mid-infrared region decreases significantly compared to those in the 1000–400 cm^{-1} region, indicating that OH groups and even small amounts of carbonate are still present in both series of samples. This fact confirms the literature data claiming that the calcination of the solids at 460 °C does not allow total dehydroxylation and decarbonation [49].

The spectra of **RHT Mg$_x$Al**, shown in Figure 7A, present the maximum of the band corresponding to ν_{OH} shifted by 100 cm^{-1} to higher wavelengths due to the larger number of OH$^-$ introduced by the memory effect compared to the spectra of the **RHT Zn$_x$Al** samples, shown in Figure 7B, a fact that suggests an increased basicity of the Mg-containing samples. After reconstruction, in the spectra of Zn-containing samples, the band for ν_{OH} is less shifted to higher wavenumbers compared to its position in the spectra of corresponding HT samples.

The results in Table 2 point out that the samples in the **Zn$_x$Al** series have a smaller specific surface, smaller specific pore volume and larger mesopores compared to those in the **Mg$_x$Al** series. The same trend is noticed for the total number of base sites and the weak and medium-strength base sites. As for the number of strong base sites, **HT Zn$_4$Al** and **HT Zn$_5$Al** have more base sites than the related **Mg$_x$Al** samples due to the presence of the zincite phase that has O^{2-} strong base sites.

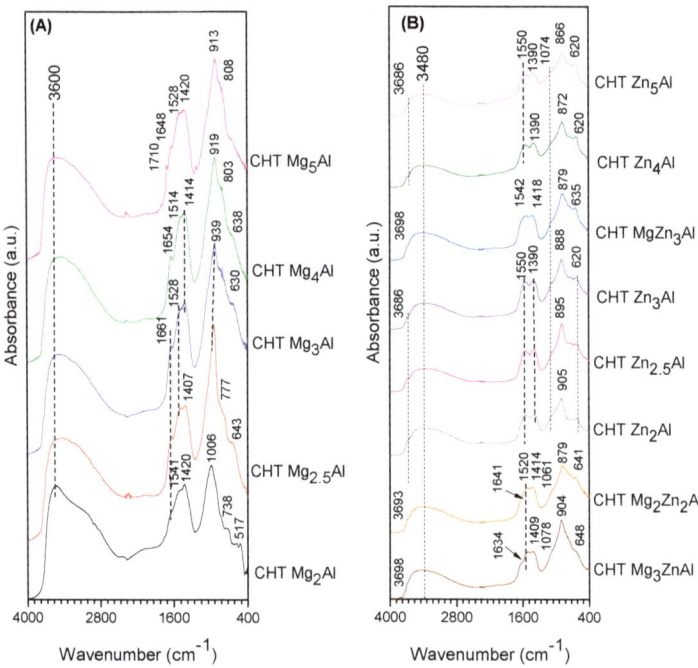

Figure 6. DRIFT spectra of mixed oxide CHT samples, (**A**) Mg$_x$Al, (**B**) Zn$_x$Al and Mg$_y$Zn$_z$Al.

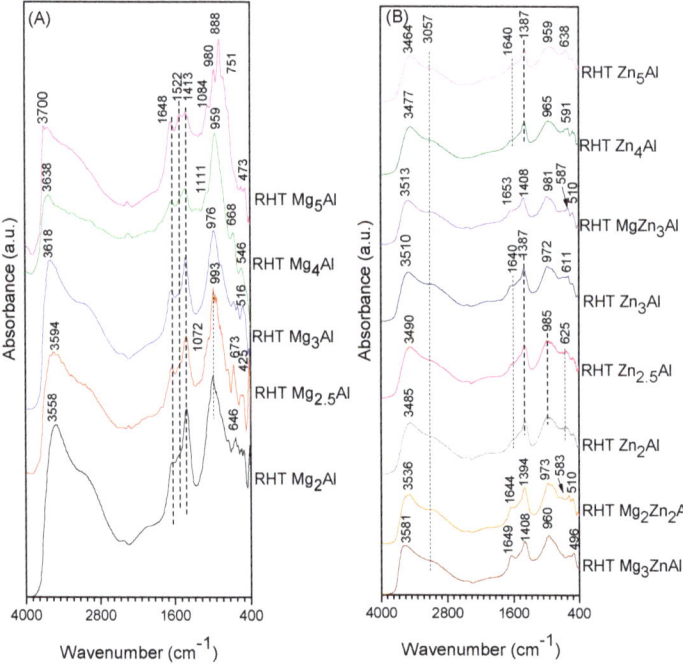

Figure 7. DRIFT spectra of reconstructed RHT samples, (**A**) Mg$_x$Al, (**B**) Zn$_x$Al and Mg$_y$Zn$_z$Al.

Table 2. Textural properties and basicities of HT samples.

Sample	S_{sp} $(m^2 \cdot g^{-1})$	Pore Volume $(cm^3 \cdot g^{-1})$	Pore Radius (Å)	Total Basicity $(mmol\ AA \cdot g^{-1})$	Strong Basic Sites $(mmol\ PhOH \cdot g^{-1})$	Weak + Medium Basic Sites * $(mmol \cdot g^{-1})$
HT Mg$_2$Al	141	0.734	207	4.05	0.25	3.80
HT Mg$_{2.5}$Al	144	0.741	204	4.92	0.32	4.60
HT Mg$_3$Al	89	0.619	251	6.56	0.26	6.30
HT Mg$_4$Al	84	0.592	260	6.95	0.35	6.60
HT Mg$_5$Al	52	0.418	188	7.00	0.10	6.90
HT Zn$_2$Al	103	0.650	239	3.65	0.15	3.50
HT Zn$_{2.5}$Al	110	0.665	233	3.94	0.24	3.70
HT Zn$_3$Al	83	0.606	256	4.70	0.30	4.40
HT Zn$_4$Al	75	0.549	273	5.47	0.37	5.10
HT Zn$_5$Al	43	0.517	290	6.42	0.42	6.00
HT Mg$_3$ZnAl	136	0.723	211	7.25	0.50	6.75
HT Mg$_2$Zn$_2$Al	102	0.648	240	6.54	0.44	6.10
HT MgZn$_3$Al	76	0.588	263	5.60	0.40	5.20

* The number of weak + medium-strength base sites is given by subtracting the number of strong base sites from the total basicity.

The samples with ternary composition, **HT Mg$_3$ZnAl** and **HT Mg$_2$Zn$_2$Al**, have a larger specific surface area and greater porosity than the reference binary compositions. The total basicity of ternary samples is higher than that of HT Zn$_4$Al due to the presence of Mg that has a lower Pauling electronegativity than Zn (e.g., 1.31 < 1.65).

The data in Table 3 show a trend of the textural properties and the basicity variation for the calcined samples in the **Zn$_x$Al** series compared to those in the **Mg$_x$Al** series, similar to that suggested by the data in Table 2 concerning HT samples.

Table 3. Textural properties and basicities of CHT samples.

Sample	S_{sp} $(m^2 \cdot g^{-1})$	Pore Volume $(cm^3 \cdot g^{-1})$	Pore Radius (Å)	Total Basicity $(mmol\ AA \cdot g^{-1})$	Strong Basic Sites $(mmol\ PhOH \cdot g^{-1})$	Weak + Medium Basic Sites * $(mmol \cdot g^{-1})$
CHT Mg$_2$Al	192	0.518	108	6.12	0.16	5.96
CHT Mg$_{2.5}$Al	196	0.646	110	8.10	0.24	7.86
CHT Mg$_3$Al	188	0.475	150	8.36	0.38	7.98
CHT Mg$_4$Al	230	0.754	120	9.17	0.47	8.70
CHT Mg$_5$Al	198	0.653	109	11.43	0.78	10.65
CHT Zn$_2$Al	140	0.732	206	4.88	0.22	4.66
CHT Zn$_{2.5}$Al	150	0.784	192	6.41	0.39	6.02
CHT Zn$_3$Al	132	0.433	209	7.40	0.40	7.00
CHT Zn$_4$Al	202	0.662	137	8.59	0.42	8.17
CHT Zn$_5$Al	164	0.537	169	10.50	0.70	9.80
CHT Mg$_3$ZnAl	262	0.859	105	10.50	0.51	9.99
CHT Mg$_2$Zn$_2$Al	248	0.813	118	9.04	0.44	8.60
CHT MgZn$_3$Al	238	0.78	116	8.64	0.43	8.21

* The number of weak + medium-strength base sites is given by subtracting the number of strong base sites from the total basicity.

The textural data displayed in Table 4 show that in the case of the reconstructed samples, the specific surface areas are significantly smaller than those of the HT samples, as was also noticed by other authors who investigated the memory effect [50,51]. However, the trends of the textural properties and the basicity variation for the **Zn$_x$Al** series compared to those in the **Mg$_x$Al** series are different than those noticed for HT samples (Table 2) and CHT samples (Table 3).

Table 4. Textural properties and basicities of RHT samples.

Sample	S_{sp} (m^2·g^{-1})	Pore Volume (cm^3·g^{-1})	Pore Radius (Å)	Total Basicity (mmol AA·g^{-1})	Strong Basic Sites (mmol PhOH·g^{-1})	Weak + Medium Basic Sites * (mmol·g^{-1})
RHT Mg$_2$Al	15	0.117	285	4.87	0.17	4.70
RHT Mg$_{2.5}$Al	26	0.183	236	5.96	0.26	5.70
RHT Mg$_3$Al	7	0.025	640	7.21	0.34	6.87
RHT Mg$_4$Al	5	0.018	895	7.42	0.52	6.90
RHT Mg$_5$Al	18	0.141	238	8.47	0.57	7.90
RHT Zn$_2$Al	21	0.164	204	4.68	0.18	4.50
RHT Zn$_{2.5}$Al	39	0.239	196	5.64	0.24	5.40
RHT Zn$_3$Al	42	0.257	182	5.80	0.30	5.50
RHT Zn$_4$Al	18	0.149	209	6.22	0.42	5.80
RHT Zn$_5$Al	19	0.154	208	7.11	0.51	6.60
RHT Mg$_3$ZnAl	28	0.197	219	8.78	0.58	8.20
RHT Mg$_2$Zn$_2$Al	42	0.296	146	8.25	0.55	7.70
RHT MgZn$_3$Al	34	0.239	180	8.04	0.54	7.50

* The number of weak + medium-strength base sites is given by subtracting the number of strong base sites from the total basicity.

The fact that Zn-containing samples have larger specific surface areas than the corresponding ones containing Mg may be the consequence of the presence of the zincite secondary phase (indicated in Figure 3) which remains between the reconstructed HT-phase crystallites.

2.2. Catalytic Activity

According to Scheme 1, besides the epoxide compound which is the main product, other by-products appear after some successive and parallel reactions.

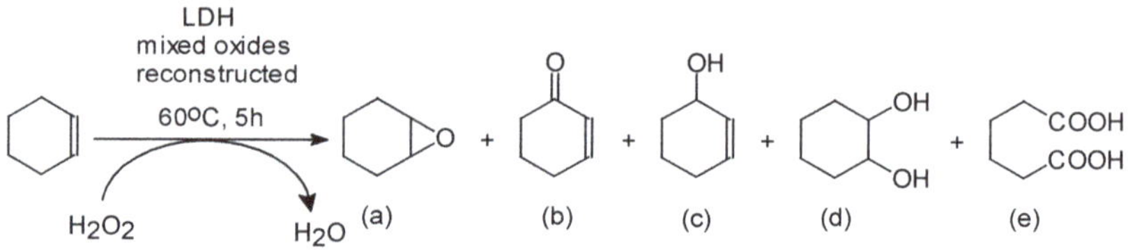

Scheme 1. Possible products for the oxidation of cyclohexene (epoxycyclohexane (**a**); 2-cyclohexene-1-one (**b**); 2-cyclohexene-1-ol (**c**); cyclohexane-1,2-diol (**d**); adipic acid (**e**)).

The blank test performed without catalyst at 60 °C for 5 h leads to a conversion of cyclohexene of 9.2%, and yield to cyclohexene oxide of 2.5%. The decomposition of hydrogen peroxide in the presence of the catalysts under similar conditions in the absence of cyclohexene and acetonitrile was found to be in the 4–5% range and therefore it was considered that the influence of the solid base catalysts on the decomposition of hydrogen peroxide is low and does not significantly influence the catalytic oxidation of cyclohexene.

The results of the catalytic tests are presented in Figure 8A–C. Since all catalysts were extremely selective for epoxidation, and the total concentration of by-products in the reaction mixture was in the range of 0.5–1 mol%, we have presented the yields to cyclohexeneoxide after 5 h.

The increasing of the Mg/Al ratio from 2 to 5, shown in Figure 8A, led to the increase in the yield to epoxide from 38%, for **HT Mg$_2$Al**, up to 82% for **HT Mg$_5$Al**. The same trend, but with values lower by around 10%, is noticed for the increasing of the Zn/Al ratio in the same range (e.g., 26% for **HT Zn$_2$Al** and 69% for **HT Zn$_5$Al**). The lower yield values obtained are a consequence of both the smaller surface areas and the inferior basicity of **HT Zn$_x$Al** samples compared to **HT Mg$_x$Al** considering the slightly higher electronegativity of Zn (1.65) compared to Mg (1.31). It could also be due to the presence in these samples of the zinc hydroxynitrate impurity. The yield to cyclohexene oxide from the samples with tricationic compositions of Mg$_y$Zn$_z$Al varies similarly to Mg content, and it is higher than the one obtained from HT Zn$_4$Al. Only the tricationic **HT Mg$_3$ZnAl** sample leads to higher yields than both **HT Zn₄Al** and **HT Mg₄Al**. This fact may be a consequence of its larger specific surface area compared to the reference bi-cationic samples. The mixed oxides show the highest activity of cyclohexeneoxide formation compared to the homologue materials HT and RHT, mostly due to their larger specific surface area and more emphasized basic character, as shown in Figure 8B. The variation of the yield as a function of the chemical composition follows the same trend as the one noticed for the freshly prepared HT materials.

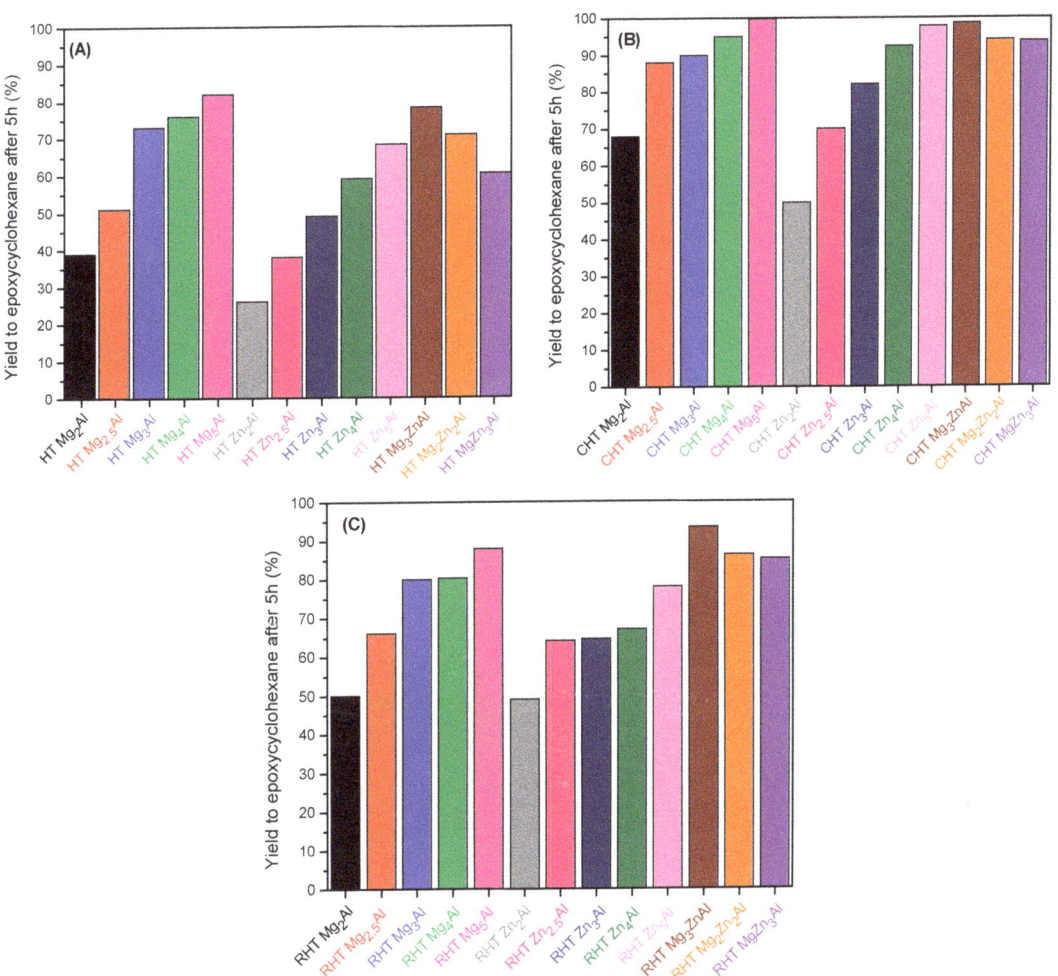

Figure 8. The variation of yield to epoxycyclohexane after 5 h on the investigated catalysts, (**A**) freshly prepared HT samples; (**B**) calcined CHT samples; (**C**) reconstructed RHT samples.

This trend is not noticed for Zn-containing reconstructed samples, shown in Figure 8C. In this case, the sample with the highest activity is **RHT Mg₃ZnAl** instead of **RHT Mg₅Al**. All the RHT samples have higher activity than the corresponding freshly prepared HT samples. As shown in Table 4, the overall basicity of these samples is higher than that of HT (Table 2), even though the specific surface area is smaller. The type of base site is also very important. Di Cosimo et al. [52] showed that the strongest base sites are O^{2-} anions, followed by OH^-, bridge cation-O and hydrogen carbonate or carbonate anions, which are considered as being medium-strength and weak base sites. In several of our early works, we found that the epoxidation is favored mostly according to the number of weak and medium-strength base sites [33,34]. A correlation of the yield to cyclohexeneoxide and the number of weak and medium-strength base sites of the investigated catalysts is presented in Figure 9.

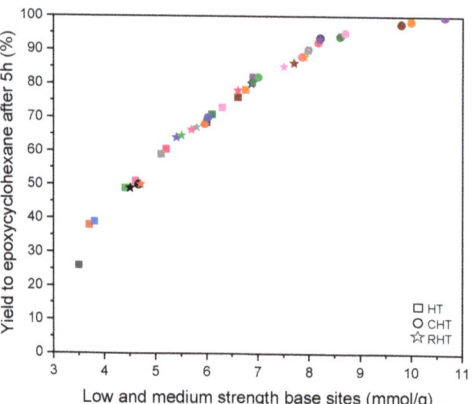

Figure 9. Variation of the yield to epoxide vs. the number of low- and medium-strength base sites (the colors of the symbols represent the sample compositions: Mg$_2$Al, Mg$_{2.5}$Al, Mg$_3$Al, Mg$_4$Al, Mg$_5$Al, Zn$_2$Al, Zn$_{2.5}$Al, Zn$_3$Al, Zn$_4$Al, Zn$_5$Al, Mg$_3$ZnAl, Mg$_2$Zn$_2$Al, MgZn$_3$Al).

The possibility of recycling was investigated only for the catalysts that gave yields to cyclohexeneoxide higher than 95% (e.g., **CHT Mg$_4$Al**, **CHT Zn$_5$Al**, **CHTMg$_3$ZnAl**, **CHT Mg$_5$Al**). The results obtained in five consecutive cycles are presented in Table 5. For the **CHT Mg$_5$Al** sample, the yield decreases after the first cycle with 10%, and falls as low as 75% after the fifth cycle. For **CHT Mg$_4$Al**, the yield decreases from 95% to 82% after the first cycle and finally to 64% after the fifth reaction cycle. This behavior is due, on one hand, to the structural modification by reconstruction under the influence of the water and, on the other hand, to the partial dissolution of Mg species by the peroxymidic acid intermediate. The analysis by AAS of the cations from the liquid reaction mixtures obtained after five reaction cycles revealed that the amount of Mg leached from CHT Mg$_5$Al was around 60 ppm, which corresponded to 2.5% w/w Mg loss from the catalyst, while for CHT Mg$_4$Al, the leached Mg was 40 ppm, which corresponded to 1.8% w/w Mg loss from the catalyst. The level of Al leached in the reaction mixture was under the detection limit. On the contrary, for **CHT Zn$_5$Al**, the yield decreases only by 4.1% after the first reaction cycle (93.6%) and drops to 85.2% after the fifth cycle. This fact is due to the better stability towards the action of the acid intermediate and poor reconstruction ability of the solid. A similar trend is noticed for **CHT Mg$_3$ZnAl** which loses only 10% of the yield to cyclohexeneoxide after five reaction cycles, suggesting a synergetic effect between Mg and Zn in this ternary composition. In the case of Zn-containing samples, the levels of cation concentration in the reaction mixtures recovered after five reaction cycles were all under the detection limit. Hence, it seems that the presence of Zn stabilizes the Mg in the framework. This fact was also indicated by the XRD patterns of the materials used in the catalytic tests before and after their regeneration by re-calcination (Figure 10).

Table 5. Results obtained in repeated reaction cycles.

Catalyst	Yield to Epoxycyclohexane After 5 h (%)				
	Cycle 1	Cycle 2	Cycle 3	Cycle 4	Cycle 5
CHT Mg$_5$Al	99.8	89.8	84.3	79.2	75.0
CHT Mg$_4$Al	95.0	82.0	76.3	69.7	64.0
CHT Zn$_5$Al	97.7	93.6	89.2	87.5	85.2
CHT Mg$_3$ZnAl	98.5	95.3	92.4	90.5	88.5

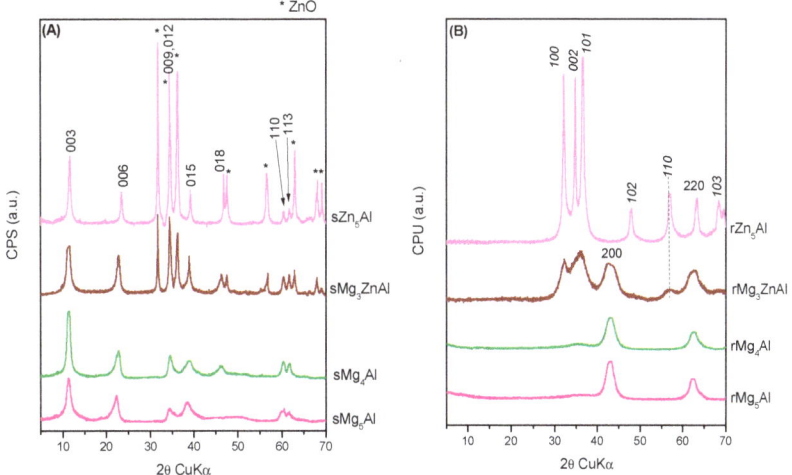

Figure 10. XRD patterns of the spent (A) (the prefix "s" added to the names of the samples) and recycled samples (B) (the prefix "r" added to the names of the samples, diffraction lines corresponding to Mg(Al)O—normal text, diffraction lines of Zn(Al)O—*italics*).

The XRD patterns of the solid samples recovered after the five reaction cycles presented in Figure 10A show that the LDH structure is recovered to some extent during the reaction. Meanwhile, the XRD patterns of the re-calcined samples regenerated after the reaction (Figure 10B) exhibit similar features to those of the parent CHT samples (see Figure 2), showing that the structure is maintained after regeneration.

3. Materials and Methods

3.1. Catalyst Preparation

The catalysts were synthesized using chemicals of analytical purity grade, $Mg(NO_3)_2 \cdot 6H_2O$, $Zn(NO_3)_2 \cdot 6H_2O$, $Al(NO_3)_3 \cdot 9H_2O$, NaOH and Na_2CO_3 were purchased from Merck, Darmstadt, Germany. Three sets of LDH materials were prepared by co-precipitation at pH 10 under low supersaturation conditions [53,54]: (i) Mg$_x$Al; (ii) Zn$_x$Al (with x = 2; 2.5; 3; 4; 5) and (iii) Mg$_y$Zn$_z$Al (y + z = 4, y = 1, 2, 3). Two aqueous solutions were prepared using the metal nitrate precursors (solution MN) and a mixture of NaOH and Na_2CO_3 (solution B). Solution MN contained the required amounts of dissolved metal nitrates for the desired values of x, y and z, at a total concentration of cations of 1.5 M. Solution B contained a molar amount of Na_2CO_3 1.5 times higher than the molar amount of Al in the MN solution while the amount of NaOH was adjusted in order to reach a total Na concentration of 3 M. Both solutions were simultaneously fed into the precipitation reactor with a feeding flow of 60 mL·h^{-1} at room temperature under vigorous stirring at 600 rpm. The obtained gel was aged for 18 h at 75 °C in mother liquor, and cooled afterwards to room temperature. Then, the solid was separated by vacuum filtration and washed with bi-distilled water until the conductivity of the washing water dropped below 100 µS·cm^{-1}. The washed solids were dried at 90 °C for 24 h under air flow. The freshly prepared solid samples were called

HT Mg$_x$Al, HT Zn$_x$Al and HT Mg$_y$Zn$_z$Al. These materials were calcined at 460 °C for 18 h in air flow to obtain the mixed oxides (**CHT Mg$_x$Al, CHT Zn$_x$Al, CHT Mg$_y$Zn$_z$Al**). The reconstruction of the hydrotalcite-like structure was performed by immersion of these calcined solids in bi-distilled water for 24 h at 25 °C, followed by drying at 90 °C for 24 h in air flow. The reconstructed materials were named **RHT Mg$_x$Al, RHT Zn$_x$Al** and **RHT Mg$_y$Zn$_z$Al**.

3.2. Catalyst Characterization

The chemical composition of the samples was determined by atomic absorption spectrometry (AAS) for the determination of Mg, Zn and Al using a Thermoelemental Solar AAS spectrophotometer (ThermoFischer SCIENTIFIC, Waltham, MA, USA). The inorganic carbon analysis (TIC) was performed with the UV-persulfate oxidation method using a HiPerTOC carbon analyzer (ThermoFischer Scientific, Waltham, MA, USA), which measures the IR absorbance of carbon dioxide produced, and determining weight loss between 105 and 200 °C to determine the amount of hydration water. The total nitrogen content was also determined with the HiPerTOC analyzer using a method based on the catalytic combustion of the samples when the nitrogen is converted into nitric oxide (NO). The NO in the gas outlet from the combustion reactor was further oxidized with ozone in a reaction chamber, generating excited nitrogen dioxide (NO$_2$*). The emitted radiation by the nitrogen dioxide when returning to the ground state was then quantified by a photomultiplier tube detector.

Powder X-ray diffraction (XRD) patterns were collected on a PANalytical X'Pert MPD system (Almelo, Netherlands) equipped with monchromatic CuKα radiation (λ = 1.5418 Å) in a continuous scan mode (counting 2 s/ 0.02° 2θ) over a 5–70° 2θ range. The PANalytical HighScore software package was used for the analysis of the XRD patterns.

DR-UV–Vis spectra were recorded with a Shimadzu 3600 UV–Vis NIR spectrometer (Shimadzu, Kyoto, Japan) equipped with an integration sphere using BaSO$_4$ as white reference.

DRIFTS spectra obtained from accumulation of 400 scans in the domain 400–4000 cm^{-1} were recorded with a NICOLET 4700 spectrometer (ThermoFisher Scientific, Waltham, MA, USA). A KBr spectrum was used as background.

N$_2$ adsorption–desorption isotherms were determined using a Micromeritics ASAP 2010 instrument (Micromeritics, Eindhoven, Netherlands). Prior to nitrogen adsorption, samples were degassed under vacuum for 24 h.

The basic character of the catalysts was determined using a method based on the irreversible adsorption of organic acids of different pK$_a$ values corresponding to the total number of base sites, e.g., acrylic acid, pK$_a$ = 4.2, and strong base sites, e.g., phenol pK$_a$ = 9.9 [55–58]. The number of weak and medium base sites is given by the difference between the amounts of adsorbed acrylic acid and phenol. The acrylic acid and phenol amounts that remained in solution were determined by UV–Vis spectrometry using the Jasco V650 UV–Vis spectrometer (Jasco, Tokyo, Japan). Three sets of determinations were performed for each solid and the error domaine was +/− 1%.

3.3. Catalytic Tests

The catalytic activity of the solids was tested in the oxidation of cyclohexene with hydrogen peroxide using acetonitrile as a reductant, as shown in Scheme 2 [30,31]. The reactions were performed in a stirred flask (50 mL) at 60 °C, for a 5 h reaction time. In a typical experiment, cyclohexene (4 mmol) and acetonitrile (32 mmol) were dissolved in 20 mL of solvent (mixture of equal volumes of water and acetone). The amount of H$_2$O$_2$ was calculated in order to reach a molar ratio of cyclohexene/H$_2$O$_2$ of 1/32 and was added dropwise during the reaction. All reagents were purchased from Merck (Darmstadt, Germany). In all reactions, the catalyst concentration was 1% (*w/w*) in the reaction mixture. The reaction was monitored in time using a GC K072320 Thermo Quest Chromatograph (ThermoFisher Scientific, Waltham, MA, USA) equipped with an FID detector and a capillary column of 30 m in length with DB5 stationary phase. The oxidation products were identified by comparison with a standard sample (retention time in GC). The reaction products were also identified by mass spectrometer coupled chromatography, using a GC/MS/MS VARIAN SATURN 2100 T equipped with a CP-SIL 8 CB Low Bleed/MS column of 30 m in length and 0.25 mm in diameter (Varian, Palo Alto, CA, USA). H_2O_2 consumption was determined by an iodometric titration at the end of the reaction. The oxidation process can lead to several products, Scheme 1, such as: epoxycyclohexane (a); 2-cyclohexene-1-one (b); 2-cyclohexene-1-ol (c); cyclohexane-1,2-diol (d); adipic acid (e).

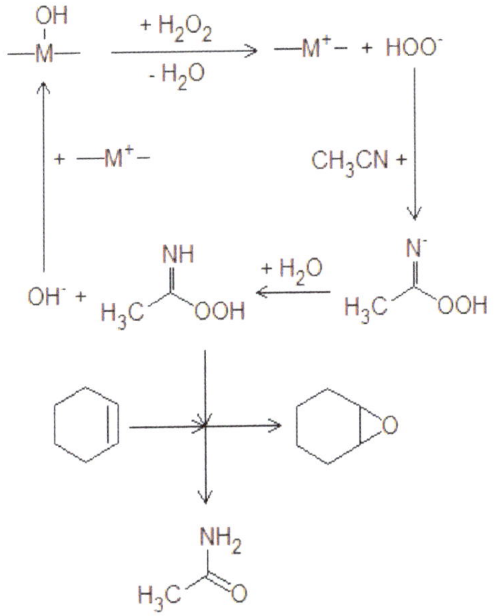

Scheme 2. Reaction mechanism on Brönsted base sites of HT solids.

3.4. Catalyst Recycling

The most active calcined catalysts were selected for recycling tests. The solid catalysts were separated from the reaction mixture by centrifugation and were dried for 5 h at 90 °C and re-calcined at 460 °C for 5 h before being used in the consecutive cycle. XRD patterns of the dried and re-calcined samples after the fifth reaction cycle were recorded. After being analyzed by GC at the end of each cycle, the liquid reaction mixtures recovered for each catalyst were mixed in order to determine the content of cations leached in the liquid using AAS. In order to perform the analysis by AAS, the cations from the liquid samples were extracted in distilled water acidified with HNO_3 (pH 5), and the volume of the aqueous extract was brought to 100 mL in volumetric flasks.

4. Conclusions

The freshly prepared HT Zn_xAl samples contain zinc hydroxynitrate $Zn_3(OH)_4(NO_3)_2$ as an impurity. This impurity is not present in the samples that contain both Mg and Zn. The calcination of dried samples leads to partial removal of OH groups under H_2O and CO_3^{2-} under the CO_2 form. The reconstruction of the initial structure is not total for Zn-containing LDHs since there are still Zn(Al)O species. The reconstruction is also poor for hydrotalcite samples at a Mg/Al ratio higher than 3. All LDH-derived solids investigated were very selective for cyclohexene epoxidation with H_2O_2 under mild reaction conditions. The catalytic activity and the basicity of the samples varies in the order HT (OH^-/CO_3^{2-}) < RHT (majority OH^-) < CHT (O^{2-}). The yield to epoxycyclohexane increases almost linearly when the number of weak and medium-strength base sites in the brucite-type layer rises in the range 4.5–8.5 mmol·g^{-1}. The basicity of Zn-containing samples decreases proportionally with the amount of Zn incorporated, but Zn improves the stability of the solids towards the action of the acid reaction intermediate and hence the maintenance of the catalytic activity. The strength of the base sites is influenced by the electronegativity of the elements in the brucite-type layer (Mg-containing LDHs have stronger base sites than Zn-containing LDHs) (e.g., Mg 1.31 < Zn 1.65). Based on the obtained results, it could be concluded that the optimal composition of the catalyst enabling a high activity and stability would be that with the ratio $Mg/Zn = 3/1$ and $(Mg + Zn)/Al = 4/1$.

Supplementary Materials: The following supporting information can be downloaded at: https://www.mdpi.com/article/10.3390/catal12020145/s1, Table S1. Structural parameters of freshly prepared samples; Table S2. Structural parameters of calcined samples; Table S3. Structural parameters of reconstructed samples.

Author Contributions: Conceptualization, R.Z. and O.D.P.; methodology, R.Z.; investigation, C.B., R.B., O.D.P., R.Z., M.F. and A.C.; resources, R.Z. and M.F.; data curation, R.B.; writing—original draft preparation, R.Z. and O.D.P.; writing—review and editing, R.Z. and O.D.P.; visualization, A.C. and M.F.; supervision, R.B. All authors have read and agreed to the published version of the manuscript.

Funding: APC was sponsored by MDPI.

Data Availability Statement: Not applicable.

Conflicts of Interest: The authors declare no conflict of interest.

References

1. Romero, M.D.; Calles, J.A.; Ocaña, M.A.; Gómez, J.M. Epoxidation of cyclohexene over basic mixed oxides derived from hydrotalcite materials: Activating agent, solvent and catalyst reutilization. *Microporous Mesoporous Mater.* **2008**, *111*, 243–253. [CrossRef]
2. Schmidt, F. New catalyst preparation technologies- observed from an industrial viewpoint. *Appl. Catal. A Gen.* **2001**, *221*, 15–21. [CrossRef]
3. Blaser, H.-U.; Studer, M. The role of catalysis for the clean production of fine chemicals. *Appl. Catal. A Gen.* **1999**, *189*, 191–204. [CrossRef]
4. Cavani, F.; Trifiro, F.; Vaccari, A. Hydrotalcite-type anionic clays: Preparation, properties and applications. *Catal. Today* **1991**, *11*, 173–301. [CrossRef]
5. Costantino, U.; Curini, M.; Montanari, F.; Nocchetti, M.; Rosati, O. Hydrotalcite-like compounds as catalysts in liquid phase organic synthesis: I. Knoevenagel condensation promoted by $[Ni_{0.73}Al_{0.27}(OH)_2](CO_3)_{0.135}$. *J. Mol. Catal. A Chem.* **2003**, *195*, 245–252. [CrossRef]
6. Mantilla, A.; Tzompantzi, F.; Manríquez, M.; Mendoza, G.; Fernández, J.L.; Gómez, R. ZnAlFe mixed oxides obtained from LDH type materials as basic catalyst for the gas phase acetone condensation. *Adv. Mat. Res.* **2010**, *132*, 55–60. [CrossRef]
7. Liu, Y.; Lotero, E.; Goodwin, J.G., Jr.; Mo, X. Transesterification of poultry fat with methanol using Mg–Al hydrotalcite derived catalysts. *Appl. Catal. A Gen.* **2007**, *331*, 138–148. [CrossRef]
8. Pavel, O.D.; Bîrjega, R.; Che, M.; Costentin, G.; Angelescu, E.; Şerban, S. The activity of Mg/Al reconstructed hydrotalcites by "memory effect" in the cyanoethylation reaction. *Catal. Commun.* **2008**, *9*, 1974–1978. [CrossRef]
9. Pavel, O.D.; Cojocaru, B.; Angelescu, E.; Pârvulescu, V.I. The activity of yttrium-modified Mg, Al hydrotalcites in the epoxidation of styrene with hydrogen peroxide. *Appl. Catal. A Gen.* **2011**, *403*, 83–90. [CrossRef]
10. Lavikainen, L.P.; Hirvi, J.T.; Kasa, S.; Pakkanen, T.A. Interaction of octahedral Mg(II) and tetrahedral Al(III) substitutions in aluminium-rich dioctahedral smectites. *Theor. Chem. Acc.* **2016**, *135*, 85. [CrossRef]
11. Sels, B.F.; De Vos, D.E.; Jacobs, P.A. Hydrotalcite-like anionic clays in catalytic organic reactions. *Catal. Rev. Sci. Eng.* **2001**, *43*, 443–488. [CrossRef]
12. Corma, A.; Iborra, S.; Miquel, S.; Primo, J. Catalysts for the production of fine chemicals: Production of food emulsifiers, monoglycerides, by glycerolysis of fats with solid base catalysts. *J. Catal.* **1998**, *173*, 315–321. [CrossRef]
13. Ono, Y. Solid base catalysts for the synthesis of fine chemicals. *J. Catal.* **2003**, *216*, 406–415. [CrossRef]
14. Zhang, F.; Xiang, X.; Li, F.; Duan, X. Layered double hydroxides as catalytic materials: Recent development. *Catal. Surv. Asia* **2008**, *12*, 253–265. [CrossRef]
15. Tichit, D.; Lutic, D.; Coq, B.; Durand, R.; Teissier, R. The aldol condensation of acetaldehyde and heptanal on hydrotalcite-type catalysts. *J. Catal.* **2003**, *219*, 167–175. [CrossRef]
16. Jiao, N.; Stahl, S.S. (Eds.) *Green Oxidation in Organic Synthesis*; John Wiley & Sons Ltd.: Hoboken, NJ, USA, 2019. [CrossRef]
17. Angelescu, E.; Ionescu, R.; Pavel, O.D.; Zăvoianu, R.; Bîrjega, R.; Luculescu, C.R.; Florea, M.; Olar, R. Epoxidation of cyclohexene with O_2 and isobutyraldehyde catalysed by cobalt modified hydrotalcites. *J. Mol. Catal. A Chem.* **2010**, *315*, 178–186. [CrossRef]
18. Litter, M.I.; Candal, R.J.; Meichtry, J.M. (Eds.) *Advanced Oxidation Technologies Sustainable Solutions for Environmental Treatments*; CRC Press: Boca Raton, FL, USA, 2014; ISBN 9781138072886.
19. Philip, R.M.; Radhika, S.; Abdulla, C.M.A.; Anilkumar, G. Recent trends and prospects in homogeneous manganese-catalysed epoxidation. *Adv. Synth. Catal.* **2021**, *363*, 1272–1289. [CrossRef]

20. Dusi, M.; Mallat, T.; Baiker, A. Epoxidation of functionalized olefins over solid catalysts. *Catal. Rev. Sci. Eng.* **2007**, *42*, 213–278. [CrossRef]

21. Hauser, S.A.; Cokoja, M.; Kühn, F.E. Epoxidation of olefins with homogeneous catalysts–Quo vadis? *Catal. Sci. Technol.* **2013**, *3*, 552–561. [CrossRef]

22. Ouidri, S.; Guillard, C.; Caps, V.; Khalaf, H. Epoxidation of olefins on photoirradiated TiO_2-pillared clays. *Appl. Clay Sci.* **2010**, *48*, 431–437. [CrossRef]

23. Cai, X.; Wang, H.; Zhang, Q.; Tong, J.; Lei, Z. Magnetically recyclable core–shell Fe_3O_4@chitosan-Schiff base complexes as efficient catalysts for aerobic oxidation of cyclohexene under mild conditions. *J. Mol. Catal. A Chem.* **2014**, *383–384*, 217–224. [CrossRef]

24. El-Korso, S.; Khaldi, I.; Bedrane, S.; Choukchou-Braham, A.; Thibault-Starzyk, F.; Bachir, R. Liquid phase cyclohexene oxidation over vanadia based catalysts with tert-butyl hydroperoxide: Epoxidation versus allylic oxidation. *J. Mol. Catal. A Chem.* **2014**, *394*, 89–96. [CrossRef]

25. Alfayate, A.; Márquez-Álvarez, C.; Grande-Casas, M.; Sánchez-Sánchez, M.; Pérez-Pariente, J. Ti(III)APO-5 materials as selective catalysts for the allylic oxidation of cyclohexene: Effect of Ti source and Ti content. *Catal. Today* **2014**, *227*, 57–64. [CrossRef]

26. Habibia, D.; Faraji, A.R.; Arshadi, M.; Fierro, J.L.G. Characterization and catalytic activity of a novel Fe nano-catalyst as efficient heterogeneous catalyst for selective oxidation of ethylbenzene, cyclohexene, and benzylalcohol. *J. Mol. Catal. A Chem.* **2013**, *372*, 90–99. [CrossRef]

27. Bujak, P.; Bartczak, P.; Polanski, J. Highly efficient room-temperature oxidation of cyclohexene and d-glucose over nanogold Au/SiO_2 in water. *J. Catal.* **2012**, *295*, 15–21. [CrossRef]

28. Khare, S.; Chokhare, R. Synthesis, characterization and catalytic activity of Fe(Salen) intercalated α-zirconium phosphate for the oxidation of cyclohexene. *J. Mol. Catal. A Chem.* **2011**, *344*, 83–92. [CrossRef]

29. Chagas, P.; Oliveira, H.S.; Mambrini, R.; Le Hyaric, M.; de Almeida, M.V.; Oliveira, L.C.A. A novel hydrofobic niobium oxyhydroxide as catalyst: Selective cyclohexane oxidation to epoxide. *Appl. Catal. A Gen.* **2013**, *454*, 88–92. [CrossRef]

30. Payne, G.B. Reactions of Hydrogen Peroxide. VII. Alkali-catalyzed epoxidation and oxidation using a nitrile as co-reactant. *J. Org. Chem.* **1961**, *26*, 659–663. [CrossRef]

31. Kirm, I.; Medina, F.; Rodriguez, X.; Cesteros, Y.; Salagre, P.; Sueiras, J. Epoxidation of styrene with hydrogen peroxide using hydrotalcites as heterogeneous catalysts. *Appl. Catal. A Gen.* **2004**, *272*, 175–185. [CrossRef]

32. Mureşeanu, M.; Georgescu, I.; Bibire, L.E.; Cârjă, G. CU^{II}(Sal-Ala)/MgAlLDH and CU^{II}(Sal-Phen)/MgAlLDH as novel catalytic systems for cyclohexene oxidation by H_2O_2. *Catal. Commun.* **2014**, *54*, 39–44. [CrossRef]

33. Zăvoianu, R.; Bîrjega, R.; Pavel, O.D.; Cruceanu, A.; Alifanti, M. Hydrotalcite like compounds with low Mo-loading active catalysts for selective oxidation of cyclohexene with hydrogen peroxide. *Appl. Catal. A Gen.* **2005**, *286*, 211–220. [CrossRef]

34. Angelescu, E.; Pavel, O.D.; Bîrjega, R.; Florea, M.; Zăvoianu, R. The impact of the "memory effect" on the catalytic activity of Mg/Al; Mg,Zn/Al; Mg/Al,Ga hydrotalcite-like compounds used as catalysts for cycloxene epoxidation. *Appl. Catal. A Gen.* **2008**, *341*, 50–57. [CrossRef]

35. Palomeque, J.; Figueras, F.; Gelbard, G. Epoxidation with hydrotalcite-intercalated organotungstic complexes. *Appl. Catal. A Gen.* **2006**, *300*, 100–108. [CrossRef]

36. Zăvoianu, R.; Ionescu, R.; Pavel, O.D.; Bîrjega, R.; Angelescu, E. Comparison between Me^{II}Mg/Al hydrotalcites and hydrotalcite-supported Me(II) acetylacetonates (Me(II) = Co, Cu or Ni) catalysts for the epoxidation of cyclohexene with molecular oxygen. *Appl. Clay Sci.* **2011**, *52*, 1–10. [CrossRef]

37. Bhattacharjee, S.; Anderson, J.A. Comparison of the epoxidation of cyclohexene, dicyclopentadiene and 1,5-cyclooctadiene over LDH hosted Fe and Mn sulfonato-salen complexes. *J. Mol. Catal. A Chem.* **2006**, *249*, 103–110. [CrossRef]

38. Zhou, W.; Zhou, J.; Chen, Y.; Cui, A.; Sun, F.; He, M.; Xu, Z.; Chen, Q. Metallophthalocyanine intercalated layered double hydroxides as an efficient catalyst for the selective epoxidation of olefin with oxygen. *Appl. Catal. A Gen.* **2017**, *542*, 191–200. [CrossRef]

39. Carriazo, D.; Lima, S.; Martín, C.; Pillinger, M.; Valente, A.A.; Rives, V. Metatungstate and tungstoniobate-containing LDHs: Preparation, characterisation and activity in epoxidation of cyclooctene. *J. Phys. Chem. Solids* **2007**, *68*, 1872–1880. [CrossRef]

40. Seftel, E.M.; Popovici, E.; Mertens, M.; De Witte, K.; Van Tendeloo, G.; Cool, P.; Vansant, E.F. Zn–Al layered double hydroxides: Synthesis, characterization and photocatalytic application. *Microporous Mesoporous Mater.* **2008**, *113*, 296–304. [CrossRef]

41. Tzompantzi, F.; Mantilla, A.; Bañuelos, F.; Fernández, J.L.; Gómez, R. Improved photocatalytic degradation of phenolic compounds with znal mixed oxides obtained from LDH Materials. *Top. Catal.* **2011**, *54*, 257–263. [CrossRef]

42. Krężel, A.; Maret, W. The biological inorganic chemistry of zinc ions. *Arch. Biochem. Biophys.* **2016**, *611*, 3–19. [CrossRef]

43. Scholz, F.; Kahlert, H. The calculation of the solubility of metal hydroxides, oxide-hydroxides, and oxides, and their visualisation in logarithmic diagrams. *ChemTexts* **2015**, *1*, 7. [CrossRef]

44. Moezzi, A.; Lee, P.-S.; McDonagh, A.M.; Cortie, M.B. On the thermal decomposition of zinc hydroxide nitrate, $Zn_5(OH)_8(NO_3)_2 \cdot 2H_2O$. *J. Solid State Chem.* **2020**, *286*, 121311. [CrossRef]

45. Angelescu, E.; Pavel, O.D.; Zavoianu, R.; Birjega, R. Cyanoethylation of ethanol over mixed oxides obtained from hydrotalcite precursors. *Rev. Roum. Chim.* **2004**, *49*, 367–375.

46. Starukh, G.; Rozovik, O.; Oranska, O. Organo/Zn-Al LDH nanocomposites for cationic dye removal from aqueous media. *Nanoscale Res. Lett.* **2016**, *11*, 228. [CrossRef]

47. Navajas, A.; Arzamendi, G.; Romero-Sarria, F.; Centeno, M.A.; Odriozola, J.A.; Gandía, L.M. DRIFTS study of methanol adsorption on Mg–Al hydrotalcite catalysts for the transesterification of vegetable oils. *Catal. Commun.* **2012**, *17*, 189–193. [CrossRef]
48. Kocík, J.; Hájek, M.; Tišler, Z.; Strejcová, K.; Velvarská, R.; Bábelová, M. The influence of long-term exposure of Mg–Al mixed oxide at ambient conditions on its transition to hydrotalcite. *J. Solid State Chem.* **2021**, *304*, 122556. [CrossRef]
49. Yi, H.; Zhao, S.; Tang, X.; Ning, P.; Wang, H.; He, D. Influence of calcination temperature on the hydrolysis of carbonyl sulfide over hydrotalcite-derived Zn–Ni–Al catalyst. *Catal. Commun.* **2011**, *12*, 1492–1495. [CrossRef]
50. Takehira, K.; Kawabata, T.; Shishido, T.; Murakami, K.; Ohi, T.; Shoro, D.; Honda, M.; Takaki, K. Mechanism of reconstitution of hydrotalcite leading to eggshell-type Ni loading on Mg single bondAl mixed oxide. *J. Catal.* **2005**, *231*, 92–104. [CrossRef]
51. Palomeque, J.; Lopez, J.; Figueras, F. Epoxydation of activated olefins by solid bases. *J. Catal.* **2002**, *211*, 150–156. [CrossRef]
52. Di Cosimo, J.I.; Díez, V.K.; Xu, M.; Iglesia, E.; Apesteguía, C.R. Structure and surface and catalytic properties of Mg-Al basic oxides. *J. Catal.* **1998**, *178*, 499–510. [CrossRef]
53. Corma, A.; Fornes, V.; Rey, F. Hydrotalcites as Base Catalysts: Influence of the Chemical Composition and Synthesis Conditions on the Dehydrogenation of Isopropanol. *J. Catal.* **1994**, *148*, 205–212. [CrossRef]
54. Jyothi, T.M.; Raja, T.; Sreekumar, K.; Talawar, M.B.; Rao, B.S. Influence of acid–Base properties of mixed oxides derived from hydrotalcite-like precursors in the transfer hydrogenation of propiophenone. *J. Mol. Catal. A Chem.* **2000**, *157*, 193–198. [CrossRef]
55. Debecker, D.; Gaigneaux, E.M.; Busca, G. Exploring, tuning, and exploiting the basicity of hydrotalcites for applications in heterogeneous catalysis. *Chem. Eur. J.* **2009**, *15*, 3920–3935. [CrossRef] [PubMed]
56. Parida, K.; Das, J. Mg/Al hydrotalcites: Preparation, characterisation and ketonisation of acetic acid. *J. Mol. Catal. A Chem.* **2000**, *151*, 185–192. [CrossRef]
57. Ionescu, R.; Pavel, O.D.; Bîrjega, R.; Zăvoianu, R.; Angelescu, E. Epoxidation of cyclohexene with H_2O_2 and acetonitrile catalyzed by Mg–Al hydrotalcite and cobalt modified hydrotalcites. *Catal. Lett.* **2010**, *134*, 309–317. [CrossRef]
58. Pavel, O.D.; Zăvoianu, R.; Bîrjega, R.; Angelescu, E. The effect of ageing step elimination on the memory effect presented by $Mg_{0.75}Al_{0.25}$ hydrotalcites (HT) and their catalytic activity for cyanoethylation reaction. *Catal. Commun.* **2011**, *12*, 845–850. [CrossRef]

used (traditional inorganic/non-traditional organic) in tailoring the physico-chemical properties of MgZn/Al LDH. In this work, we focus on using tetramethylammonium hydroxide (TMAH), an organic alkali that is cheaper than hexamethylenetetramine, is easily soluble in water, and allows the induction of precipitation at atmospheric pressure. Additionally, the optimum activity of LDH-type materials, mixed oxides, and reconstructed LDH are highlighted in the Claisen-Schmidt Condensation between benzaldehyde and cyclohexanone and in the cyclohexanone self-condensation.

2. Results and Discussion

2.1. Characterization of Catalysts

The XRD pattern for the LDH synthesized through the traditional route of co-precipitation in the presence of inorganic alkalis, LDH-MgZnAl-CO_3^{2-}/OH^--CP, presented sharp and symmetric reflections at small angles for the (003), (006), (009), (012), (015), and (018) planes, while at higher angles there were broad and weak reflections for the (110) and (113) planes, which were representative of LDH-type materials (ICDD 70-2151), Figure 1A [23].

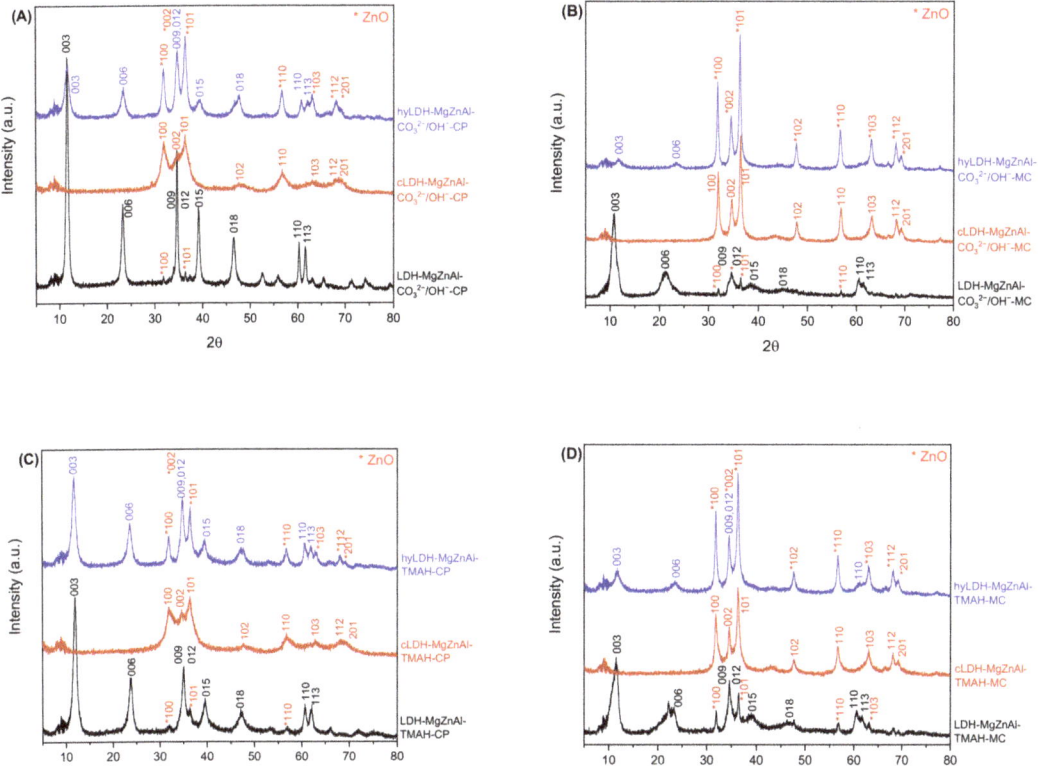

Figure 1. XRD patterns of the materials synthesized though (**A**) co-precipitation and (**B**) mechanochemical method, both in the presence of Na_2CO_3/NaOH, and (**C**) co-precipitation and (**D**) mechanochemical method, both in the presence of TMAH; (*—ZnO).

These reflections were indexed in a hexagonal lattice with an *R3m* rhombohedral symmetry. Aside from these, some additional fine diffraction lines corresponding to zincite phase appeared in the domain of 2θ = 31–38° (ICDD 005-0664). Its presence was noticeable in the XRD pattern of the calcined material, cLDH-MgZnAl-CO$_3^{2-}$/OH$^-$-CP, where diffraction lines corresponding to ZnO ((100), (002), and (101)) were more intense. Reconstruction by memory effect led to a mixture of stable zincite phase and Mg/Al LDH. At the same time, the IFS parameter (i.e., the interlayer distance) decreased from 2.82 Å to 2.80 Å, and 2θ$_{003}$ shifted towards a higher value from 11.6259° to 11.6421°, as shown in Table 1, thus denoting the presence of smaller species in the interlayer space, which, according to the literature [20], are OH$^-$ groups that partially replace the CO$_3^{2-}$ groups following the calcination–reconstruction process. This behavior was also observed in the DRIFT spectra, as shown in Figure 2.

Table 1. The network parameter of samples.

Hydrotalcite Samples	Lattice Parameters		IFS * (Å)	2θ$_{003}$ (°)	I$_{003}$/I$_{006}$	I$_{003}$/I$_{110}$	FWHM$_{003}$	D ** (Å)
	a (Å)	*c* (Å)						
LDH-MgZnAl-CO$_3^{2-}$/OH$^-$-CP	3.0717	22.8689	2.82	11.6259	2.96	5.54	0.6064	131.8
hyLDH-MgZnAl-CO$_3^{2-}$/OH$^-$-CP	3.0497	22.8017	2.80	11.6421	1.97	3.42	1.1043	72.4
LDH-MgZnAl-CO$_3^{2-}$/OH$^-$-MC	3.0504	22.4314	2.68	11.8652	2.54	5.33	0.8885	90.0
hyLDH-MgZnAl-CO$_3^{2-}$/OH$^-$-MC	3.0497	22.6427	2.75	11.7234	2.20	4.48	0.9709	82.3
LDH-MgZnAl-TMAH-CP	3.0617	24.7662	3.46	10.8279	3.57	5.08	1.1644	68.6
hyLDH-MgZnAl-TMAH-CP	2.9757	22.7965	2.80	11.6610	1.59	1.13	0.9020	88.6
LDH-MgZnAl-TMAH-MC	3.0580	23.6969	3.10	11.3150	2.83	3.45	1.6050	49.8
hyLDH-MgZnAl-TMAH-MC	3.0535	22.7515	2.78	11.6700	2.18	3.45	1.1800	67.7
Mixed oxides samples	***a* (Å)**			**2θ$_{101}$ (°)**	**I$_{003}$**		**FWHM$_{101}$**	**D *** (Å)**
cLDH-MgZnAl-CO$_3^{2-}$/OH$^-$-CP	4.9536			36.2400	279		1.8266	45.8
cLDH-MgZnAl-CO$_3^{2-}$/OH$^-$-MC	4.9483			36.3628	257		1.5500	53.9
cLDH-MgZnAl-TMAH-CP	4.9332			36.3949	736		0.4925	169.9
cLDH-MgZnAl-TMAH-MC	4.9449			36.3055	175		0.6023	138.9

* IFS represents the interlayer free distance; 4.8Å brucite sheet thickness [42]. ** D represents the mean crystallite size (derived from the Debye–Scherrer equation) determined from the FWHM of the (003) reflection for LDH samples. *** D represents the mean crystallite size (derived from the Debye–Scherrer equation) determined from the FWHM of the (101) reflection for mixed oxides.

The easily noticeable decrease in *a* network parameter (i.e., the distance between the network cations) was due to the extraction of Zn from the LDH network followed by the stable phase of zincite synthesis made by calcination. The mechano-chemical LDH, LDH-MgZnAl-CO$_3^{2-}$/OH$^-$-MC, presented particularities similar to the above-mentioned, as shown in Figure 1B, but the amount of zincite increased and that of the LDH phase decreased. Moreover, the number of interlayer species with small size, i.e., OH$^-$, is higher compared to LDH obtained by co-precipitation, a fact highlighted by an IFS value of 2.68 Å compared to 2.82 Å, and a shifting from 2θ$_{003}$ to 11.8652°. After the reconstruction of the layered structure hyLDH-MgZnAl-CO$_3^{2-}$/OH$^-$-MC, a large amount of zincite remained as the stable phase. This behavior simultaneously led to an increase in the IFS value of 2.75 Å, as well as a shift to lower values of 11.7234 for 2θ$_{003}$. The mean crystallite size derived from the Debye–Scherrer equation shows an insignificant variation (from 90.0 Å to 82.3 Å) compared to that noticed for co-precipitated materials (from 131.8 Å to 72.4 Å).

The use of TMAH in LDH synthesis by both co-precipitation and mechano-chemical methods produced a pure LDH structure mixed with small amounts of zincite phase, as shown in Figure 1C,D. The IFS values of LDH obtained by both methods were increased (LDH-MgZnAl-TMAH-CP of 3.46 Å and LDH-MgZnAl-TMAH-MC of 3.10 Å), because, besides OH$^-$ and CO$_3^{2-}$ groups, in the interlayer space there were also small amounts of TMAH as well as tri-methyl amine (a ubiquitous impurity from TMAH; its existence being also demonstrated by DRIFT). The calcination processes eliminated both organic compounds as dimethyl ether and methanol [43]. The LDH structure reconstruction through the memory effect was present for all materials synthesized in the presence of TMAH. The diffraction lines of hyLDH-MgZnAl-TMAH-CP and hyLDH-MgZnAl-TMAH-

MC were more intense compared to those of the materials synthesized with inorganic alkalis. All mixed oxide samples prepared by calcination of parent LDH showed only ZnO lines with no important differences in a network parameter and $2\theta_{101}$, except for the mean crystallite size, where organic alkalis generated values three times higher compared with the inorganic one, as shown in Table 1. Because the diffraction lines corresponding to the Mg(Al^{3+})O solid solution of MgO-periclase-type (JCPDS-45-0946) phases were not obvious, this phase was highly dispersed.

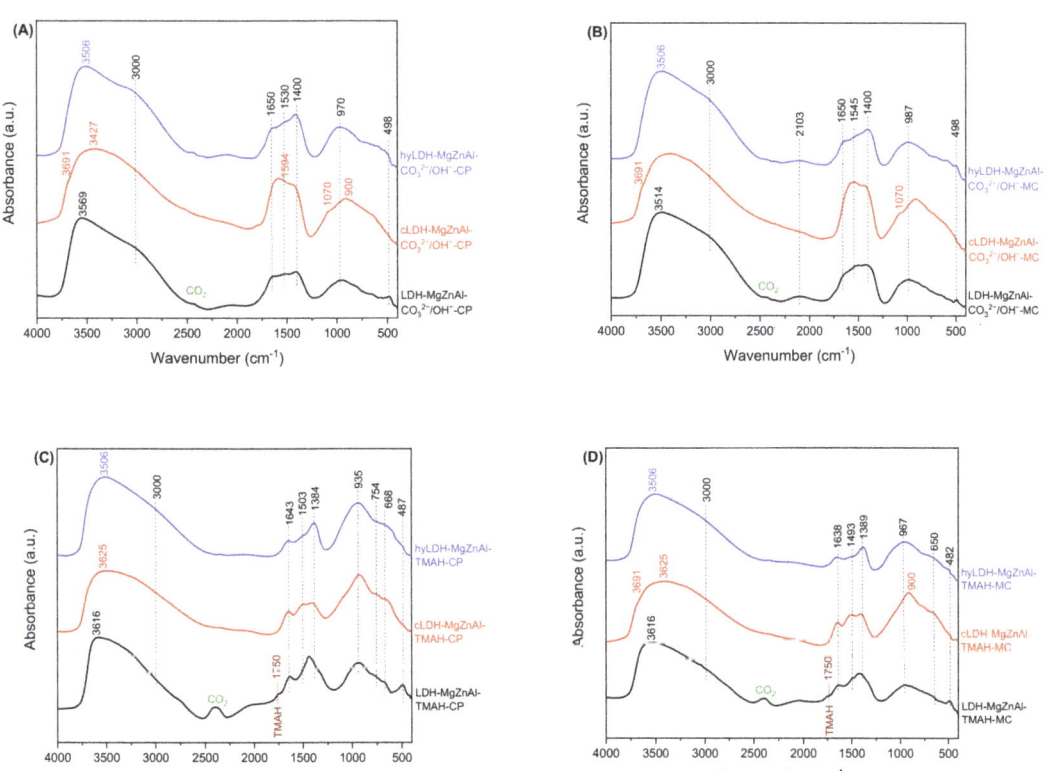

Figure 2. DRIFT spectra of the materials synthesized though (**A**) co-precipitation and (**B**) mechano-chemical method, both in the presence of Na$_2$CO$_3$/NaOH, as well as (**C**) co-precipitation and (**D**) mechano-chemical method, both in the presence of TMAH.

DRIFT spectra of the investigated LDH, as shown in Figure 2A–D, presented a large band in the 3700–3400 cm^{-1} domain corresponding to the vibration of hydroxyl groups, ν_{O-H}, which at 3000 cm^{-1} was assigned to hydrogen bonds between carbonate anion and water molecules, both situated in the interlayer space [44], a band at 1638–1650 cm^{-1} characteristic of the H$_2$O bending vibration of interlayer LDH structure, and a band at 1200–600 cm^{-1} assigned to the CO$_3$$^{2-}$ group vibration, while that below 600 cm^{-1} was assigned to Mg–O, Zn–O, and Al–O bonds.

The band at 1750 cm^{-1} for samples synthesized with TMAH was assigned to this hydrolysis agent, which was present in small quantities in the pores of LDH. This remnant compound, due to its pronounced base character, generated a higher physisorption of atmospheric CO$_2$, exhibiting a band at 2450 cm^{-1} (stronger than that appearing in the spectra of materials prepared with inorganic alkalis). It is noteworthy that in the case of the TMAH prepared samples, the bands from 1650 cm^{-1} and 1400 cm^{-1} shifted to

1640–1638 cm^{-1} and 1389–1384 cm^{-1}, respectively. The calcination at 460 °C eliminated the remnant TMAH and adsorbed CO$_2$, leading to the total disappearance of their corresponding IR adsorption bands. In addition, there was also a partial removal of OH$^-$ and CO$_3{}^{2-}$ from the network, leading to a decreased intensity of the corresponding bands, as also remarked by other authors working on LDH prepared with inorganic alkalis, even when the calcination was performed at 650–700 °C [19]. The spectra of the calcined samples also present a band of characteristic stretching vibrations of structural hydroxyl groups, which were coordinated to Mg or Al octahedral at 3691 cm^{-1} [45]. This phase was evidently highly dispersed in the solid matrix since its reflections were absent in the XRD patterns. The bands at 3000 cm^{-1} and that at 1638–1650 cm^{-1} were well restored in the spectra of the reconstructed samples. Simultaneously, the band at 1400–1384 cm^{-1} became stronger.

The UV–VIS spectra of the samples obtained in the presence of inorganic alkalis using both preparation methods, as depicted in Figure 3A,B, showed a large absorption band in the wavelength range of 240–380 nm, with maxima at 348 nm and 359 nm for the LDH. This shifting was due to the presence of zincite phase in higher amounts for the samples obtained through mechano-chemical method.

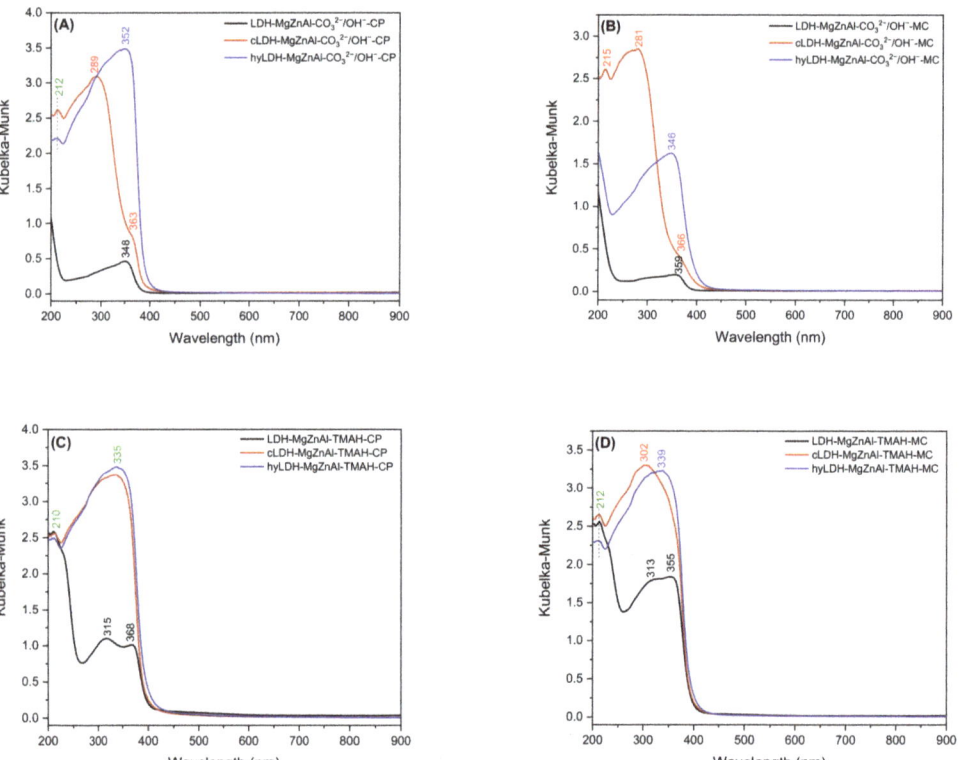

Figure 3. UV–VIS spectra of the materials synthesized though (**A**) co-precipitation and (**B**) mechano-chemical method, both in the presence of Na$_2$CO$_3$/NaOH, as well as (**C**) co-precipitation and (**D**) mechano-chemical method, both in the presence of TMAH.

In mixed oxide samples, the maxima identified at 363 nm and 366 nm corresponded to zinc oxide nanoparticles as impurities in the materials [46], while those at 289 nm and 281 nm were attributed to the band gap absorption in ZnO/MgO nanocomposites, which present larger band gaps as compared to ZnO [47]. The bands at 212 and 215 nm indicated the presence of Mg(OH)$_2$ and MgO [48]. For samples obtained in the presence of TMAH by both routes, as shown in Figure 3C,D, the common band at 210 nm indicated the presence of the organic base in their structure, while those at 313 nm and 315 nm were due to the presence of zincite phase inside the layered structure. Mixed oxides and reconstructed LDH also showed bands for zincite phase shifted to higher values for layered materials.

The basicity of samples, as shown in Table 2, decreased in the order mixed oxides > reconstructed LDH > parent LDH samples, a trend valid for both strong base sites as well as the sum of weak and medium base sites, regardless of the synthesis route. However, there was a slight increase in basicity for the samples prepared by mechano-chemical route. Additionally, the presence of small TMAH in LDH provided a pronounced base character compared to the LDH prepared with inorganic alkalis. In the meantime, the reconstructed LDH showed a decreased specific surface area comparative to those of the parent ones, due to the tendency of hexagonal lamellar crystals to cluster together in large conglomerate systems under vermiculate form, involving less defined platelets, which affect the access of reactants as well as the probe molecule N$_2$ to active sites [49,50]. However, the partial replacement of carbonate anions with hydroxyl groups following the calcination–hydration process leads to an increase in the basicity of the reconstructed samples due to the pronounced base character of these groups, as shown in Table 2. Regarding the textural properties of mixed oxides, they are in the proper range of these types of materials [4].

Table 2. The surface area and basicity of materials.

Hydrotalcite Samples	Surface Area $(m^2 g^{-1})$	Pore Volume $(cm^3 \cdot g^{-1})$	Average Pore Width (Å)	Total Number of Base Sites $(mmol \cdot g^{-1})$ *	Distribution of Base Sites	
					Strong Base Sites $(mmol \cdot g^{-1})$ **	Weak and Medium Base Sites $(mmol \cdot g^{-1})$ ***
LDH-MgZnAl-CO$_3$$^{2-}$/OH$^-$-CP	69	0.387	222	7.23	0.48	6.75
cLDH-MgZnAl-CO$_3$$^{2-}$/OH$^-$-CP	258	0.843	124	10.38	0.53	9.85
hyLDH-MgZnAl-CO$_3$$^{2-}$/OH$^-$-CP	25	0.186	238	8.93	0.57	8.36
LDH-MgZnAl-CO$_3$$^{2-}$/OH$^-$-MC	150	0.498	132	7.42	0.51	6.91
cLDH-MgZnAl-CO$_3$$^{2-}$/OH$^-$-MC	266	0.845	121	10.48	0.57	9.91
hyLDH-MgZnAl-CO$_3$$^{2-}$/OH$^-$-MC	28	0.203	208	8.99	0.61	8.38
LDH-MgZnAl-TMAH-CP	2	0.005	115	7.71	0.50	7.21
cLDH-MgZnAl-TMAH-CP	235	0.765	112	10.64	0.55	10.09
hyLDH-MgZnAl-TMAH-CP	1	0.004	118	9.36	0.59	8.77
LDH-MgZnAl-TMAH-MC	44	0.118	106	7.85	0.54	7.31
cLDH-MgZnAl-TMAH-MC	241	0.798	124	10.66	0.56	10.10
hyLDH-MgZnAl-TMAH-MC	9	0.104	118	9.40	0.64	8.76

* mmol of acrylic acid. ** mmol of phenol. *** the difference between total number of base sites—strong base sites.

TMAH used as organic alkali in the synthesis of LDH by both the co-precipitation and mechano-chemical methods acted also as template molecule, as shown in Figure 4. The mechano-chemical method compared to the co-precipitation method in the presence of traditional inorganic alkali led to different pore widths (i.e., with maxima at 40 Å, 57 Å, and 150 Å, respectively). While preparation with traditional inorganic alkali led to a large pore width, with maxima of 358 Å for LDH-MgZnAl-CO$_3$$^{2-}$/OH$^-$-CP and 150 Å for LDH-MgZnAl-CO$_3$$^{2-}$/OH$^-$-MC, the preparation in the presence of organic alkali tended to decrease the pore width, leading to two well-defined maxima, as shown in Figure 4 C,D: one at 36 Å for LDH-MgZnAl-TMAH-CP and 34 Å for LDH-MgZnAl-TMAH-MC, and a second one at 132 Å for LDH-MgZnAl-TMAH-CP and 90 Å for LDH-MgZnAl-TMAH-MC. The appearance of these two domains was not unusual, because the porosity was due to the size of the organic compounds, tri-methyl amine having the lower value and TMAH having the higher one.

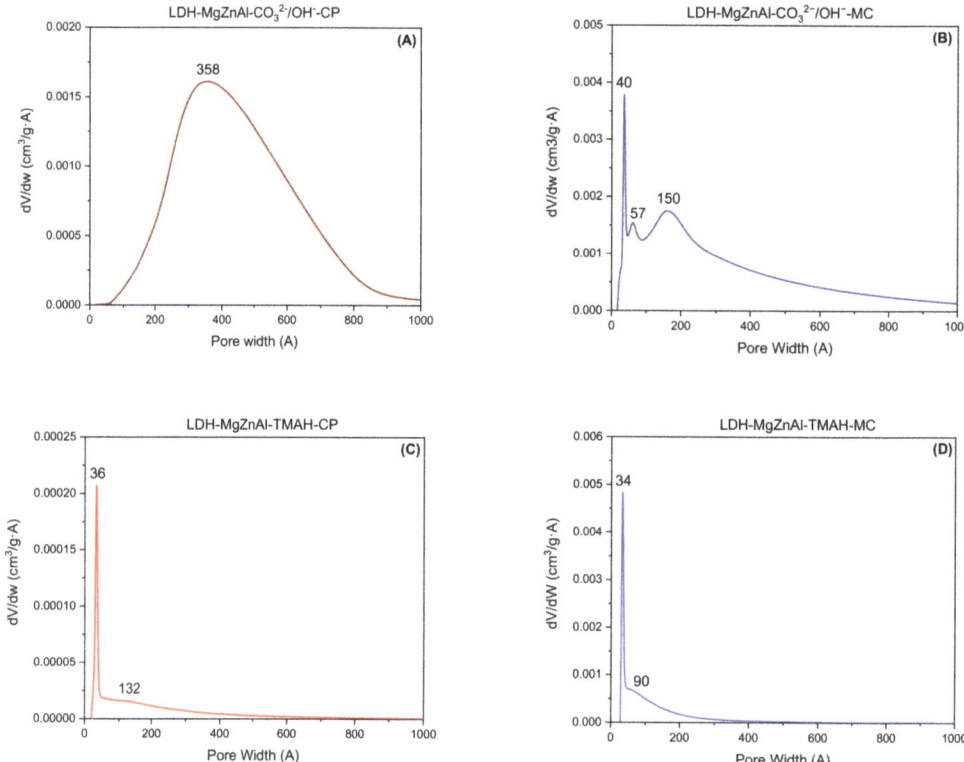

Figure 4. The representation of LDH sample pore widths; (**A**) LDH-MgZnAl-CO_3^{2-}/OH^--CP; (**B**) LDH-MgZnAl-CO_3^{2-}/OH^--MC; (**C**) LDH-MgZnAl-TMAH-CP; (**D**) LDH-MgZnAl-TMAH-MC.

2.2. Catalytic Activity

2.2.1. Claisen-Schmidt Condensation

The blank test at room temperature after 5 h was in the range of experimental errors of 0.97% in cyclohexanone conversion, increasing at 9.13% at 120 °C and 2 h, with total selectivity toward 2,6-DBCHO (Scheme 1). The cyclohexanone conversion followed the same order: mixed oxides > reconstructed LDH > parent LDH samples, regardless of the preparation method, with total selectivity toward 2,6-DBCHO, as shown in Figure 5. Additionally, there was an improvement in the catalytic activities of materials prepared in the presence of organic alkalis compared to those prepared with inorganic alkalis, regardless of the preparation methods. However, a linear dependence between total basicity and conversion values was determined, as shown in Figure 6. Notably, at the end of the reactions, no products from self-cyclohexanone condensations or benzoic acid were found in the analyzed reaction mixtures. The total selectivity towards 2,6-DBCHO was also explained by the ability of 2-BCHO to adsorb onto active sites from pores with no steric hindrance, due to the non-planar shape of this molecule. On the same note, the presence of zincite phase to a different extent may also play a role in increasing the yield of 2,6-DBCHO. Lower activity was noticed for the samples prepared with inorganic alkalis by mechano-chemical method. The comparison of the method types as well as of the hydrolysis agents used reveals the benefits presented by the use of organic instead of inorganic alkalis, but also of the advantages of co-precipitation over the mechano-chemical method.

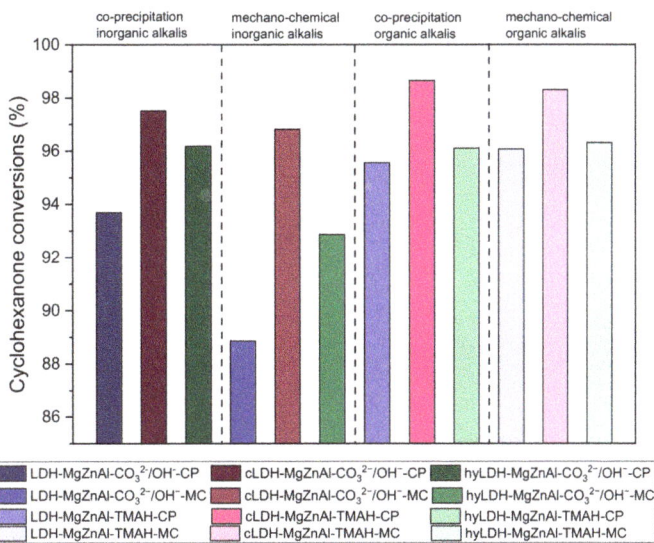

Figure 5. Cyclohexanone conversion after aldol condensation for 2 h, 120 °C, 20 mg of catalyst.

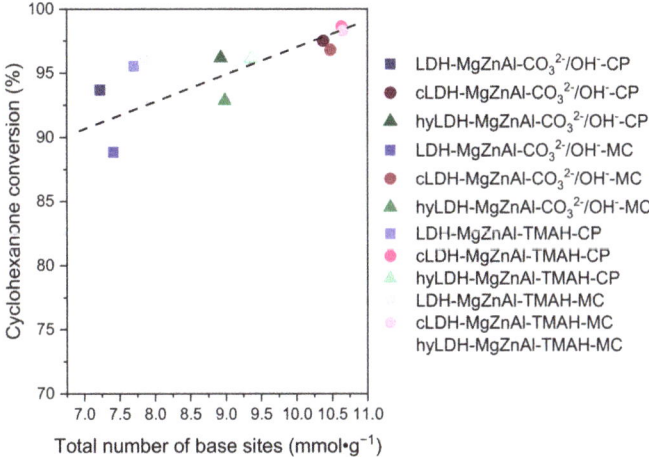

Figure 6. The cyclohexanone conversion vs. total number of base sites for investigated catalysts.

2.2.2. Cyclohexanone Self-Condensation

Blank reaction at room temperature or reflux after 5 h did not provide more than 0.19% of cyclohexanone conversion. The cyclohexanone conversions were significantly lower than those reached in Claisen-Schmidt Condensation, and their magnitude follows the same variation trend: mixed oxides > reconstructed LDH > parent LDH samples, regardless of the preparation method. There was also a high selectivity toward the *mono*-condensed product (Scheme 2), as shown in Table 3.

Table 3. Experimental data gathered for cyclohexanone conversion after 5 h, reflux, 5 wt.% catalyst, solvent-free.

Catalysts	Conv. $C_6H_{10}O$ (%)	Sel. A (%)	Sel. A1 (%)	Sel. B (%)	Sel. B1 (%)
LDH-MgZnAl-CO_3^{2-}/OH^--CP	1.01	69.00	27.31	1.93	1.76
cLDH-MgZnAl-CO_3^{2-}/OH^--CP	1.09	75.31	24.31	0.22	0.16
hyLDH-MgZnAl-CO_3^{2-}/OH^--CP	0.70	69.74	29.64	0.44	0.17
LDH-MgZnAl-CO_3^{2-}/OH^--MC	0.45	66.39	33.61	0.00	0.00
cLDH-MgZnAl-CO_3^{2-}/OH^--MC	0.80	73.74	26.26	0.00	0.00
hyLDH-MgZnAl-CO_3^{2-}/OH^--MC	0.49	69.87	30.13	0.00	0.00
LDH-MgZnAl-TMAH-CP	2.65	73.35	11.36	8.32	6.97
cLDH-MgZnAl-TMAH-CP	7.29	82.48	15.08	1.70	0.74
hyLDH-MgZnAl-TMAH-CP	6.23	75.73	12.20	1.53	0.55
LDH-MgZnAl-TMAH-MC	0.78	76.82	19.75	1.98	1.45
cLDH-MgZnAl-TMAH-MC	1.03	80.30	19.70	0.00	0.00
hyLDH-MgZnAl-TMAH-MC	0.67	74.21	25.79	0.00	0.00

This changing in paradigm compared to that presented for Claisen-Schmidt Condensation, where the selectivity toward the *di*-condensate compound was total, was due to the rigidity of the double bond connecting cyclohexane moieties in the *mono*-condensate compound, which led to steric hindrances in accessing the porous structure of the catalyst. This fact was confirmed also by the dependence of cyclohexanone conversion on total number of base sites Figure 7.

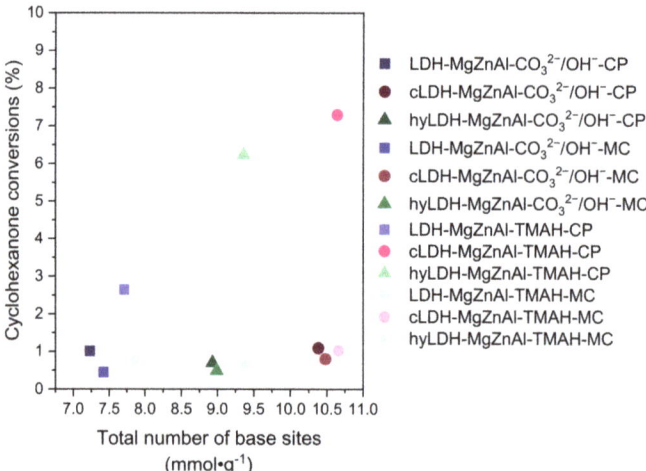

Figure 7. Cyclohexanone conversion vs. total number of base sites of investigated catalysts in self-cyclohexanone condensation.

The selectivities in the *di*-condensation product did not exceed 10%, as their occurrence was related only to the active sites on the external surface of the solid catalysts. Therefore, the best catalytic activities were presented by the materials obtained in the presence of organic alkalis and the co-precipitation method.

2.3. Catalyst Reusability

The stability of catalysts (LDH-MgZnAl-TMAH-CP, cLDH-MgZnAl-TMAH-CP and hyLDH-MgZnAl-TMAH-CP) was checked in three consecutive Claisen-Schmidt Condensation runs. After that, the conversion decreased to less than 4%, with no modification of diffraction lines in XRD patterns, as shown in Supplementary Materials Figure S1, thus confirming the stability of these materials in this reaction, Figure 8.

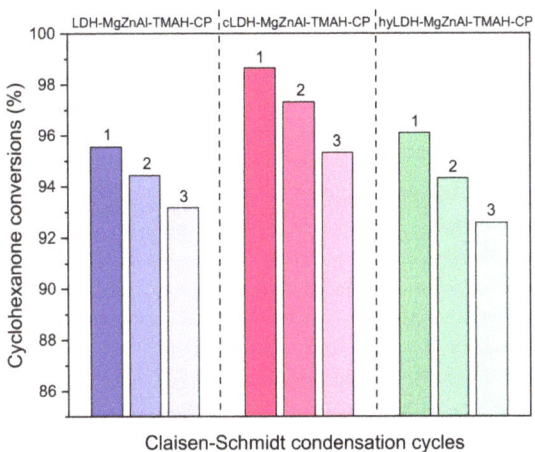

Figure 8. The catalyst reusability after 3 cycles in Claisen-Schmidt Condensation for catalysts prepared by co-precipitation with organic alkalis.

3. Materials and Methods

3.1. Catalyst Preparation

The layered double hydroxide $Mg_{0.325}Zn_{0.325}Al_{0.25}$ was synthesized through both co-precipitation and the mechano-chemical method using the traditional inorganic alkalis, but also using a non-conventional organic base. Co-precipitation was carried out with inorganic alkalis according to a methodology already reported [44], mixing solutions of $Mg(NO_3)_2 \cdot 6H_2O$, $Zn(NO_3)_2 \cdot 6H_2O$, and $Al(NO_3)_3 \cdot 9H_2O$ (0.325/0.325/0.25 molar ratio and 1.5M) in the presence of NaOH and Na_2CO_3 ($NaOH/Na_2CO_3$ of 2.5 molar ratio and 1M Na_2CO_3) with a pH of 10, at room temperature and at a rate of 600 rpm. The TIM854, NB pH/EP/Stat pH-STAT Titrator was used to add both solutions at a feed rate of 60 mL·h^{-1}. The suspension was then aged for 18 h at 80 °C in an air atmosphere, cooled at room temperature and filtered, washed with bi-distillated water until a pH of 7 was attained, and dried for 24 h in air at 120 °C (**LDH-MgZnAl-CO$_3$$^{2-}$/OH$^-$-CP**). After that, calcination of this sample at 460 °C for 18 h in air led to the corresponding mixed oxide (**cLDH-MgZnAl-CO$_3$$^{2-}$/OH$^-$-CP**). The reconstruction of the layered structure was completed through memory effect by the immersion of cLDH-MgZnAl-CO$_3$$^{2-}$/OH$^-$-CP in bi-distilled water for 24 h at room temperature, followed by drying for 24 h at 120 °C in air (**hyLDH-MgZnAl-CO$_3$$^{2-}$/OH$^-$-CP**). The nontraditional mechano-chemical method was carried out by a direct milling of the precursors in a Mortar Grinder RM 200 for 1 h at 100 rpm (**LDH-MgZnAl-CO$_3$$^{2-}$/OH$^-$-MC**) at a pH of approx. 10, with no water addition or additional aging process. Further protocols for mixed oxides and reconstructed layered samples were identical to the co-precipitation method (**cLDH-MgZnAl-CO$_3$$^{2-}$/OH$^-$-MC** and **hyLDH-MgZnAl-CO$_3$$^{2-}$/OH$^-$-MC**). Both the co-precipitation and mechano-chemical routes were also used to generate LDH in the presence of a nontraditional organic alkali represented by TMAH. (TetraMethylAmonium Hydroxide; 25 wt.% in water) maintaining the same operational parameters (**LDH-MgZnAl-TMAH-CP**; **cLDH-MgZnAl-TMAH-CP**; **hyLDH-MgZnAl-TMAH-CP**). Notably, the bi-distilled water used in the washing step was 10 times lower compared to when inorganic alkalis were used as precipitation agents, where 2500 mL bi-distilled water was used to obtain 10 g of catalyst. The pH values during the co-precipitation and the washing stages were monitored using a Consort 853 Multiparameter equipped with pH electrode (Fisher Scientific, Turnhout, Belgium). A volume of TMAH identical to the one employed in the co-precipitation route was applied to obtain the LDH sample by mechano-chemical route (**LDH-MgZnAl-TMAH-MC**) under similar operating conditions (**cLDH-MgZnAl-TMAH-MC**; **hyLDH-MgZnAl-TMAH-MC**).

Index: LDH—layered double hydroxides; cLDH—mixed oxides; hyLDH—reconstructed layered structure; MgZnAl—the involved cations; CO_3^{2-}/OH^-—inorganic alkalis ($Na_2CO_3/NaOH$); TMAH—tetramethylammonium hydroxide; CP—co-precipitation; MC—mechano-chemical.

3.2. Catalyst Characterization

Powder X-ray diffraction patterns were recorded using a Shimadzu XRD 7000 diffractometer with Cu Kα radiation (λ = 1.5418 Å, 40 kV, 40 mA) at a scanning speed of $0.10°\cdot min^{-1}$ in the 2θ range of 5–80°. DRIFTS spectra were recorded with JASCO FT/IR-4700 spectrometer by an accumulation of 128 scans in 400–4000 cm^{-1} domain. DR UV–VIS spectra were recorded in the range 900–200 nm on a Jasco V-650 UV–VIS spectrophotometer with integration sphere, using Spectralon as a white reference. N_2 adsorption–desorption isotherms were determined using a Micromeritics ASAP 2010 instrument, where prior to nitrogen adsorption, samples were outgassed under vacuum for 24 h at 120 °C. The distribution of the pore sizes was determined by BJH desorption dV/dw pore volume using Faas correction. The original N_2 sorption isotherms are included in the Supplementary Materials, Figure S2. The base character of the catalysts was determined by an irreversible adsorption of organic molecules of different *pKa* method [51–53] (e.g., acrylic acid, *pKa* = 4.2, for total base sites and phenol, *pKa* = 9.9, for strong base sites), where the number of weak and medium base sites was calculated as the difference between the amounts of adsorbed acrylic acid and phenol.

3.3. Catalytic Tests

3.3.1. Claisen-Schmidt Condensation

The Claisen-Schmidt Condensation was carried out in a thermo-stated glass reactor with a water-cooled condenser, where a mixture of benzaldehyde (0.002 moles ReagentPlus >99%, Sigma-Aldrich, Darmstadt, Germany), cyclohexanone (0.001 moles >99%, Sigma-Aldrich) and 20 mg of catalyst (at a benzaldehyde/catalyst ratio of 10/1) was stirred under solvent-free conditions for 2 h at 120 °C [37]. After that the catalyst was removed by filtration, the mixture was washed with 1 mL of toluene and analyzed by Thermo-Quest GC with a FID detector and a capillary column 30 m in length with a DB5 stationary phase. The compounds were identified by mass spectrometer coupled chromatography, using a GC/MS/MS Varian Saturn 2100 T equipped with a CP-SIL 8 CB Low Bleed/MS column 30 m in length and 0.25 mm in diameter.

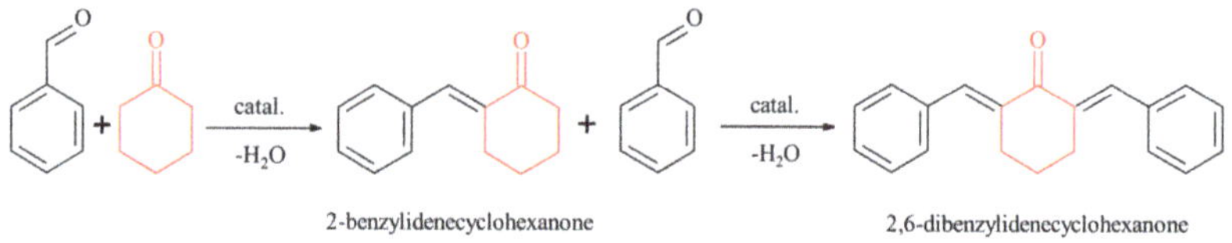

Scheme 1. Claisen-Schmidt Condensation between benzaldehyde and cyclohexanone.

3.3.2. The Aldol Cyclohexanone Self-Condensation

The aldol cyclohexanone self-condensation was investigated in a similar reactor to the one utilized for the Claisen-Schmidt Condensation, mixing 0.01 moles of cyclohexanone with 5 wt.% catalyst under solvent-free conditions [54]. After 5 h at reflux, the catalyst was removed from the mixture by filtration and the liquid reaction mixture was analyzed by GC-FID. Additionally, mass spectrometer coupled chromatography was used for compound identification.

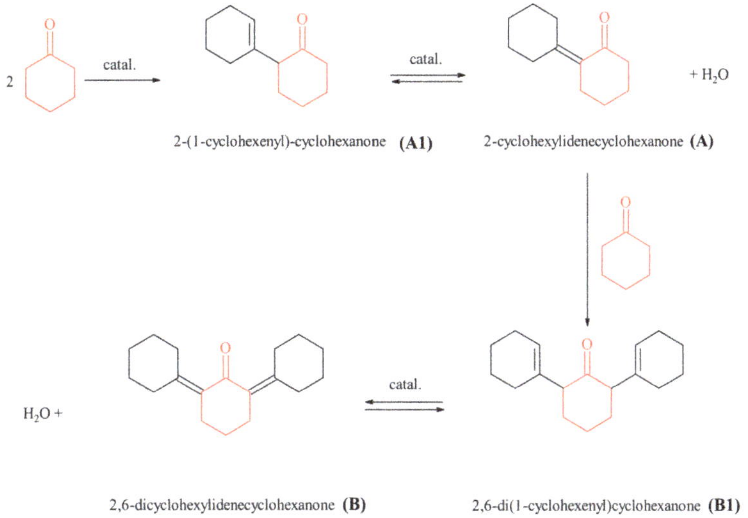

Scheme 2. The aldol self-condensation of cyclohexanone; (**A1**)—2-(1-cyclohexenyl)-cyclohexanone; (**A**)—2 –cyclohexylidenecyclohexanone; (**B**)—2,6-dicyclohexylidenecyclohexanone; (**B1**)—2,6-di (1-cyclohexenyl)cyclohexanone.

3.4. Catalyst Recycling

The LDH-MgZnAl-TMAH-CP, cLDH-MgZnAl-TMAH-CP, and hyLDH-MgZnAl-TMAH-CP, the most active catalysts in their class, were selected for recycling tests. The layered catalysts were separated from the reaction mixture by filtration, washed with 1 mL of toluene and dried for 5 h at 120 °C in air before being used in the consecutive cycles. The same parameters were used for the mixed oxide with the specification that the thermal treatment was carried out at 460 °C.

4. Conclusions

Both preparation methods employed led to the obtaining of LDH materials spiked with zincite phase. However, the amount of additional phase was higher in mechanochemically prepared catalysts. All solids exhibited the memory effect of the Mg/Al LDH phase, while Zn was conserved as stable zincite phase. Using TMAH as an organic alkali for LDH preparation brings a number of advantages: *(i)* a smaller quantity of water involved in the washing step, *(ii)* preventing contamination with alkali metal cations, and *(iii)* tailoring the LDH texture. Mechano-chemically prepared materials show a pronounced basicity compared than that of co-precipitated ones, while LDH materials prepared with TMAH present a higher basicity than that of LDH prepared with inorganic alkalis. Regardless of the preparation methods, organic/inorganic alkalis or reactions, the activity of the catalysts decreased in the order: mixed oxides > reconstructed LDH > parent LDH samples. For Claisen-Schmidt Condensation, the conversions were higher than 90% after 2 h with a total selectivity toward 2,6-dibenzylidenecyclohexanone, while in the self-condensation of cyclohexanone, the conversion did not exceed 7.29% after 5 h. The yields reached for 2,6-dibenzylidenecyclohexanone after 2 h reaction were comparable to the highest yields reported by other authors who worked with more complex catalytic systems. These catalytic behaviors were a consequence of the cooperation between a number of active base sites, which promoted the activation of the reagent molecules and the shape selectivity due to the particular porous structure of the samples prepared with organic alkalis. The catalytic materials presented a good stability after three cycles in Claisen-Schmidt Condensation.

Supplementary Materials: The following supporting information can be downloaded at: https://www.mdpi.com/article/10.3390/catal12070759/s1, Figure S1: XRD patterns of the catalysts recycled; Figure S2: Isotherm linear plot of LDH samples.

Author Contributions: Conceptualization, O.D.P.; Data curation, R.Z.; Funding acquisition, E.E.J.; Investigation, R.Z., S.-D.M. and O.D.P.; Methodology, B.C. and O.D.P.; Resources, M.T. and E.E.J.; Supervision, V.I.P. and S.O.; Writing—original draft, R.Z., O.D.P. and E.E.J.; Writing—review & editing, V.I.P., O.D.P. and S.O. All authors have read and agreed to the published version of the manuscript.

Funding: This work was financially supported by The Education, Scholarship, Apprenticeships and Youth Entrepreneurship Programmer—EEA Grants 2014-2021, Project No. 18-Cop-0041.

Data Availability Statement: The data are available on request from the corresponding author.

Conflicts of Interest: The authors declare no conflict of interest.

References

1. Basahel, S.N.; Al-Thabaiti, S.A.; Narasimharao, K.; Ahmed, N.S.; Mokhtar, M. Nanostructured Mg–Al Hydrotalcite as Catalyst for Fine Chemical Synthesis. *J. Nanosci. Nanotechnol.* **2014**, *14*, 1931–1946. [CrossRef] [PubMed]
2. Wang, K.; Wang, T.; Islam, Q.A.; Wu, Y. Layered double hydroxide photocatalysts for solar fuel production. *Chin. J. Catal.* **2021**, *42*, 1944–1975. [CrossRef]
3. Yadav, G.D.; Wagh, D. Claisen-Schmidt Condensation using Green Catalytic Processes: A Critical Review. *ChemistrySelect* **2020**, *5*, 9059–9085. [CrossRef]
4. Cavani, F.; Trifirò, F.; Vaccari, A. Hydrotalcite-type anionic clays: Preparation, properties and applications. *Catal. Today* **1991**, *11*, 173–301. [CrossRef]
5. Huang, Y.; Liu, C.; Rad, S.; He, H.; Qin, L. A Comprehensive Review of Layered Double Hydroxide-Based Carbon Composites as an Environmental Multifunctional Material for Wastewater Treatment. *Processes* **2022**, *10*, 617. [CrossRef]
6. Johnston, A.-L.; Lester, E.; Gomes, R.L. Understanding Layered Double Hydroxide properties as sorbent materials for removing organic pollutants from environmental waters. *J. Environ. Chem. Eng.* **2021**, *9*, 105197. [CrossRef]
7. Dias, A.C.; Fontes, M.P.F. Arsenic (V) removal from water using hydrotalcites as adsorbents: A critical review. *Appl. Clay Sci.* **2020**, *191*, 105615. [CrossRef]
8. Yang, H.; Xiong, C.; Liu, X.; Liu, A.; Li, T.; Ding, R.; Shah, S.P.; Li, W. Application of layered double hydroxides (LDHs) in corrosion resistance of reinforced concrete-state of the art. *Constr. Build. Mater.* **2021**, *307*, 124991. [CrossRef]
9. Mohapi, M.; Sefadi, J.S.; Mochane, M.J.; Magagula, S.I.; Lebelo, K. Effect of LDHs and Other Clays on Polymer Composite in Adsorptive Removal of Contaminants: A Review. *Crystals* **2020**, *10*, 957. [CrossRef]
10. Shirin, V.A.; Sankar, R.; Johnson, A.P.; Gangadharappa, H.; Pramod, K. Advanced drug delivery applications of layered double hydroxide. *J. Control Release* **2020**, *330*, 398–426. [CrossRef]
11. Lauermannová, A.-M.; Paterová, I.; Patera, J.; Skrbek, K.; Jankovský, O.; Bartůněk, V. Hydrotalcites in Construction Materials. *Appl. Sci.* **2020**, *10*, 7989. [CrossRef]
12. Ho, P.H.; Ambrosetti, M.; Groppi, G.; Tronconi, E.; Palkovits, R.; Fornasari, G.; Vaccari, A.; Benito, P. Structured Catalysts-Based on Open-Cell Metallic Foams for Energy and Environmental Applications. *Stud. Surf. Sci. Catal.* **2019**, *178*, 303–327. [CrossRef]
13. Sikander, U.; Sufian, S.; Salam, M. A review of hydrotalcite based catalysts for hydrogen production systems. *Int. J. Hydrog. Energy* **2017**, *42*, 19851–19868. [CrossRef]
14. Bravo-Suárez, J.J.; Páez-Mozo, E.A.; Oyama, S.T. Review of the synthesis of layered double hydroxides: A thermodynamic approach. *Quim. Nova* **2004**, *27*, 601–614. [CrossRef]
15. Gonçalves, J.M.; Martins, P.R.; Angnes, L.; Araki, K. Recent advances in ternary layered double hydroxide electrocatalysts for the oxygen evolution reaction. *New J. Chem.* **2020**, *44*, 9981–9997. [CrossRef]
16. Kühl, S.; Schumann, J.; Kasatkin, I.; Hävecker, M.; Schlögl, R.; Behrens, M. Ternary and quaternary Cr or Ga-containing ex-LDH catalysts—Influence of the additional oxides onto the microstructure and activity of Cu/ZnAl2O4 catalysts. *Catal. Today* **2015**, *246*, 92–100. [CrossRef]
17. Ulibarri, M.A.; Pavlovic, I.; Barriga, C.; Hermosın, M.C.; Cornejo, J. Adsorption of anionic species on hydrotalcite-like compounds: Effect of interlayer anion and crystallinity. *J. Appl. Clay Sci.* **2001**, *18*, 17–27. [CrossRef]
18. Takehira, K.; Shishido, T.; Shouro, D.; Murakami, K.; Honda, M.; Kawabata, T.; Takaki, K. Novel and effective surface enrichment of active species in Ni-loaded catalyst prepared from Mg–Al hydrotalcite-type anionic clay. *Appl. Catal. A: Gen.* **2005**, *279*, 41–51. [CrossRef]
19. Marchi, A.; Apesteguía, C. Impregnation-induced memory effect of thermally activated layered double hydroxides. *Appl. Clay Sci.* **1998**, *13*, 35–48. [CrossRef]
20. Teodorescu, F.; Pălăduță, A.-M.; Pavel, O. Memory effect of hydrotalcites and its impact on cyanoethylation reaction. *Mater. Res. Bull.* **2013**, *48*, 2055–2059. [CrossRef]
21. Barbosa, C.A.; Ferreira, A.M.D.; Constantino, V.; Coelho, A.C.V. Preparation and Characterization of Cu(II) Phthalocyanine Tetrasulfonate Intercalated and Supported on Layered Double Hydroxides. *J. Incl. Phenom. Macrocycl. Chem.* **2002**, *42*, 15–23. [CrossRef]

22. He, J.; Wei, M.; Li, B.; Kang, Y.; Evans, D.G.; Duan, X. Preparation of Layered Double Hydroxides. *Struct. Bond.* **2005**, *119*, 89–119. [CrossRef]
23. Pavel, O.D.; Stamate, A.-E.; Zăvoianu, R.; Bucur, I.C.; Bîrjega, R.; Angelescu, E.; Pârvulescu, V.I. Mechano-chemical versus co-precipitation for the preparation of Y-modified LDHs for cyclohexene oxidation and Claisen-Schmidt condensations. *Appl. Catal. A: Gen.* **2020**, *605*, 117797. [CrossRef]
24. Chen, L.; Sun, B.; Wang, X.; Qiao, F.; Ai, S. 2D ultrathin nanosheets of Co–Al layered double hydroxides prepared in l-asparagine solution: Enhanced peroxidase-like activity and colorimetric detection of glucose. *J. Mater. Chem. B* **2013**, *1*, 2268–2274. [CrossRef]
25. Wang, Y.; Zhang, Y.; Liu, Z.; Xie, C.; Feng, S.; Liu, D.; Shao, M.; Wang, S. Layered Double Hydroxide Nanosheets with Multiple Vacancies Obtained by Dry Exfoliation as Highly Efficient Oxygen Evolution Electrocatalysts. *Angew. Chem. Int. Ed.* **2017**, *56*, 5867–5871. [CrossRef]
26. Musella, E.; Gualandi, I.; Scavetta, E.; Rivalta, A.; Venuti, E.; Christian, M.; Morandi, V.; Mullaliu, A.; Giorgetti, M.; Tonelli, D. Newly developed electrochemical synthesis of Co-based layered double hydroxides: Toward noble metal-free electro-catalysis. *J. Mater. Chem. A* **2019**, *7*, 11241–11249. [CrossRef]
27. Rong, F.; Zhao, J.; Yang, Q.; Li, C. Nanostructured hybrid NiFeOOH/CNT electrocatalysts for oxygen evolution reaction with low overpotential. *RSC Adv.* **2016**, *6*, 74536–74544. [CrossRef]
28. Ciotta, E.; Pizzoferrato, R.; Di Vona, M.L.; Ferrari, I.V.; Richetta, M.; Varone, A. Increasing the Electrical Conductivity of Layered Double Hydroxides by Intercalation of Ionic Liquids. *Mater. Sci. Forum* **2018**, *941*, 2209–2213. [CrossRef]
29. Pavel, O.D.; Stamate, A.-E.; Bacalum, E.; Cojocaru, B.; Zăvoianu, R.; Pârvulescu, V.I. Catalytic behavior of Li-Al-LDH prepared via mechanochemical and co-precipitation routes for cyanoethylation reaction. *Catal. Today* **2021**, *366*, 227–234. [CrossRef]
30. Iyi, N.; Matsumoto, T.; Kaneko, Y.; Kitamura, K. A Novel Synthetic Route to Layered Double Hydroxides Using Hexamethylenete-tramine. *Chem. Lett.* **2004**, *33*, 1122–1123. [CrossRef]
31. Schmidt, J.G. Ueber die Einwirkung von Aldehyd auf Furfurol. *Ber. Dtsch. Chem. Ges.* **1880**, *13*, 2342–2345. [CrossRef]
32. Claisen, L. Condensationen der Aldehyde mit Acetessig- und Malonsäureäther. *Ber. Dtsch. Chem. Ges.* **1881**, *14*, 345–349. [CrossRef]
33. Sallum, L.O.; Duarte, V.S.; Custodio, J.M.F.; Faria, E.C.M.; da Silva, A.M.; Lima, R.S.; Camargo, A.J.; Napolitano, H.B. Cyclohexanone-Based Chalcones as Alternatives for Fuel Additives. *ACS Omega* **2022**, *7*, 11871–11886. [CrossRef] [PubMed]
34. Eagon, S.; Hammill, J.T.; Fitzsimmons, K.; Sienko, N.; Nguyen, B.; Law, J.; Manjunath, A.; Wilkinson, S.P.; Thompson, K.; Glidden, J.E.; et al. Antimalarial activity of 2,6-dibenzylidenecyclohexanone derivatives. *Bioorganic Med. Chem. Lett.* **2021**, *47*, 128216. [CrossRef]
35. Tang, Y.; Xu, J.; Gu, X. Modified calcium oxide as stable solid base catalyst for Aldol condensation reaction. *J. Chem. Sci.* **2013**, *125*, 313–320. [CrossRef]
36. Sluban, M.; Cojocaru, B.; Parvulescu, V.I.; Iskra, J.; Korošec, R.C.; Umek, P. Protonated titanate nanotubes as solid acid catalyst for aldol condensation. *J. Catal.* **2017**, *346*, 161–169. [CrossRef]
37. Jain, D.; Khatri, C.; Rani, A. Synthesis and characterization of novel solid base catalyst from fly ash. *Fuel* **2011**, *90*, 2083–2088. [CrossRef]
38. Motiur Rahman, A.F.M.; Ali, R.; Jahng, Y.; Kadi, A.A. A Facile Solvent Free Claisen-Schmidt Reaction: Synthesis of α,α′-bis-(Substituted-benzylidene)cycloalkanones and α,α′-bis-(Substituted-alkylidene)cycloalkanones. *Molecules* **2012**, *17*, 571–583. [CrossRef]
39. Mortezaei, Z.; Zendehdel, M.; Bodaghifard, M.A. Synthesis and characterization of functionalized NaP Zeolite@CoFe2O4 hybrid materials: A micro–meso-structure catalyst for aldol condensation. *Res. Chem. Intermed.* **2020**, *46*, 2169–2193. [CrossRef]
40. Vashishtha, M.; Mishra, M.; Undre, S.; Singh, M.; Shah, D.O. Molecular mechanism of micellar catalysis of cross aldol reaction: Effect of surfactant chain length and surfactant concentration. *J. Mol. Catal. A: Chem.* **2015**, *396*, 143–154. [CrossRef]
41. Hu, Z.G.; Liu, J.; Zeng, P.L.; Dong, Z.B. Synthesis of α, α′-bis(Substituted Benzylidene)Ketones Catalysed by a SOCl₂/EtOH reagent. *J. Chem. Res.* **2004**, *2004*, 55–56. [CrossRef]
42. Miyata, S. The Syntheses of Hydrotalcite-Like Compounds and Their Structures and Physico-Chemical Properties I: The Systems Mg²⁺-Al³⁺-NO₃⁻, Mg²⁺-Al³⁺-Cl⁻, Mg²⁺-Al³⁺-ClO₄⁻, Ni²⁺-Al³⁺-Cl⁻ and Zn²⁺-Al³⁺-Cl⁻. *Clays Clay Miner.* **1975**, *23*, 369–375. [CrossRef]
43. Musker, W.K. A Reinvestigation of the Pyrolysis of Tetramethylammonium Hydroxide. *J. Am. Chem. Soc.* **1964**, *86*, 960–961. [CrossRef]
44. Pavel, O.; Zăvoianu, R.; Bîrjega, R.; Angelescu, E.; Pârvulescu, V.I. Mechanochemical versus co-precipitated synthesized lanthanum-doped layered materials for olefin oxidation. *Appl. Catal. A: Gen.* **2017**, *542*, 10–20. [CrossRef]
45. Frost, R.L.; Cash, G.A.; Kloprogge, J. 'Rocky Mountain leather', sepiolite and attapulgite—an infrared emission spectroscopic study. *Vib. Spectrosc.* **1998**, *16*, 173–184. [CrossRef]
46. Singh, D.K.; Pandey, D.K.; Yadav, R.R.; Singh, D. A study of nanosized zinc oxide and its nanofluid. *Pramana* **2012**, *78*, 759–766. [CrossRef]
47. Anandan, K.; Siva, D.; Rajesh, K. Structural and optical properties of (ZnO/MgO) nanocomposites. *Int. J. Eng. Res. Technol.* **2018**, *7*, 493–499. [CrossRef]
48. Yousefi, S.; Ghasemi, B.; Tajally, M.; Asghari, A. Optical properties of MgO and Mg(OH) 2 nanostructures synthesized by a chemical precipitation method using impure brine. *J. Alloys Compd.* **2017**, *711*, 521–529. [CrossRef]
49. Winter, F.; Xia, X.; Hereijgers, B.P.C.; Bitter, J.H.; van Dillen, A.J.; Muhler, A.M.; de Jong, K.P. On the Nature and Accessibility of the Brønsted-Base Sites in Activated Hydrotalcite Catalysts. *J. Phys. Chem. B* **2006**, *110*, 9211–9218. [CrossRef]

50. Pavel, O.; Bîrjega, R.; Che, M.; Costentin, G.; Angelescu, E.; Şerban, S. The activity of Mg/Al reconstructed hydrotalcites by "memory effect" in the cyanoethylation reaction. *Catal. Commun.* **2008**, *9*, 1974–1978. [CrossRef]
51. Debecker, D.P.; Gaigneaux, E.M.; Busca, G. Exploring, Tuning, and Exploiting the Basicity of Hydrotalcites for Applications in Heterogeneous Catalysis. *Chem. Eur. J.* **2009**, *15*, 3920–3935. [CrossRef]
52. Parida, K.; Das, J. Mg/Al hydrotalcites: Preparation, characterisation and ketonisation of acetic acid. *J. Mol. Catal. A Chem.* **2000**, *151*, 185–192. [CrossRef]
53. Pavel, O.; Zăvoianu, R.; Bîrjega, R.; Angelescu, E. The effect of ageing step elimination on the memory effect presented by $Mg_{0.75}Al_{0.25}$ hydrotalcites (HT) and their catalytic activity for cyanoethylation reaction. *Catal. Commun.* **2011**, *12*, 845–850. [CrossRef]
54. Angelescu, E.; Bîrjega, R.; Pavel, O.; Che, M.; Costentin, G.; Popoiu, S. Hydrotalcites (HTs) and mesoporous mixed oxides obtained from HTs, basic solid catalysts for cyclohexanone condensation. *Stud. Surf. Sci. Catal.* **2005**, *156*, 257–264. [CrossRef]

catalysts

Article

On the Effect of the M^{3+} Origin on the Properties and Aldol Condensation Performance of MgM^{3+} Hydrotalcites and Mixed Oxides

Valeriia Korolova [1], Oleg Kikhtyanin [2], Martin Veselý [3], Dan Vrtiška [1], Iva Paterová [3], Vlastimil Fíla [4], Libor Čapek [5] and David Kubička [1,2,*]

1   Department of Petroleum Technology and Alternative Fuels, University of Chemistry and Technology Prague, Technická 5, 166 28 Prague, Czech Republic; korolovv@vscht.cz (V.K.); vrtiskad@vscht.cz (D.V.)
2   Technopark Kralupy, University of Chemistry and Technology Prague, nám. G. Karse 7/2, 278 01 Kralupy nad Vltavou, Czech Republic; oleg.kikhtyanin@vscht.cz
3   Department of Organic Technology, University of Chemistry and Technology Prague, Technická 5, 166 28 Prague, Czech Republic; veselyr@vscht.cz (M.V.); dudkovai@vscht.cz (I.P.)
4   Department of Inorganic Technology, University of Chemistry and Technology Prague, Technická 5, 166 28 Prague, Czech Republic; filav@vscht.cz
5   Department of Physical Chemistry, Faculty of Chemical Technology, University of Pardubice, Studentská 573, 532 10 Pardubice, Czech Republic; libor.capek@upce.cz
*   Correspondence: kubickad@vscht.cz

Citation: Korolova, V.; Kikhtyanin, O.; Veselý, M.; Vrtiška, D.; Paterová, I.; Fíla, V.; Čapek, L.; Kubička, D. On the Effect of the M^{3+} Origin on the Properties and Aldol Condensation Performance of MgM^{3+} Hydrotalcites and Mixed Oxides. *Catalysts* **2021**, *11*, 992. https://doi.org/10.3390/catal11080992

Academic Editors: Ioan-Cezar Marcu and Octavian Dumitru Pavel

Received: 2 August 2021
Accepted: 17 August 2021
Published: 18 August 2021

Publisher's Note: MDPI stays neutral with regard to jurisdictional claims in published maps and institutional affiliations.

Abstract: Hydrotalcites (HTCs) are promising solid base catalysts to produce advanced biofuels by aldol condensation. Their main potential lies in the tunability of their acid-base properties by varying their composition. However, the relationship between the composition of hydrotalcites, their basicity, and their catalytic performance has not yet been fully revealed. Here, we investigate systematically the preparation of HTCs with the general formula of Mg$_6$M$^{3+}_2$(OH)$_{16}$CO$_3$·4H$_2$O, where M^{3+} stands for Al, Ga, Fe, and In, while keeping the Mg/M^{3+} equal to 3. We use an array of analytical methods including XRD, N$_2$ physisorption, CO$_2$-TPD, TGA-MS, FTIR-ATR, and SEM to assess changes in the properties and concluded that the nature of M^{3+} affected the HTC crystallinity. We show that the basicity of the HTC-derived mixed oxides decreased with the increase in atomic weight of M^{3+}, which was reflected by decreased furfural conversion in its aldol condensation with acetone. We demonstrate that all MgM^{3+} mixed oxides can be fully rehydrated, which boosted their activity in aldol condensation. Taking all characterization results together, we conclude that the catalytic performance of the rehydrated HTCs is determined by the "host" MgO component, rather than the nature of M^{3+}.

Keywords: hydrotalcites; mixed oxides; aldol condensation; basic catalysts

1. Introduction

The Green Chemistry principles aroused a strong interest in the development of solid base catalysts for important base-catalyzed reaction, such as transesterification, aldol condensation, or alkylation [1–9]. The use of solid base catalysts would allow for reducing the amount of produced waste waters significantly along with simplifying the product separation and improving the product quality. Magnesium-containing mixed oxides are among the most studied materials as they can be prepared rather easily by calcination of the so-called hydrotalcites (HTCs), i.e., hydroxycarbonates with layered structure and the general formula [M$^{2+}_{1-x}$M$^{3+}_x$(OH)$_2$]$^{b+}$[A^{n-}]$_{b/n}$·mH$_2$O, where M^{2+} and M^{3+} are divalent and trivalent metal cations, respectively, and A is an interlayer anion, typically carbonate [1,10,11]. Besides Mg, Zn or Ni are used as divalent cations, whereas the trivalent cations are represented primarily by Al that can be replaced e.g., by Ga or Fe [1,9,12–17]. Due to the so-called memory effect, the mixed oxides having a Lewis-base character can be rehydrated to form hydrotalcites in hydroxide form (instead of the carbonate form) with a Brønsted-base character [2,3,18]. This opens not only a range of the application possibilities, but also of the fundamental questions relating to the relationship between their structure and basicity as well as between their physico-chemical properties and catalytic performance.

The acid–base properties of HTC-based catalysts for organic reactions can be tailored by isomorphous substitution of Mg and Al cations with various other di- and trivalent cations [12,13,19]. The basicity of M^{2+}Al hydrotalcites decreased when Mg was replaced by Zn or Ni which was reflected in the catalytic performance of the HTC-derived materials [19–22]. However, the effect of Al substitution by another M^{3+} element in the MgM^{3+} hydrotalcite-like materials on their acid-base properties and catalytic performance has not yet been systematically described.

Aldol condensation is currently used to synthesize fine chemicals [5,23,24], and it holds a promising potential for development of sustainable aviation fuels [25–28] and bio-based monomers [29,30]. In addition, it is a suitable reaction for probing the acid-base character of catalysts as only the accessible, i.e., catalytically active basic sites can be reached by the reactants in both mixed oxides and reconstructed hydrotalcites. In contrast, the common CO_2-TPD method is suitable for probing the basic sites only in mixed oxides as the rehydrated hydrotalcites react with CO_2 and are transformed into hydroxycarbonates [16]. Moreover, CO_2 can also reach those basic sites that are not accessible to the reactants, i.e., sites relevant to CO_2 adsorption and storage, but not to catalysis.

The wide variability of the typically used hydrotalcite synthesis protocols including their activation by calcination yielding mixed oxides and by calcination followed by re-hydration affording reconstructed hydrotalcites and their use in different reactions under varying reaction conditions thwart the fundamental evaluation of the specific influence of the M^{3+} cation on the structure and activity of the synthesized materials. Therefore, we have synthesized a series of MgM^{3+} hydrotalcites with M^{3+} being Al, Ga, Fe, and In using the same synthesis procedure and evaluated the derived mixed oxides and reconstructed hydrotalcites with respect to their physico-chemical properties and catalytic performance in aldol condensation of furfural and acetone. This allowed us to elucidate the role of the trivalent cation on the structure, properties, and catalytic performance (including its stability) of the synthesized materials.

2. Results and Discussion

2.1. Catalyst Synthesis and Characterization

The ICP and XRD results confirmed that the coprecipitation of magnesium nitrate and the corresponding M^{3+} nitrate (M^{3+} being Al, Ga, In, and Fe, respectively) by the alkaline solution of Na_2CO_3 and NaOH resulted in the formation of the respective hydrotalcite with the targeted Mg/M^{3+} ratio of 3. The ICP results also proved that sodium that could interfere in the catalytic studies was removed almost completely during washing of the precipitates. The Na/M^{3+} atomic ratio was in the range 0.025–0.033 and did not reveal any dependence on the M^{3+} nature.

The measured Mg/M^{3+} atomic ratios are reported in Table 1 together with the calculated formula of the as prepared hydrotalcites. The calculation was possible as phase pure hydrotalcite phases characterized by the diffraction lines at $2\theta \approx 11.3°$, $22.5°$, $34.5°$, $39°$, $46°$, $60°$, and $61.5°$ were formed (Figure 1A) without any admixtures being detectable. The formation of the phase-pure HTC materials suggested an effective substitution of Mg^{2+} cations in the brucite-like layers by M^{3+} cations (even in case of In^{3+} with the largest ionic radius of 0.8 Å) being in line with the previous studies [16,17,31,32].

Table 1. Mg/M^{3+} atomic ratios in the as-prepared samples according to ICP results and the proposed HTC formula.

Sample	Ionic Radius of M^{3+}, Å *	Mg^{2+}/M^{3+} by ICP	Proposed Chemical Formula of Prepared Solids	Theoretical Weight Loss under Calcination
MgAl-AP	0.535	3.0	Mg$_6$Al$_2$(OH)$_{16}$CO$_3$·4H$_2$O	43.1
MgGa-AP	0.62	2.8	Mg$_{5.6}$Ga$_2$(OH)$_{15.2}$CO$_3$·3.8H$_2$O	34.6
MgFe-AP	0.645	2.9	Mg$_{5.8}$Fe$_2$(OH)$_{15.6}$CO$_3$·3.9H$_2$O	38.0
MgIn-AP	0.8	2.9	Mg$_{5.8}$In$_2$(OH)$_{15.6}$CO$_3$·3.9H$_2$O	32.2

*—from http://abulafia.mt.ic.ac.uk/shannon/ptable.php (Accessed on 2 July 2021).

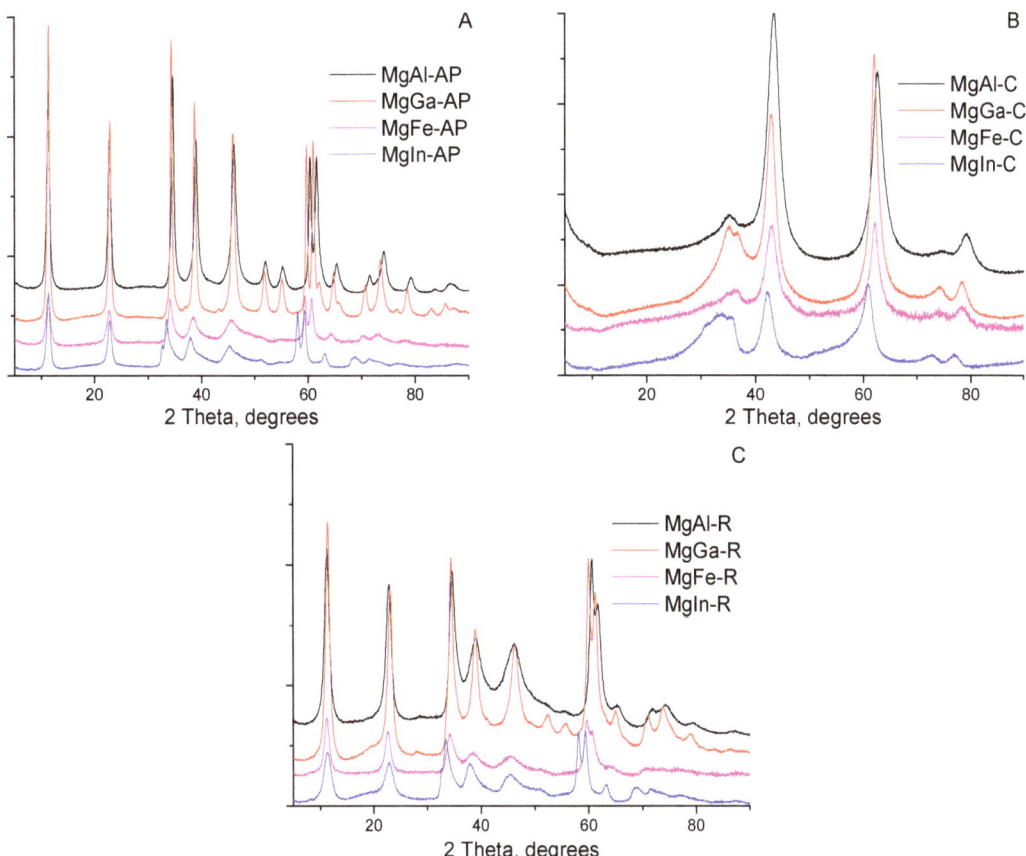

Figure 1. XRD patterns of as-prepared (**A**), calcined (**B**), and rehydrated (**C**) MgAl, MgGa, MgFe, and MgIn samples.

Although all as-prepared samples exhibited the hydrotalcite structure, the intensity of the reflections in the XRD patterns differed significantly (Figure 1) indicating difference in crystallinity. The crystallinity was evaluated using the area of the signals attributed to two different basal reflections (003 and 110) and is reported as relative crystallinity (Table 2). MgGa-AP and MgAl-AP were the most crystalline among the samples with the crystallinity close to 100%, followed by MgFe-AP and MgIn-AP with the crystallinity of 32–41% (Figure 1A and Table 2). Such a difference in the crystallinity could either reflect the impact of M^{3+} cations on the dimensions of HTC platelets and the existence of local defects in the layered structure or suggest the presence of X-ray amorphous impurities.

Table 2. Crystallinity, lattice parameters, and crystallite size (L(Å)) of the prepared samples.

Sample	Relative CrystalLinity (%)	HTC Basal Spacing d_{003}, (Å)	Unit Cell a, (Å)	MgO d_{200}, (Å)	Crystallite Size, L (Å)			
					d_{003}	d_{110}	d_{200}	d_{220}
MgAl-AP	98	7.8	3.063		116	188		
MgGa-AP *	100	7.81	3.096		153	231		
MgFe-AP	32	7.85	3.111		86	143		
MgIn-AP	41	7.83	3.178		98	157		
MgAl-C				2.089			30	31
MgGa-C				2.107			40	46
MgFe-C				2.119			33	40
MgIn-C				2.141			39	45
MgAl-R	88	7.83	3.061		57	85		
MgGa-R	95	7.78	3.088		78	127		
MgFe-R	29	7.85	3.103		62	78		
MgIn-R	38	7.81	3.172		54	114		

*—the sample with the largest crystallinity taken as a reference.

The diffraction lines assigned to (003) and (110) reflections (i.e., at $\approx$11.3° and 60°) were used to calculate the basal spacing between the layers (d_{003}) and the unit cell dimension a (as $a = 2d_{110}$), respectively. The d_{003} value increased from 7.80–7.81 Å to 7.83–7.85 Å with the increase in the ionic radius of M^{3+} cations (Table 2), which reflected the change in the thickness of each hydroxide layer due to the change in the charge density of the layers [33]. The unit cell dimension increased linearly from 3.063 Å to 3.178 Å with the increase in the M^{3+} ionic radius (Table 2). Both diffraction lines were used to calculate the sizes of the crystallites in the direction of the layers stacking (d_{003}) and of the layers plane (d_{110}). The dimensions are given in Table 2; they correlate well with the relative crystallinity values, i.e., the larger the crystallinity, the larger the crystallite size (i.e., the size of the coherent domain).

The hydrotalcite structure of all samples was completely decomposed because of the calcination at 450 °C and the corresponding mixed oxide phase was formed as evidenced by the diffraction lines at 2θ $\approx$ 43.0° and 62.5° (Figure 1B). These reflections are characteristic for the periclase structure (MgO). Nonetheless, the presence of M^{3+} cations in the MgO structure can be deduced from the increase in the MgO basal spacing in the calcined samples, d_{200}, from 2.089 Å to 2.141 Å with the increasing ionic radius of M^{3+}. It is noteworthy that the average crystallite size (L) of the MgM-C samples calculated from d_{200} and d_{220} basal reflections was considerably lower than the crystallite size of the corresponding as-prepared hydrotalcites (Table 2). Unfortunately, the presence of other plausible phases, in particular spinel and isolated M^{3+} oxides, can be neither excluded nor confirmed by the XRD data.

It is well documented that the HTC-derived mixed oxides, when rehydrated, can be reconstructed to the original hydrotalcite structure. This has been demonstrated mainly for MgAl materials [18,34–39], but some examples of successful transformation of MgGa, MgFe, and MgIn mixed oxides to hydrotalcites exist as well [16,17,31,40]. It worth noting that rehydrated MgIn HTC was previously obtained only by the hydrothermal treatment of a corresponding MgIn mixed oxide (140 °C, 24 h) in Na_2CO_3 solution [31]. By using the same rehydration procedure, we have succeeded in transforming all MgM^{3+} mixed oxides to the corresponding hydrotalcites as evidenced by the appearance of the characteristic HTC diffraction lines at 2θ $\approx$ 11.3, 22.5, 34.5°, etc. and disappearance of the mixed oxide diffraction lines at 2θ $\approx$ 43.0° and 62.5° (Figure 1C). In particular, the reconstruction of the MgIn HTC structure by the treatment of MgIn mixed oxide with pure water at ambient temperature has not yet been reported. A comparison of the diffractograms in Figure 1A,C shows that the intensities of the XRD reflexes of the rehydrated samples are much smaller than those of the as prepared hydrotalcites. However, as demonstrated in Table 2, the relative crystallinity of the rehydrated samples was not significantly lower than that of the as-prepared hydrotalcites.

The unit cell size (Table 2) of the rehydrated materials was only slightly lower than in case of the as prepared HTC, but it preserved its linear dependence on the M^{3+} ionic radius. While the crystallinity and unit cell size were preserved, there was a dramatic decrease in the size of the coherent domains in both directions, i.e., in the stacking (d_{003}) as well as in the platelet plane (d_{110}) (Table 2). The results indicate that, although the crystalline structure was reconstructed and the unit cell size was virtually unchanged due to rehydration, there were some imperfections due to the rehydration process that resulted in the observed decrease in the crystallite size, i.e., in the coherent domain size. This can be plausibly explained by the incomplete insertion of the M^{3+} cations in their original positions in the brucite-like layers (which is supported by the observed decrease in the unit cell size) and the formation MgM^{3+} spinel or M^{3+} oxide phases that break the structure periodicity and thus reduce the coherent domain size. Apparently, the extent is rather small as neither of these phases was detected by XRD, i.e., their coherent domain sizes were too small.

The visual inspection of prepared materials using SEM revealed differences in the morphology of the samples due to the differences in the M^{3+} element and due to the calcination-rehydration treatments (Figure 2). Regardless of the M^{3+} cation in the structure, all AP-HTCs showed a well-developed layered structure with haphazardly oriented and intergrown platelets (Figure 2). These agglomerates were several micrometers large and consisted of differently sized platelets (<1 mm). While MgAl-AP and MgGa-AP had well-shaped platelets (0.3–0.6 and 0.5–0.8 mm, respectively, (Figure 2), the platelets of MgFe-AP and MgIn-AP were more heterogeneous and layered character was less developed, particularly in case of MgIn-AP. These observations are in line with the relative crystallinity data based on XRD (Table 2), i.e., the samples with higher crystallinity have larger and well-organized platelets.

Upon calcination and calcination followed by rehydration, the size of the agglomerates did not change significantly, but the shape and degree of heterogeneity increased because of the rehydration treatment. This is demonstrated on the MgGa-R, i.e., the sample with the highest crystallinity among rehydrated samples, in Figure 2 (2R) (enlarged image is available in Figure S1 in the SI). In great contrast to MgGa-AP, the MgGa-R platelets were crumbled, and their surface was cracked which supports the observed decrease in the coherent domain size (Table 2). A similar trend was observed for all the MgM^{3+} rehydrated materials. Consequently, the individual platelets of MgIn R could be barely distinguished (Figure 2).

The textural properties of the AP hydrotalcites reflected the relative crystallinity of the samples, i.e., the BET surface area decreased linearly with the increasing relative crystallinity of the AP-HTCs (Tables 2 and 3), and it also decreased with the increasing crystallite size (d_{003} and d_{110}) determined from the XRD data (Table 2). A similar trend was also observed for the total pore volume. The calcination affording mixed oxides resulted in a tremendous increase in the BET area (2 to 7 times) and total pore volume (1.5 to 3 times) in comparison with the AP-HTCs (Table 3). Again, the BET surface decreases linearly with the increasing crystallite size of the mixed oxides (d_{200} and d_{220}), proving that the crystallite size (the coherent domain size) is the decisive structural parameter affecting the specific surface area of both AP-HTCs and mixed oxides.

Figure 2. Scanning electron micrographs of the prepared samples: 1—MgAl, 2—MgGa, 3—MgFe, 4—MgIn. A—as-prepared, C—calcined, R—rehydrated for 20 min.

Table 3. BET surface of the as-prepared hydrotalcites (AP), derived mixed oxides (C), and rehydrated hydrotalcites (R), as well as the basic properties of MgM-C samples.

Material	BET Surface Area (m^2/g)			Pore Volume (cm^3/g)			Concentration of Basic Sites in MgM-C, μmol/g	Density of Basic Sites in MgM-C, μmol/m^2	Calculated Ratio $\mu mol_{CO2}/\mu mol_{Mg2+}$ in MgM-C
	AP	C	R	AP	C	R			
MgAl	31	222	5	0.094	0.343	0.013	270	1.2	0.015
MgGa	34	164	4	0.095	0.221	0.011	225	1.4	0.017
MgFe	103	185	12	0.275	0.416	0.031	171	0.9	0.012
MgIn	76	134	19	0.196	0.288	0.039	167	1.3	0.015

On the other hand, the calcination followed by rehydration resulted in very low BET surface areas (<20 m^2/g, i.e., only 12–25% of the BET value obtained for the AP-HTCs) of the MgM-R materials that is in contrast with the decreased crystallite size of the rehydrated HTCs in comparison with their AP counterparts. Thus, there has to be an additional parameter affecting the textural properties. The most plausible explanation is the presence of anions and water molecules in the interlayer space blocking access or the existence of local defects. This conclusion is supported by the trend among the MgM-R materials where the materials with larger platelets (as seen on SEM images, Figure 2) have smaller BET area and total pore volume (Table 3).

The structural changes due to calcination leading to MgM mixed oxides and due to calcination and rehydration affording MgM rehydrated hydrotalcites were further corroborated by the FTIR (Figure S2) and TGA-MS (Figure S3) characterization results that are presented in detail in the SI. The total weight loss observed in the TGA-MS experiments decreased logically with the increasing atomic weight of the M^{3+} cation (Figure S3), and it agreed with the theoretical weight loss calculated from the MgM-AP HTC composition (Table 1). It can be thus inferred that the AP hydrotalcites are indeed phase pure HTCs having the composition reported in Table 1. Moreover, the ratio of the weight loss of a MgM-R to the weight loss of the corresponding MgM-AP was close to unity (Figure S3), which provides more evidence that nearly complete reconstruction of the HTC structure by mixed oxide rehydration was achieved independently on the nature of the M^{3+} cation.

The structural changes in the MgM materials due to differences in composition as well as calcination and rehydration treatments also affect the number and character of the acid and base sites that are essential for the catalytic activity of these materials. The number of basic sites in MgM-C was determined using CO_2-TPD (Figure S4 in the SI) and the results are reported in Table 3. There is a clear increase in the number of basic sites with the decrease in ionic radius of the M^{3+} cation. Nonetheless, it has to be noted that the number of basic sites is given per one gram of the material, i.e., when keeping the atomic ratio of Mg/M^{3+} constant, the relative content of Mg in one gram of sample increases with the decreasing atomic weight of M^{3+}. As a result, the specific Mg content correlates with the number of basic sites. Thus, when normalizing the number of basic sites per 1 mol of Mg^{2+} rather than 1 g of mixed oxide, a constant value of 0.015 mmol of basic sites (adsorbed CO_2) per 1 mmol of Mg atoms is obtained (Table 3). In other words, the number of basic sites is a function of the Mg content per 1 g of material and, thus, the M^{3+} cation does not contribute to the basic character and a simple "spacer" between the Mg-related basic sites can be seen. In addition, the CO_2 desorption profiles were virtually identical for all MgM-C materials which indicated that the strength of CO_2 adsorption, i.e., the strength of the basic sites, was not influenced by the M^{3+} cation nature (Figure S3 in the SI).

2.2. Catalyst Performance

The changes in the catalyst properties resulting from the replacement of Al^{3+} by other M^{3+} elements as well as due to the calcination and rehydration treatments were further characterized by their performance in aldol condensation of furfural with acetone. The basic scheme of aldol condensation of furfural with acetone is shown in Scheme 1.

In general, furfural (F) reacts with acetone (Ac) forming a hydrated intermediate, 4-(2-furyl)-4-hydroxybutan-2-one (FAc-OH) that dehydrates giving a first condensation product, 4-(2-furyl)-3-buten-2-one (FAc). The FAc can react with another F resulting in the formation of 1,4-pentadien-3-one, 1,5-di-2-furanyl (F_2Ac) as the second condensation product. The plausible self-condensation of acetone over mixed oxides was insignificant corresponding to acetone conversion <2% and thus did not affect the aldol condensation of F with Ac. Moreover, the catalyst leaching was also excluded following an approach described previously [41]. The total carbon mass balance including the reactants and the major reaction products, i.e., FAc-OH, FAc, and F_2Ac always exceeded 95%. Heavier reaction products, most likely due to consecutive condensation reactions, were observed only at high furfural conversion. Due to their poor identification and low concentration, they were excluded from the carbon balance calculation.

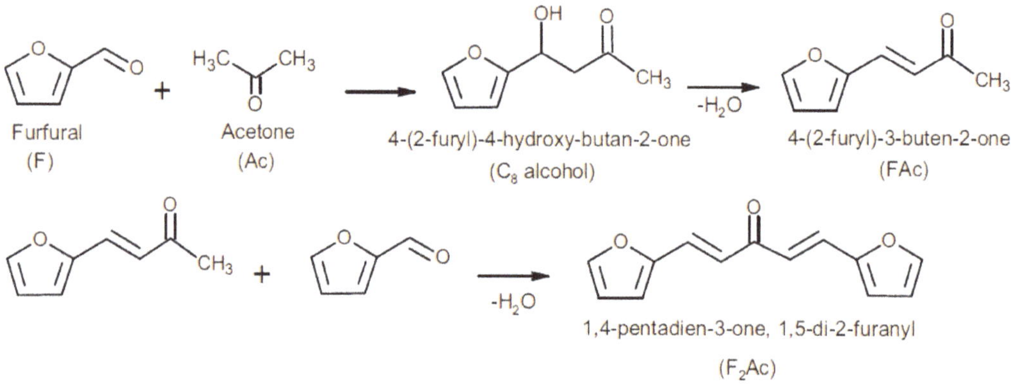

Scheme 1. Basic reaction scheme of aldol condensation between furfural with acetone (from [34]).

In preliminary experiments, the catalytic performance of the four as-prepared materials was evaluated in aldol condensation of furfural and acetone at T = 50 °C and acetone:furfural ratio of 10. Independently from the origin of M^{3+}, furfural conversion was below 1.5% in all cases, thus proving the absence of a strong basicity in the as-prepared HTCs.

The furfural conversion and selectivity to all three main products at 50 °C over MgM-C mixed oxides are summarized in Figures 3A and 3B, respectively. The initial reaction rate decreased in order MgAl-C > MgGa-C > MgIn-C > MgFe-C (Figure 3A). After the initial rapid increase in F conversion (ca. the initial 40 min), the F conversion continued to grow only moderately, which can be ascribed to the gradual catalyst reaction either due to the deposition of heavier products formed by consecutive condensation reactions or due to the specific blockage of basic sites, e.g., by furoic acid formed by Cannizzaro reaction [42]. The initial reaction rate (reflecting the rate of furfural disappearance during initial 10 min, $mmol_F \cdot g_{cat}^{-1} \cdot min^{-1}$) increased linearly with an increase in the number of basic sites per gram in a catalyst demonstrating the role of basic sites (Figure 4A). From the catalyst synthesis point of view, it follows that smaller M^{3+} cations are preferred as they allow for increasing the number of basic sites per gram of catalysts while maintaining the same Mg/M^{3+} ratio. The reaction rate data also exhibit positive correlation with the BET surface area of the calcined samples, thus indicating the importance of the accessibility of the active sites (Figure 4B). These two positive correlations at the same time suggest that the surface density of the basic sites ($\mu mol/m^2$) does not affect to a great extent in the studied range (0.9–1.4 $\mu mol/m^2$, Table 3) the reaction rate.

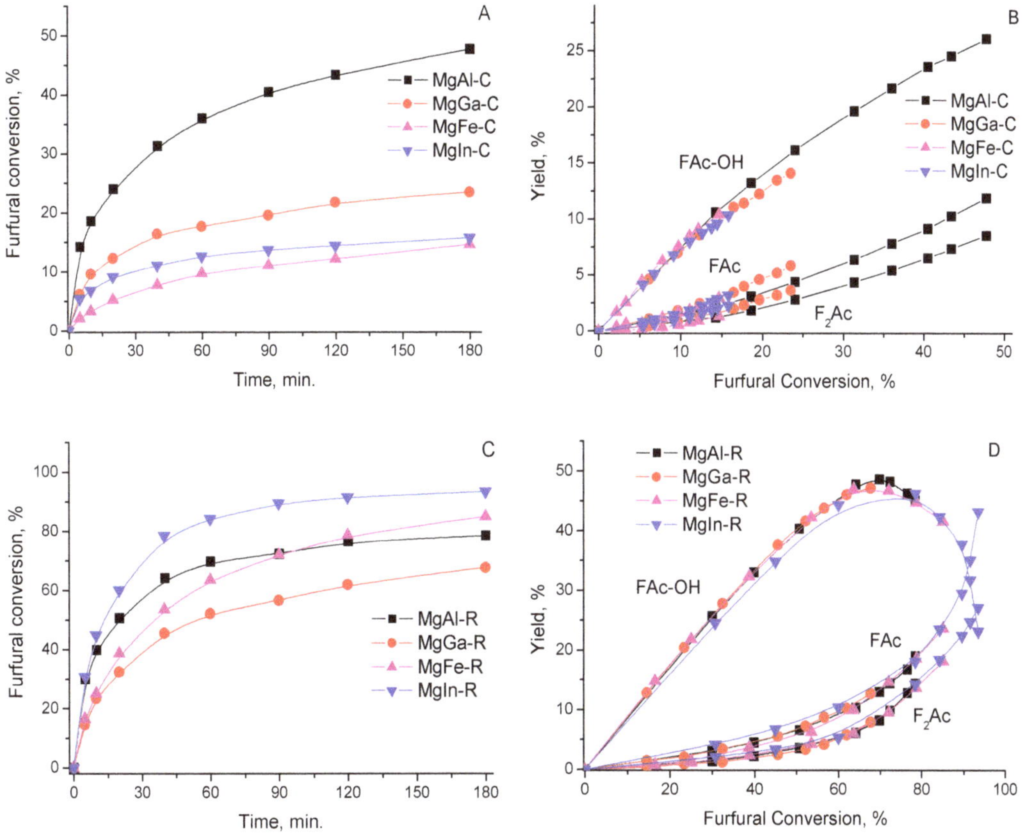

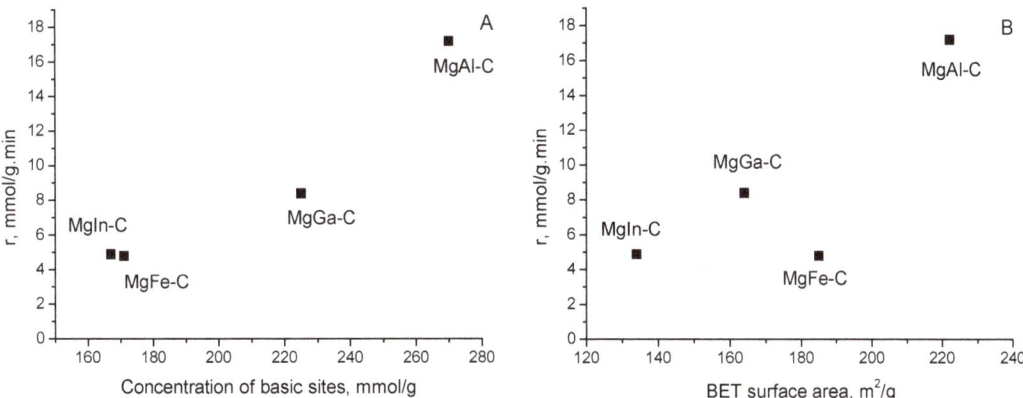

Figure 3. The dependence of the furfural conversion on the reaction time observed on MgM-C ((A), $T_{reac.} = 50$ °C, molar ratio F:Ac – 1:10) and MgM-R (C), $T_{reac.} – 25$ °C, molar ratio F:Ac = 1:5) samples. Yield for the reaction products is observed on MgM-C (**B**) and MgM-R (**D**) as a function of the furfural conversion.

Figure 4. The dependence of the initial reaction rate on the concentration of basic sites (**A**) and BET surface area (**B**) of MgM-C samples.

The selectivity to the main products was virtually identical at a given conversion regardless of the nature of the M^{3+} and the number and concentration of basic sites (Figure 3B). It can be, thus, inferred that all MgM^{3+} mixed oxides possessed the same nature of basic sites, i.e., that MgO provided the basic sites, with a similar strength as seen by TPD results (Figure S4 in the SI). In conclusion, the M^{3+} cation determines the number of available basic sites (per gram of catalyst) in MgM^{3+} mixed oxides, but it does not affect either the type or the strength of the basic sites present. Moreover, the linear relationship between the initial activity and the number of basic sites also indicates that either all basic sites determined by CO_2 desorption or their same percentage in all MgM^{3+} mixed oxides are accessible to the reactants and take part in the reaction.

The rehydration of the MgM^{3+} mixed oxides has resulted in a significant increase in the aldol condensation rate as documented in Figure 3C. Despite the aldol condensation reaction using rehydrated HTCs being performed at only 25 °C, significantly higher F conversion was obtained than over the MgM^{3+} mixed oxide catalysts at 50 °C (Figure 3A). While for MgAl materials successful rehydration resulting in the transformation of the Lewis basic sites into the more intrinsically active Bronsted basic sites is well known [2,34,43,44], it has been reported only scarcely for MgGa and MgFe materials [16,17,40]. A successful rehydration of MgIn mixed oxide is reported for the first time and the rehydrated MgIn-R HTC was even found to be a more efficient catalyst for aldol condensation of F and Ac. In line with the aldol condensation over mixed oxides reported here and the previous studies on aldol condensation, the superior performance of the MgIn-R catalyst should be related to its superior number of accessible basic sites.

Due to the chemical reaction between rehydrated hydrotalcites, i.e., materials in exclusively hydroxy form, with CO_2 resulting in the replacement of some hydroxyls and formation of the more stable and thermodynamically-favored hydroxycarbonates, TPD-CO_2 cannot be used to determine the number of basic sites. Nonetheless, the catalytic data can be used to compare the number of basic sites in different rehydrated catalysts. The data in Figure 3C show that the initial reaction rate decreased in the order MgIn ≈ MgAl > MgFe ≈ MgGa. As with the MgM^{3+} mixed oxides, the decrease in catalytic activity was observed at longer reaction times (>40 min) and the final conversion decreased in the order MgIn > MgFe > MgAl > MgGa (Figure 3C), which can be concluded to be also the order of the number of accessible basic sites. As in the case of MgM-C catalysts, all MgM-R have also exhibited a virtually identical selectivity to the main reaction products (Figure 3D). It suggests that the upon-rehydration created Bronsted basic sites had similar strength distribution, i.e., the differences in the observed conversion have to be caused by other catalyst properties than by their base-sites strength.

The inspection of the specific BET surface areas of the rehydrated MgM-R materials further supports the significantly higher intrinsic catalytic activity of the Bronsted basic sites than of the Lewis basic sites as the BET area of the rehydrated materials was 7 to 26 times lower (Table 3) than that of their mixed oxide counterparts. Interestingly, the most severe drop (20- to 26-times lower BET for MgM-R than for MgM-C) was observed for MgAl and MgGa, i.e., the most crystalline materials. In contrast, the BET decrease only 7- to 13-times for the less crystalline MgFe and MgIn rehydrated catalysts (Table 3). These results agree with the observed decrease in catalyst activity at longer reaction times (Figure 3C) as the rehydrated HTC catalysts with the largest BET surface area (MgIn-R and MgFe-R) exhibited a weaker decrease in the catalyst activity (i.e., a larger difference between the final conversion and the conversion after ca. 10 min) than MgAl-R and MgGa-R that had a similar initial activity as MgIn-R and MgFe-R, respectively (Figure 3C). As discussed above, the textural data (BET area, total pore volume) are further corroborated by the relative crystallinity results (XRD) and by the HTC platelets size (SEM). In particular, the decreasing size of the rehydrated HTC platelets corresponds well to the increasing furfural conversion (Figure 5). This is in line with the findings of Abello et al. [2] who suggested that only the active sites located at the edges of the platelets were operative in aldol condensations. As the formation of smaller platelets inherently increases the number of OH$^-$ ions near the edges, the reconstructed HTCs with a smaller size of HTC platelets should possess enhanced catalytic activity.

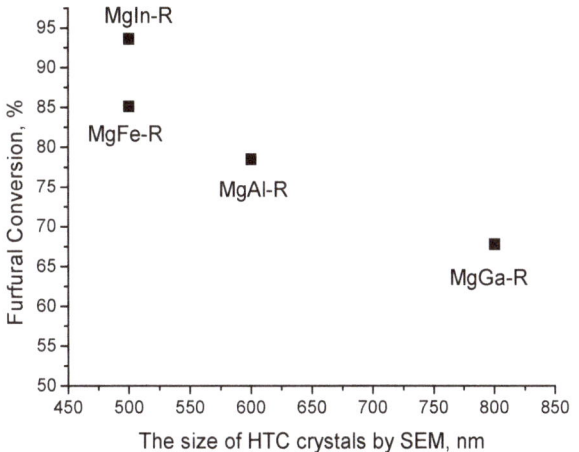

Figure 5. A correlation between the size of HTC platelets in rehydrated MgM-R20 hydrotalcites evaluated by SEM and furfural conversion over these catalysts after 180 min of the reaction.

The stability of the MgM^{3+} catalysts was assessed in the consecutive catalytic cycles consisting always of calcination, rehydration, and reaction (Figure 6A). All studied MgM-R catalysts exhibited stable catalytic performance without any obvious significant change in furfural conversion during the three catalytic cycles (Figure 6A). In fact, in the consecutive experiments with MgAl-R and MgGa-R, an increase in the conversion of furfural was observed in comparison with the first experiment, while MgFe-R and MgIn-R maintained its conversion in the second and third run on the same level (Figure 6A). The catalyst stability was further supported by the stable yields of all three main reaction products as a function of furfural conversion independently on the nature of the M^{3+} cation and regardless of the number of consecutive runs (Figure 6B). Once again, this shows the indirect catalytic role of the M^{3+} cation, i.e., it was not involved directly in the aldol condensation reaction.

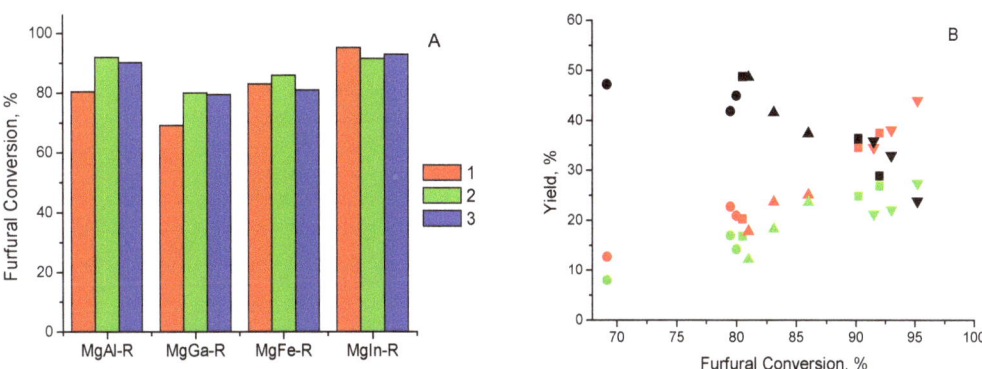

Figure 6. (**A**). Furfural conversion in three consecutive cycles [calcination-rehydration-reaction] in the presence of different MgM-R materials; (**B**) the yield of reaction products observed in three reaction cycles. Black symbols—FAc-OH, red symbols—FAc, green symbols—F_2Ac; squares—MgAl-R, circles—MgGa-R, triangle up—MgFe-R, triangle down—MgIn-R. Molar ratio furfural: acetone = 5, $T_{reac.}$ = 25 °C, reaction time—3 h.

3. Experimental Materials and Methods

3.1. Catalyst Preparation

MgM^{3+} hydrotalcite (HTC) samples (where M stands for Al, Ga, Fe, and In) with a Mg/M^{3+} atomic ratio in a reactive mixture of 3:1 were prepared based on a coprecipitation method adopted from [34,41]. Mg(NO$_3$)$_2$·6H$_2$O (99.9%, Lach:Ner, Neratovice, Czech Republic), Al(NO$_3$)$_3$·9H$_2$O (98.8%, Lach:Ner, Czech Republic), Ga(NO$_3$)$_3$·xH$_2$O (99.9%, Sigma-Aldrich, Prague, Czech Republic), Fe(NO$_3$)$_3$·9H$_2$O (98%, Sigma-Aldrich) and In(NO$_3$)$_3$·xH$_2$O (99.9%, Sigma-Aldrich) were used to prepare a salt solution, while NaOH (99.6%, Lach:Ner, Czech Republic) and Na$_2$CO$_3$ (99%, Penta, Czech Republic) were used to prepare an alkaline solution. An aqueous solution of Mg and M^{3+} nitrates (total metal ion concentration of 0.5 mol·L^{-1}) was slowly added to 200 mL of redistilled water. The flow rate of the simultaneously added alkaline solution of Na$_2$CO$_3$ (0.2 mol·L^{-1}) and NaOH (1 mol·L^{-1}) was controlled to maintain the reaction pH at the desired value (10.0 ± 0.1). The coprecipitation was carried out for 3 h under vigorous stirring (450 rpm) at 25 °C. The resulting suspension was then aged at 25 °C under stirring for 1.5 h. The precipitate was then filtered, washed several times with a plenty of distilled water (at least 2 L), and dried for 12 h at room temperature and then for 12 h at 60 °C. Finally, the as-prepared MgM^{3+} HTCs (further denoted as MgM-AP) were calcined in static air at 450 °C for 3 h in a muffle oven to produce the corresponding MgM mixed oxides (further denoted as MgM-C) and transferred into a desiccator at the end of the calcination procedure to avoid contamination by atmospheric CO$_2$ during cooling. The rehydration of the mixed oxides was performed by their stirring at 200 RPM in water (0.5 g of freshly calcined material per 100 mL of redistilled water) at room temperature for 20 min, followed by filtration of the resulting solid (further denoted as MgM-R) using a Buchner funnel equipped with a vacuum pump. The MgM-R samples were immediately transferred into a reactor loaded with furfural and acetone to initiate aldol condensation.

3.2. Physico-Chemical Characterization

The phase composition of the prepared catalysts was determined by X-ray diffraction using a diffractometer PANanalytical X'Pert3 Powder and CuKα radiation. The XRD patterns were recorded in a range of 2θ = 5–70°. The relative crystallinity (%) of the as-prepared MgM HTCs and rehydrated HTCs was estimated from the area of signals at the diffraction angle 2θ ≈ 11.5° and 23° (basal reflections (003) and (006)), and 2θ ≈ 60° (basal reflection (110)). In this case, the sample with the highest crystallinity was used as the reference. The average size of coherent domains, further referred to as crystallite size (L), of the prepared materials was estimated from the X-ray line broadening (the values of the fullwidth at half-maximum) using the Scherrer equation (L = 0.9λ/β cosθ), considering different basal reflections: (003) and (110) reflections for the as-prepared and rehydrated HTCs, or (200) and (220) reflections for mixed oxides. The content of Mg^{2+}, M^{3+}, and Na$^+$ in the samples were analyzed by ICP using Agilent 5100 ICP OES.

Nitrogen physisorption was measured at 77 K for calcined catalysts using a static volumetric adsorption system (TriFlex analyzer, Micromeritics, Norcross, GA, USA). The samples were degassed at 473 K (12 h) prior to N$_2$ adsorption analysis to obtain a clean surface. The adsorption isotherms were fitted using the Brunauer–Emmett–Teller (BET) method for the specific surface area.

All samples were characterized by the FTIR-ATR (Attenuated Total Reflection) technique. An infrared spectrometer IRAffinity-1 (Shimadzu, Kyoto, Japan) with Quest ATR accessory with a diamond crystal (Specac, Fort Washington, PA, USA) was used to record the FTIR spectra. LabSolution IR software (Shimadzu, Kyoto, Japan) was used as an interface between the spectrometer and the control computer. The spectra were recorded in the 4000–400 cm^{-1} region using the spectral resolution of 2 cm^{-1}.

TGA-MS examination of the as-prepared and rehydrated HTC samples in an N$_2$ atmosphere was performed using TG-DTA Setsys Evolution (Setaram, Caluire, France) instrument in the temperature range of 25–700 °C (the molecular ions m/z = 18 and 44 were analyzed).

Temperature programmed desorption (TPD) of CO_2 was carried out using a Micromeritics Instrument, AutoChem II 2920. For desorbed CO_2 detection, both a thermal conductivity detector (TCD) and a quadrupole mass spectrometer (MKS Cirrus 2 Analyzer) with a capillary coupling system were used. A catalyst sample (0.06 g) was placed in a quartz U-shaped tube. Prior to adsorption of CO_2, the catalyst was heated under a helium flow (30 mL·min^{-1}) up to 450 °C and kept at 450 °C for 60 min to remove impurities from the sample. In the following step, the sample was cooled down to an adsorption temperature of 40 °C. Measured pulses of CO_2 (pulse volume, 5 mL) were injected into the helium gas and carried through the catalyst sample until adsorption saturation. Then, the sample was purged with helium for 60 min to remove physisorbed CO_2. Afterwards, the linear temperature program (10 °C min^{-1}) was started at a temperature of 40 °C, and the sample was heated up to a temperature of 600 °C. The amount of the desorbed CO_2 was determined by calibration.

The sample morphology was investigated using scanning electron microscopy (SEM) with an FEG electron gun (FIB-SEM TESCAN LYRA3GMU, Brno, Czech Republic) at the acceleration voltage of 10 kV. Prior to the SEM measurement, the sample was placed on a carbon conductive tape and coated by a 5 nm thin gold layer in a sputter coater (Quorum Q150R S, Emitech SC7640 Sputter Coater, Polaron, Laughton, UK).

3.3. Catalytic Tests

Acetone (99.98%, Penta, Czech Republic) and furfural (99%, Sigma-Aldrich) were used as reactants in all catalytic experiments.

It was shown in our recent article [45] that acid impurities in furfural could influence the performance of solids with basic properties in aldol condensation, and that catalyst was partially spent for the neutralization of the acid impurities. Moreover, a routine distillation of an as-received furfural had only a temporary effect on the properties of a furfural source, which was readily re-oxidized and became acidic even if stored in a dark place at decreased temperature. Therefore, it was difficult to maintain the stability and reproducibility of the performance of basic catalysts in aldol condensation. Thus, in the present study, an as-received furfural was first distilled using a vacuum rotator-evaporator and then stabilized by 2,6-di-tert-butyl-4-methylphenol (DBMP, 99%, Sigma Aldrich) using weight ratio DBMP/Furfural = 0.04. In separate experiments, it was established that the addition DBMP to furfural preserved the properties of furfural as a reagent for aldol condensation for at least 3–4 weeks, which was visually observed by the absence of darkening of this compound during storage. On the other hand, DBMP did not influence the performance of neither mixed oxides nor rehydrated HTCs catalysts in aldol condensation.

Aldol condensation of the stabilized furfural with acetone was carried out in a 100 mL stirred batch reactor (a glass flask reactor) at a temperature of either 50 °C in case of mixed oxides or 25 °C in case of rehydrated HTCs at ambient pressure. The mixture of either 37.9 g of acetone and 6.27 g of furfural (acetone to furfural molar ratio 10:1) or 18.95 g of acetone and 6.27 g of furfural (acetone to furfural molar ratio 5:1) was used for experiments with the mixed oxides or the rehydrated HTCs correspondingly. The catalyst used in the experiments was prepared on the basis of 1 g of an as-prepared hydrotalcite, which corresponded to 0.6 to 0.7 g of calcined catalyst (mixed oxide). Prior to the catalytic tests, the reaction mixture was stirred at 400 rpm and stabilized at reaction temperature. The loss of acetone during the catalytic experiments was prevented by using a cooler–condenser above the reactor. After that, the necessary amount of MgM-C (a mixed oxide powder, freshly calcined in a muffle oven at 450 °C and stored in a desiccator) or MgM-R (a freshly rehydrated HTC, filtered with Buchner funnel) was added, and the reaction was carried out for 180 min at 400 rpm. It was previously established that the reaction was limited neither by external nor internal mass transfer under the chosen reaction conditions (in tests with changing stirring rate and catalyst particle size [34]). Liquid reaction products were periodically withdrawn from the reactor during the experiment, centrifugated, filtered, diluted with methanol (1:25 by volume), and analyzed by Agilent 7820 GC unit equipped with a flame ionization detector (FID), using a HP-5 capillary column (30 m/0.32 mm ID/0.25 μm). Catalytic results on aldol condensation of furfural and acetone were described by conversion and selectivity parameters that were calculated as follows [34,41]:

Reactant conversion (t) (mol%) = 100 × (reactant$_{t=0}$ − reactant$_{t=t}$)/reactant$_{t=0}$, where t stands for reaction time

Selectivity to product i = 100 × (mole of reactant converted to product i)/(total moles of reactant converted)

Carbon balance was monitored in all experiments as the total number of carbon atoms detected in each organic compound with Cn atoms (where n = 3, 5, 8, . . . , etc.) divided by the initial number of carbon atoms in the F + Ac feed:

C balance (%) = (3 mol C3 + 5 mol C5 . . . + nmol Cn)/(3 mol C3(t = 0) + 5 mol C5(t = 0)).

Each catalytic experiment was repeated several (at least 3) times to prove the reproducibility of the obtained results (with experimental error evaluated as ±5%).

4. Conclusions

To elucidate the impact of the nature of M^{3+} cation in MgM^{3+} hydrotalcites (HTCs) on the physico-chemical properties and the catalytic performance of the corresponding mixed oxides and rehydrated HTCs, four samples of MgM^{3+} HTCs (M = Al, Ga, Fe, In) were prepared by a precipitation method using the same synthesis protocol. XRD data evidenced that the as-prepared materials were phase-pure HTCs, nevertheless with varied crystallinity. The characterization of the as-prepared samples indicated that the observed difference in the crystallinity could be explained by the difference in their crystallite size, defectiveness, and intergrowth of the layered structure, rather than by the presence of an amorphous phase. The CO_2-TPD study showed that the concentration of the basic sites in MgM^{3+} mixed oxides differed, and it gradually decreased with the growth in the atomic weight of the M^{3+} element. Nevertheless, all mixed oxides had almost the same distribution of the basic site strengths. Interestingly, when the total amount of adsorbed CO_2 was expressed per MgO site, it was the same within the experimental error for all studied mixed oxides. Accordingly, the catalytic performance of the mixed oxides in aldol condensation of furfural and acetone was determined by their basic characteristics. The used 20 min rehydration of the mixed oxides resulted in almost complete (>90%) recovery of the HTC structure with crystallinity values close to those for the as-prepared HTCs. To the best of our knowledge, the successful formation of rehydrated MgIn HTC using pure water at ambient temperature was reported here for the first time. The SEM study evidenced that the rehydration of the mixed oxides resulted in the formation of crystals with irregular and defective lamellar structure, which could explain the slightly decreased crystallinity and the decreased crystallite size of the rehydrated HTCs. The activity of all rehydrated MgM^{3+} HTCs in aldol condensation of furfural and acetone was much higher compared to that of the mixed oxides. Moreover, the difference in furfural conversion observed between the four rehydrated HTCs was obvious, but the observed trend in the activity of the catalysts correlated neither with their crystallinity nor with the atomic weight of the M cation nor with the concentration of the basic sites in the mixed oxides used as precursors. This pointed to the existence of additional factors, for example, the influence of the rehydration conditions that could determine the amount of accessible active sites in these materials and, as a consequence, their catalytic performance. A series of three consecutive experiments with the same catalyst loading demonstrated that the activity of the rehydrated MgM^{3+} catalysts in aldol condensation was fully restored after their separation from the reaction mixture, followed by re-calcination and re-rehydration steps. The performed experiments suggest that the rehydrated MgM^{3+} materials (M = Al, Ga, Fe, In) possess comparable catalytic performance in aldol condensation of furfural and acetone in terms of activity, selectivity, and stability, being apparently determined by the properties of the "host" component, i.e., magnesium oxide.

Supplementary Materials: The following are available online at https://www.mdpi.com/article/10.3 390/|c}atal11080992/s1.

Author Contributions: Conceptualization, D.K. and O.K.; methodology, O.K.; investigation, V.K., M.V., D.V., I.P. and V.F.; data curation, V.K.; writing—original draft preparation, V.K. and O.K.; writing—review and editing, D.K. and L.Č.; visualization, V.K.; supervision, O.K.; project administration, D.K.; funding acquisition, O.K. and L.Č. All authors have read and agreed to the published version of the manuscript.

Funding: This research was funded by the Czech Science Foundation, grant number 19-22978S.

Data Availability Statement: Not applicable.

Acknowledgments: The authors are grateful to the Czech Science Foundation for the financial support (Project No. 19-22978S).

Conflicts of Interest: The authors declare no conflict of interest.

References

1. Debecker, D.P.; Gaigneaux, E.M.; Busca, G. Exploring, Tuning, and Exploiting the Basicity of Hydrotalcites for Applications in Heterogeneous Catalysis. *Chem. Eur. J.* **2009**, *15*, 3920–3935. [CrossRef]
2. Abelló, S.; Medina, F.; Tichit, D.; Pérez-Ramírez, J.; Groen, J.C.; Sueiras, J.E.; Salagre, P.; Cesteros, Y. Aldol Condensations Over Reconstructed Mg–Al Hydrotalcites: Structure–Activity Relationships Related to the Rehydration Method. *Chem. Eur. J.* **2005**, *11*, 728–739. [CrossRef] [PubMed]
3. Tichit, D.; Coq, B. Catalysis by Hydrotalcites and Related Materials. *Cattech* **2003**, *7*, 206–217. [CrossRef]
4. Tichit, D.; Bennani, M.N.; Figueras, F.; Tessier, R.; Kervennal, J. Aldol condensation of acetone over layered double hydroxides of the meixnerite type. *Appl. Clay Sci.* **1998**, *13*, 401–415. [CrossRef]
5. Climent, M.J.; Corma, A.; Fornés, V.; Guil-Lopez, R.; Iborra, S. Aldol Condensations on Solid Catalysts: A Cooperative Effect between Weak Acid and Base Sites. *Adv. Synth. Catal.* **2002**, *344*, 1090–1096. [CrossRef]
6. Guida, A.; Lhouty, M.H.; Tichit, D.; Figueras, F.; Geneste, P. Hydrotalcites as base catalysts. Kinetics of Claisen-Schmidt condensation, intramolecular condensation of acetonylacetone and synthesis of chalcone. *Appl. Catal. A-Gen.* **1997**, *164*, 251–264. [CrossRef]
7. Kikhtyanin, O.; Hora, L.; Kubička, D. Unprecedented selectivities in aldol condensation over Mg–Al hydrotalcite in a fixed bed reactor setup. *Catal. Commun.* **2015**, *58*, 89 92. [CrossRef]
8. Červený, J.; Šplíchalová, J.; Kačer, P.; Kovanda, F.; Kuzma, M.; Červený, L. Molecular shape selectivity of hydrotalcite in mixed aldol condensations of aldehydes and ketones. *J. Mol. Catal. A Chem.* **2008**, *285*, 150–154. [CrossRef]
9. Othman, M.R.; Helwani, Z.; Martunus; Fernando, W.J.N. Synthetic hydrotalcites from different routes and their application as catalysts and gas adsorbents: A review. *Appl. Organomet. Chem.* **2009**, *23*, 335–346. [CrossRef]
10. Sideris, P.J.; Nielsen, U.G.; Gan, Z.; Grey, C.P. Mg/Al ordering in layered double hydroxides revealed by multinuclear NMR spectroscopy. *Science* **2008**, *321*, 113–117. [CrossRef]
11. Nishimura, S.; Takagaki, A.; Ebitani, K. Characterization, synthesis and catalysis of hydrotalcite-related materials for highly efficient materials transformations. *Green Chem.* **2013**, *15*, 2026–2042. [CrossRef]
12. Sels, B.F.; de Vos, D.E.; Jacobs, P.A. Hydrotalcite-like anionic clays in catalytic organic reactions. *Catal. Rev. Sci. Eng.* **2011**, *43*, 443–488. [CrossRef]
13. Cavani, F.; Trifiro, F.; Vaccari, A. Hydrotalcite-type anionic clays: Preparation, properties and applications. *Catal. Today* **1991**, *11*, 173–301. [CrossRef]
14. Debek, R.; Motak, M.; Grzybek, T.; Galvez, M.E.; da Costa, P. A Short Review on the Catalytic Activity of Hydrotalcite-Derived Materials for Dry Reforming of Methane. *Catalysts* **2017**, *7*, 32. [CrossRef]
15. Takehira, K.; Shishido, T. Preparation of supported metal catalysts starting from hydrotalcites as the precursors and their improvements by adopting "memory effect". *Catal. Surv. Asia* **2007**, *11*, 1–30. [CrossRef]
16. Kikhtyanin, O.; Capek, L.; Tisler, Z.; Velvarska, R.; Panasewicz, A.; Diblikova, P.; Kubicka, D. Physico-Chemical Properties of MgGa Mixed Oxides and Reconstructed Layered Double Hydroxides and Their Performance in Aldol Condensation of Furfural and Acetone. *Front. Chem.* **2018**, *6*, 176. [CrossRef] [PubMed]
17. Kocík, J.; Frolich, K.; Perková, I.; Horáček, J. Pyroaurite-based Mg-Fe mixed oxides and their activity in aldol condensation of furfural with acetone: Effect of oxide composition. *J. Chem. Technol. Biotechnol.* **2019**, *94*, 435–445. [CrossRef]
18. Tichit, D.; Lutic, D.; Coq, B.; Durand, R.; Teissier, R. The aldol condensation of acetaldehyde and heptanal on hydrotalcite-type catalysts. *J. Catal.* **2003**, *219*, 167–175. [CrossRef]

19. Smolakova, L.; Frolich, K.; Kocik, J.; Kikhtyanin, O.; Capek, L. Surface Properties of Hydrotalcite-Based Zn(Mg)Al Oxides and Their Catalytic Activity in Aldol Condensation of Furfural with Acetone. *Ind. Eng. Chem. Res.* **2017**, *56*, 4638–4648. [CrossRef]

20. Sánchez-Cantú, M.; Pérez-Díaz, L.M.; Rubio-Rosas, E.; Abril-Sandoval, V.H.; Merino-Aguirre, J.G.; Reyes-Cruz, F.M.; Orea, L. MgZnAl hydrotalcite-like compounds preparation by a green method: Effect of zinc content. *Chem. Pap.* **2014**, *68*, 638–649. [CrossRef]

21. Hernández, W.Y.; Aliç, F.; Verberckmoes, A.; van der Voort, P. Tuning the acidic–basic properties by Zn-substitution in Mg–Al hydrotalcites as optimal catalysts for the aldol condensation reaction. *J. Mater. Sci.* **2017**, *52*, 628–642. [CrossRef]

22. Reichle, W.T. Catalytic reactions by thermally activated, synthetic, anionic clay minerals. *J. Catal.* **1985**, *94*, 547–557. [CrossRef]

23. Fakhfakh, N.; Cognet, P.; Cabassud, M.; Lucchese, Y.; de los Ríos, M.D. Stoichio-Kinetic Modeling and Optimization of Chemical Synthesis: Application to the Aldolic Condensation of Furfural on Acetone. *Chem. Eng. Process.* **2008**, *47*, 349–362. [CrossRef]

24. Climent, M.J.; Corma, A.; Garcia, H.; Guil-Lopez, R.; Iborra, S.; Fornes, V. Acid-base bifunctional catalysts for the preparation of fine chemicals: Synthesis of jasminaldehyde. *J. Catal.* **2001**, *197*, 385–393. [CrossRef]

25. Li, X.; Sun, J.; Shao, S.; Hu, X.; Cai, Y. Aldol condensation/hydrogenation for jet fuel from biomass-derived ketone platform compound in one pot. *Fuel Process. Technol.* **2021**, *215*, 106768. [CrossRef]

26. Han, F.; Xu, J.; Li, G.; Xu, J.; Wang, A.; Cong, Y.; Zhang, T.; Li, N. Synthesis of renewable aviation fuel additives with aromatic aldehydes and methyl isobutyl ketone under solvent-free conditions. *Sustain. Energy Fuels* **2021**, *5*, 556–563. [CrossRef]

27. Sacia, E.R.; Balakrishnan, M.; Deaner, M.H.; Goulas, K.A.; Toste, F.D.; Bell, A.T. Highly Selective Condensation of Biomass-Derived Methyl Ketones as a Source of Aviation Fuel. *ChemSusChem* **2015**, *8*, 1726–1736. [CrossRef] [PubMed]

28. Kikhtyanin, O.; Kadlec, D.; Velvarská, R.; Kubička, D. Using Mg-Al Mixed Oxide and Reconstructed Hydrotalcite as Basic Catalysts for Aldol Condensation of Furfural and Cyclohexanone. *ChemCatChem* **2018**, *10*, 1464–1475. [CrossRef]

29. Stadler, B.M.; Wulf, C.; Werner, T.; Tin, S.; de Vries, J.G. Catalytic Approaches to Monomers for Polymers Based on Renewables. *ACS Catal.* **2019**, *9*, 8012–8067. [CrossRef]

30. Chang, H.; Gilcher, E.B.; Huber, G.W.; Dumesic, J.A. Synthesis of performance-advantaged polyurethanes and polyesters from biomass-derived monomers by aldol-condensation of 5-hydroxymethyl furfural and hydrogenation. *Green Chem.* **2021**. [CrossRef]

31. Thomas, G.S.; Kamath, P.V. Reversible thermal behavior of the layered double hydroxides (LDHs) of Mg with Ga and In. *Mater. Res. Bull.* **2005**, *40*, 671–681. [CrossRef]

32. Aramendía, M.A.; Avilés, Y.; Benítez, J.A.; Borau, V.; Jiménez, C.; Marinas, J.M. Comparative study of Mg/Al and Mg/Ga layered double hydroxides. *Microporous Mesoporous Mater.* **1999**, *29*, 319–328. [CrossRef]

33. Li, F.; Jiang, X.; Evans, D.G.; Duan, X. Structure and Basicity of Mesoporous Materials from Mg/Al/In Layered Double Hydroxides Prepared by Separate Nucleation and Aging Steps Method. *J. Porous Mater.* **2005**, *12*, 55–63. [CrossRef]

34. Kikhtyanin, O.; Tišler, Z.; Velvarská, R.; Kubička, D. Reconstructed Mg-Al hydrotalcites prepared by using different rehydration and drying time: Physico-chemical properties and catalytic performance in aldol condensation. *Appl. Catal. A Gen.* **2017**, *536*, 85–96. [CrossRef]

35. Climent, M.J.; Corma, A.; Iborra, S.; Velty, A. Synthesis of methylpseudoionones by activated hydrotalcites as solid base catalysts. *Green Chem.* **2002**, *4*, 474–480. [CrossRef]

36. Roelofs, J.C.A.A.; Lensveld, D.J.; van Dillen, A.J.; de Jong, K.P. On the Structure of Activated Hydrotalcites as Solid Base Catalysts for Liquid-Phase Aldol Condensation. *J. Catal.* **2001**, *203*, 184–191. [CrossRef]

37. Figueras, F.; Lopez, J.; Sanchez-Valente, J.; Vu, T.T.H.; Clacens, J.-M.; Palomeque, J. Isophorone Isomerization as Model Reaction for the Characterization of Solid Bases: Application to the Determination of the Number of Sites. *J. Catal.* **2002**, *211*, 144–149. [CrossRef]

38. Pérez-Ramírez, J.; Abello, S.; van der Pers, N.M. Memory Effect of Activated Mg–Al Hydrotalcite: In Situ XRD Studies during Decomposition and Gas-Phase Reconstruction. *Chem. Eur. J.* **2007**, *13*, 870–878. [CrossRef] [PubMed]

39. Sharma, S.K.; Parikh, P.A.; Jasra, R.V. Reconstructed Mg/Al hydrotalcite as a solid base catalyst for synthesis of jasminaldehyde. *Appl. Catal. A Gen.* **2010**, *386*, 34–42. [CrossRef]

40. Hibino, T.; Tsunashima, A. Calcination and rehydration behavior of Mg-Fe-CO3 hydrotalcite-like compounds. *J. Mater. Sci. Lett.* **2000**, *19*, 1403–1405. [CrossRef]

41. Kikhtyanin, O.; Čapek, L.; Smoláková, L.; Tišler, Z.; Kadlec, D.; Lhotka, M.; Diblíková, P.; Kubička, D. Influence of Mg–Al Mixed Oxide Compositions on Their Properties and Performance in Aldol Condensation. *Ind. Eng. Chem. Res.* **2017**, *56*, 13411–13422. [CrossRef]

42. Kikhtyanin, O.; Lesnik, E.; Kubička, D. The occurrence of Cannizzaro reaction over Mg-Al hydrotalcites. *Appl. Catal. A Gen.* **2016**, *525*, 215–225. [CrossRef]

43. Xu, C.; Gao, Y.; Liu, X.; Xin, R.; Wang, Z. Hydrotalcite reconstructed by in situ rehydration as a highly active solid base catalyst and its application in aldol condensations. *RSC Adv.* **2013**, *3*, 793–801. [CrossRef]

44. Abelló, S.; Vijaya-Shankar, D.; Pérez-Ramírez, J. Stability, reutilization, and scalability of activated hydrotalcites in aldol condensation. *Appl. Catal. A Gen.* **2008**, *342*, 119–125. [CrossRef]

45. Kikhtyanin, O.; Korolova, V.; Spencer, A.; Dubnová, L.; Shumeiko, B.; Kubička, D. On the influence of acidic admixtures in furfural on the performance of MgAl mixed oxide catalysts in aldol condensation of furfural and acetone. *Catal. Today* **2021**, *367*, 248–257. [CrossRef]

Article

CuAlCe Oxides Issued from Layered Double Hydroxide Precursors for Ethanol and Toluene Total Oxidation

Hadi Dib, Rebecca El Khawaja, Guillaume Rochard *, Christophe Poupin, Stéphane Siffert and Renaud Cousin *

Univ. Littoral Côte d'Opale, U.R. 4492, UCEIV, Unité de Chimie Environnementale et Interactions sur le Vivant, SFR Condorcet FR CNRS 3417, F-59140 Dunkerque, France; hadi.dib@univ-littoral.fr (H.D.); rebecca.el-khawaja@univ-littoral.fr (R.E.K.); christophe.poupin@univ-littoral.fr (C.P.); stephane.siffert@univ-littoral.fr (S.S.)
* Correspondence: guillaume.rochard@univ-littoral.fr (G.R.); renaud.cousin@univ-littoral.fr (R.C.)

Received: 1 July 2020; Accepted: 24 July 2020; Published: 3 August 2020

Abstract: CuAlCe oxides were obtained from hydrotalcite-type precursors by coprecipitation using a M^{2+}/M^{3+} ratio of 3. The collapse of the layered double hydroxide structure following the thermal treatment leads to the formation of mixed oxides (CuO and CeO_2). The catalytic performance of the copper-based catalysts was evaluated in the total oxidation of two Volatile Organic Compounds (VOCs): ethanol and toluene. XRD, SEM Energy-Dispersive X-ray Spectrometry (EDX), H_2-temperature programmed reduction (TPR) and XPS were used to characterize the physicochemical properties of the catalysts. A beneficial effect of combining cerium with CuAl-O oxides in terms of redox properties and the abatement of the mentioned VOCs was demonstrated. The sample with the highest content of Ce showed the best catalytic properties, which were mainly related to the improvement of the reducibility of the copper species and their good dispersion on the surface. The presence of a synergetic effect between the copper and cerium elements was also highlighted.

Keywords: layered double hydroxide; catalytic oxidation; ethanol; toluene; VOC

---

1. Introduction

In the last few decades, extensive efforts have been focused on searching for methods of Volatile Organic Compound (VOC) abatement. VOCs produce ozone, which contributes to the formation of smog and global warming. The emission of these pollutants in the atmosphere is today, strictly regulated. For this, one of the most promising technologies is catalytic total oxidation, which represents an environmentally friendly control technology [1,2]. Studies of VOCs' catalytic oxidation have been widely reported [3–6]. Low temperatures (generally around 250–500 °C) are required compared to thermal oxidation, which requires high temperatures (650–1100 °C). VOC oxidation has been carried out over noble metal and transition metal catalysts.

Noble metals are more active at low temperatures but are costly and often have low stability [7]. However, transition metal oxides (especially for Co-, Mn- and Cu-based catalysts) [8–10] can be cheaper alternatives to this kind of catalyst, which are known to be more resistant to poisoning. Cobalt-based catalysts have been highly reported in the literature as a great economical choice compared to noble metal catalysts [7,11,12] but cause ethical problems because of their toxicity [13,14]. Manganese oxides have also been promising for the total oxidation of VOCs. However, they are most challenging to design due to having several oxidation states and multiple oxide polymorphs being found for every manganese oxidation state, each manifesting distinct catalytic performance for VOC oxidation [15,16]. Thus, a copper catalyst could be a good candidate for VOC oxidation.

Moreover, Layered Double Hydroxides (LDH), named anionic clays or hydrotalcite-like materials, have been revealed to be interesting oxide precursors for the oxidation reaction. These consist of brucite-like $Mg(OH)_2$ positively charged layers separated by an anionic layer compensating the positive charge, along with the water molecules trapped inside these layers. The most frequently occurring hydrotalcite (HT) compound is $Mg_6Al_2(OH)_{16}CO_3,4H_2O$. A wide variety of LDH compounds are available due to the substitution of divalent and trivalent cations. After calcination, oxide materials are formed and show unique properties such as high surface area, thermal stability and well-mixed oxide homogeneity [8].

CuO-based catalysts present remarkable catalytic performance and selectivity for CO_2 concerning the VOC oxidation reaction [17–19]. The feasibility of the partial substitution or total replacement of Mg^{2+} by Cu^{2+} is already described in the literature [20–22]. In previous work, Cu-based hydrotalcite-like compounds have gained significant interest, among the transition metals, due to their high efficiency as catalysts in some important processes such as the total oxidation of toluene [8]. Cerium oxide (CeO_2) is widely used as a promoter in various redox reactions due to its reducibility and its high oxygen storage capacity (OSC) [23,24], which can improve the catalytic properties. A recent study has shown that the addition of cerium in the CuAl hydrotalcite-like material improves activity and selectivity towards NO selective catalytic reduction by lowering the temperature of carbon monoxide oxidation [25]. In addition, cerium incorporation has already been studied and tends to favor the reduction of metal cations in the catalyst based on mixed oxides prepared via the hydrotalcite route, which could lead to an increase in the catalytic activity [26]. Thus, in this work, a new and original approach is to combine cerium with Cu-Al materials using the LDH synthesis approach.

This research work aims to synthesize $Cu_6Al_{2-x}Ce_x$-O catalysts using the hydrotalcite precursor to generate efficient catalytic systems for VOC abatement. Additionally, the effect of Ce content within the materials will be studied in catalytic $Cu_6Al_{2-x}Ce_x$-O systems for the total oxidation of toluene (probe molecule of BTEX) and ethanol (probe molecule of oxygenated VOCs).

2. Results

2.1. Structural Properties

The XRD profiles of the hydrotalcite precursors before the calcination process are shown in Figure 1A. All the samples show the typical diffraction patterns of hydrotalcite-like materials, $Cu_6Al_2(OH)_{16}CO_3,4H_2O$ (JCPDS-ICDD 37-0630). This result confirms that the hydrotalcite structure is retained even after cerium addition. $Cu_6Al_{2-x}Ce_x$-HT materials also present pure ceria (JCPDS-ICDD 01-081-0792), which can be explained by the partial oxidation of Ce^{3+} cations to Ce^{4+} during the synthesis [27]. However, the increase in ceria addition leads to a gradual decrease in the intensities of each peak, indicating the loss of crystallinity of the corresponding phase. This might be due to the distortion of the brucite layers since Ce^{3+} ions have a larger ionic radius (1.01 Å) than Al^{3+} (0.54 Å) [28]. Therefore, Ce^{3+} cations are probably not or partially incorporated in the hydrotalcite structure.

After a thermal treatment under air at 500 °C of the $Cu_6Al_{2-x}Ce_x$-HT samples, XRD measurements were also performed on $Cu_6Al_{2-x}Ce_x$-O samples. Following the thermal treatment, the collapse of the lamellar structure of hydrotalcite-like compounds leads to the formation of CuO oxides (JCPDS-ICDD 48-1548) and ceria (Figure 1B). For all of the samples analyzed, no peak corresponding to Al_2O_3 is observed, thus indicating that the aluminum species are in amorphous states [25]. However, no diffractions of $Cu_xAl_yO_4$ spinel-like structures are obtained for any of the $Cu_6Al_{2-x}Ce_x$-O samples. This result could be related to the degree of crystallization of the spinel phases [21]. In fact, the evolution of X-ray diffractograms as a function of the calcination temperature (60–1000 °C) for copper-based catalysts has been studied [20]. Since high calcination temperatures (above 800 °C) are required for the formation of the $CuAl_2O_4$ spinel phase, no spinel-like structures are expected to be detected after a thermal treatment at 500 °C. Moreover, the rate of the formation of the latter is greatly reduced for hydrotalcites with high copper contents caused by the decrease in the ions' diffusion in the presence of large particles of copper oxide [20,29].

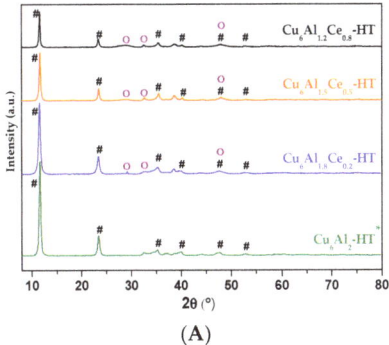
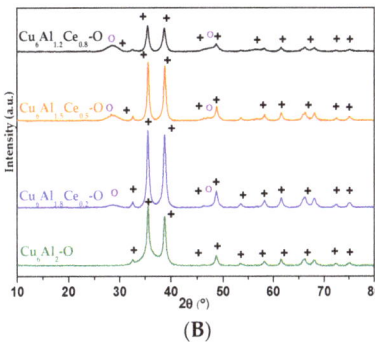

(A) **(B)**

Figure 1. Diffraction patterns of the hydrotalcite precursors (**A**) and the mixed oxide samples (**B**) #: $Cu_6Al_2(OH)_{16}CO_3,4H_2O$; +: CuO; O: CeO_2; * mitigated intensity.

In Table 1, the CuO crystallite size is reported. When adding Ce in low amounts, the CuO crystallite size is not affected, while an increase in Ce content leads to the maximum CuO crystallite size (20.7 nm) and then a decrease to 15.2 nm. Instead, the intensity of this phase decreases with an increase in the Ce content while the same amount of Cu is present in the materials. This can suggest that the addition of Ce allows the formation of amorphous Cu species instead of CuO crystals.

Table 1. Structural and textural features of the $Cu_6Al_{2-x}Ce_x$-O samples.

Samples	CuO Crystallite Size (nm) *	Experimental Atomic Ratio (Cu/Al/Ce) **	BET Surface Area (m^2/g)	Pore Diameter (nm)	Pore Volume (cm^3/g)
$Cu_6Al_{1.2}Ce_{0.8}$-O	15.2	5.5/1.2/0.8	47	17.3	0.19
$Cu_6Al_{1.5}Ce_{0.5}$-O	20.7	5.9/1.5/0.5	37	22.7	0.22
$Cu_6Al_{1.8}Ce_{0.2}$-O	17.7	5.6/1.8/0.2	33	23.9	0.19
Cu_6Al_2-O	17.5	5.6/2.0/-	17	35.4	0.14

* CuO crystallite size determined by the Debye–Scherrer equation based on (111) peak. ** Obtained via ICP analysis.

2.2. Textural Properties

To understand the catalytic behavior, physicochemical characterizations of the oxide samples were performed. Nitrogen physisorption isotherms for all the Cu_6Al_2-O and $Cu_6Al_{2-x}Ce_x$-O materials are represented in Figure 2. IV-type isotherms with H3-type hysteresis loops are observed for all samples, corresponding mostly to a mesoporous structure and narrow plate-like particles (IUPAC classification). The solids textural properties and chemical compositions are reported in Table 1. The chemical compositions are largely maintained especially the Ce and Al atomic ratio, while the Cu ratio seems slightly lower than the theoretical ratio. An increase in the Brunauer–Emmett–Teller (BET) surface area is revealed with an increase in Ce content in the materials. An increase in pore volume is observed when adding Ce to the materials, but no significant evolution is revealed upon an increase in Ce content in the materials. This observation could be explained by the progressive content of cerium oxide in the materials, which possess higher textural properties (110 $m^2 \cdot g^{-1}$ and 0.5 $cm^3 \cdot g^{-1}$ for CeO_2 reference obtained via hydrothermal synthesis) than CuO and Al_2O_3.

The aim of using hydrotalcite-like precursors is to obtain a very good dispersion of the resulting metal oxides. In this context, the surface morphology of $Cu_6Al_{1.2}Ce_{0.8}$-O was analyzed and is presented in Figure 3. A non-uniform and porous structure was observed upon SEM study, confirming the N_2 physisorption results. However, the results of Energy-Dispersive X-ray Spectrometry (EDX) suggest a good dispersion of the elements (Cu, Al and Ce) based on an analysis of seven zones, randomly selected (Figure 4). These results confirm the advantages of the synthesis-by-hydrotalcite route in order to obtain homogeneous catalytic materials. It should be noted that Zedan et al. [18] synthesized Cu-Ce-O catalysts using the combustion method but the SEM-EDX results showed that these materials had a certain heterogeneity.

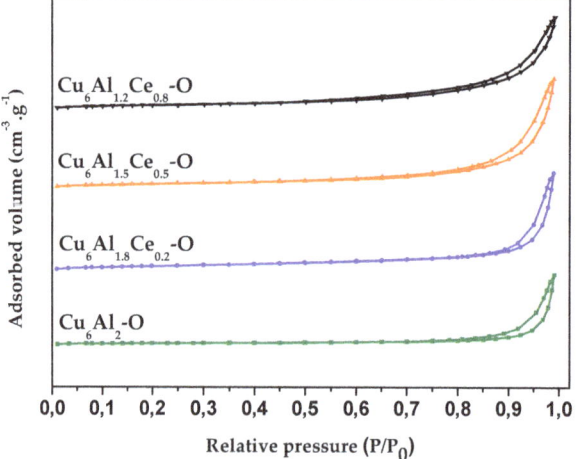

Figure 2. N_2 adsorption/desorption isotherms of $Cu_6Al_{2-x}Ce_x$-O samples.

Figure 3. SEM image of $Cu_6Al_{1.2}Ce_{0.8}$-O samples.

2.3. H_2-Temperature Programmed Reduction

The H_2-TPR profiles of simple, binary and ternary oxides are represented in Figure 5. As the samples have high copper contents, only signals obtained up to a reduction temperature of 350 °C and corresponding to the reduction of copper species were analyzed. The deconvolution of the TPR signals was carried out to assign the hydrogen reduction peaks. The H_2 consumptions of α, β and γ peaks (multi-peaks fitted according to the Gaussian method) are shown in Table 2. The reduction of copper species is only taken into consideration for the theoretical H_2 consumption calculations. Copper oxide, CuO, shows only one main peak at 290 °C corresponding to the reduction of copper (from Cu^{2+} to Cu^0) [22].

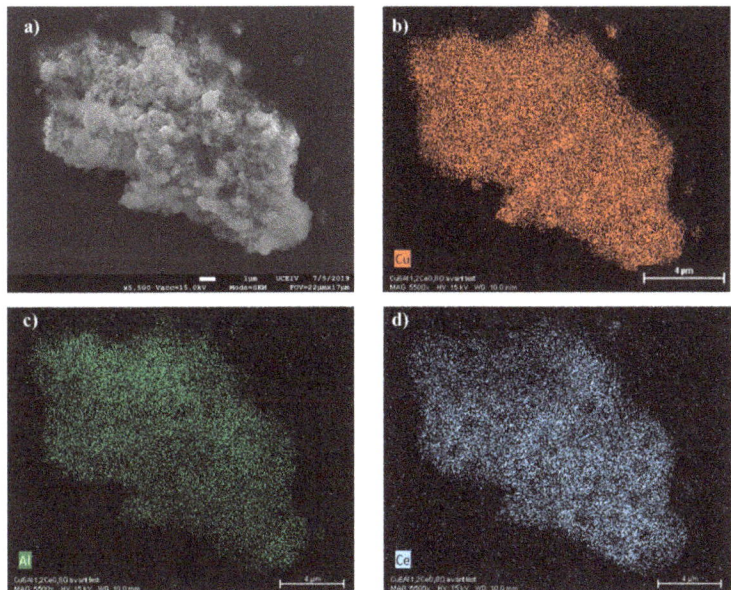

Figure 4. SEM images and element mapping of $Cu_6Al_{1.2}Ce_{0.8}$-O sample.

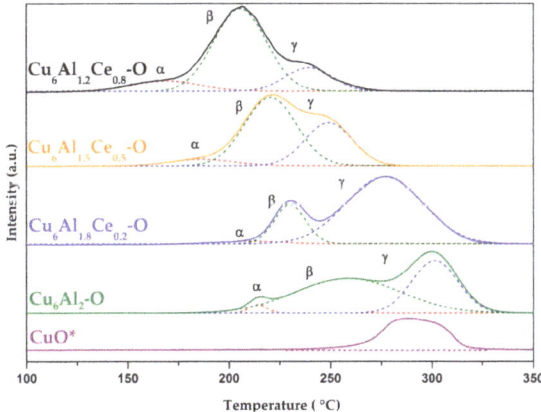

Figure 5. H_2- temperature programmed reduction (TPR) profiles of CuO and $Cu_6Al_{2-x}Ce_x$-O samples.

Table 2. Reduction peak temperatures and H_2 consumption for $Cu_6Al_{2-x}Ce_x$-O samples.

Samples	Temperature (°C)			H_2 Consumption (μmol/g) (T < 350 °C)		
	T α	T β	T γ	Peak α	Peak β	Peak γ
CuO	-	-	289	n.d *	n.d *	n.d *
$Cu6Al_{1.2}Ce_{0.8}$-O	168	206	239	942	6081	1542
$Cu6Al_{1.5}Ce_{0.5}$-O	187	221	249	590	4974	2866
$Cu_6Al_{1.8}Ce_{0.2}$-O	210	230	277	263	1666	6838
Cu_6Al_2-O	215	259	302	181	5235	3610

* not determined.

Regarding the mixed oxide Cu_6Al_2-O, three reduction peaks (α, β and γ) are observed. The α reduction peak located at a lower temperature is attributed to the reduction of Cu^{2+} ions highly dispersed on the surface. The β reduction peak corresponds to the reduction of CuO cluster species. The γ reduction peak, located at higher temperatures, is assigned to larger CuO bulk-like particles/crystalline CuO [22,30–32]. The ceria-based samples present three reduction peaks (α, β and γ), which correspond to the zones of the reduction of the copper species in interaction with cerium species. As shown in Table 2, the reduction temperature of these three peaks shifts to lower temperatures with the increase of ceria content. The redox properties improvement is attributed to the decrease in the crystallite size shown in XRD results previously. On the other hand, researchers suggest that a synergetic interaction between CuO and CeO_2 could considerably improve the reducibility of the catalysts [33–35].

The H_2 consumption (Table 2) of the α and β peaks increases with the addition of Ce in the catalyst. Therefore, the Ce addition favors copper active species dispersion on the surface and promotes their reduction [35,36]. The decreasing consumption values of the γ peak indicate the diminution of the CuO bulk-like proportion of particles. It can be concluded that Ce addition enhanced the transition from CuO bulk-like particles into CuO clusters while promoting the high dispersion of Cu^{2+} on the surface.

2.4. X-ray Photoelectron Spectroscopy

XPS was performed in order to gain more insight into the surface composition and the chemical state of the elements existing on the surface of $Cu_6Al_{2-x}Ce_x$-O catalysts.

The Cu 2p photopeaks are presented in Figure S1. The Cu 2p spectra exhibit the principal peaks of Cu $2p_{3/2}$ and Cu $2p_{1/2}$ (centered at 934.4 and 954.2 eV, respectively) and shake-up satellite bands at 938–948 eV. No significant energy shift is observed with increasing Ce content. In line with the literature, it is reported that the higher Cu $2p_{3/2}$ binding energy and the satellite peak are two most important XPS characteristics of Cu^{2+} species. Meanwhile, the lower Cu $2p_{3/2}$ binding energy and absence of the shake-satellite peak are characteristic of reduced copper species (Cu^+ and Cu^0) [26,37,38]. Thus, Cu species initially existing in these samples are mainly in divalent oxidation states (Cu^{2+}). Furthermore, the lack of band at lower apparent Cu $2p_{3/2}$ binding energy (932.4 eV) demonstrates the absence of Cu^+ or Cu^0 species in the catalysts. According to Table 3, an increase in the nCu/nM atomic ratio as function of Ce content is noticeable. The presence of cerium in catalyst can then increase the dispersion of copper species on the surface and therefore promote their reductions. This can be directly linked to the formation of smaller particles of CuO reported in the reduction part.

Table 3. Summarized XPS results from Ce 3d, O 1s spectra and atomic ratios.

Samples	B.E. * Ce 3d u'''/eV	Ce^{III}/Ce^{IV}	B.E. * O 1s/eV (%) O-I	O-II	nCu/nM **	nO/nTot ***
$Cu_6Al_{1.2}Ce_{0.8}$-O	917.1	0.11	530.2 (26.1%)	532.6 (73.9%)	0.48	0.81
$Cu_6Al_{1.5}Ce_{0.5}$-O	916.8	0.13	530.0 (23.0%)	532.0 (77.0%)	0.39	0.47
$Cu_6Al_{1.8}Ce_{0.2}$-O	916.4	0.36	529.8 (17.9%)	531.8 (82.1%)	0.33	0.57
Cu_6Al_2-O	-	-	530.3 (24.0%)	532.2 (76.0%)	0.65	0.56

* Binding energy. ** nM = nCu + nAl + nCe. *** nTot = nO + nCu + nAl + nCe.

The Ce 3d spectra are presented in Figure S2, and the corresponding binding energies are assembled in Table 3. The Ce 3d spectra can be deconvoluted into ten peaks (Figure 6) with six components for Ce^{IV} (u'''/v''/u''/v''/u/v) and four components for Ce^{III} (u'/v'/u^0/v^0). This indicates a combination of Ce^{3+} and Ce^{4+} ions within the catalysts [26,32]. The ratio Ce^{III}/Ce^{IV} (Table 3) decreased upon increasing the amount of Ce in the composition of the mixed oxide. Therefore, a partial substitution of Al^{3+} by Ce^{3+} is possible at low concentrations of Ce. This substitution can be less significant when this concentration increases, leading to the formation of CeO_2, which can explain the abundance of the Ce^{4+} proportion and the improvement of the textural properties.

The Al 2p photoemission spectra for the studied samples are shown in Figure 7. Three peaks can be distinguished in each spectrum. Two broad-centered peaks at 84.5 and 77.5 eV are attributed to the

Cu 3p signal, while the shoulder peak at lower energy (74.2 eV) represents the Al 2p signal. The Al^{3+} cations are characteristic of the octahedral species present in alumina oxides, Al_2O_3 [39]. As expected, the signal and the appropriate atomic percentage of Al decreased with the Ce content, while the Cu signal remained approximatively the same. No modifications of Al valence occurred with the addition of Ce.

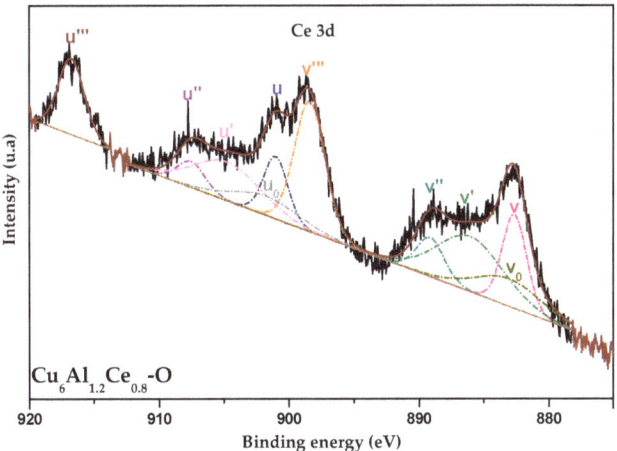

Figure 6. Fitted curve of Ce 3d spectrum for $Cu_6Al_{1.2}Ce_{0.8}$-O sample.

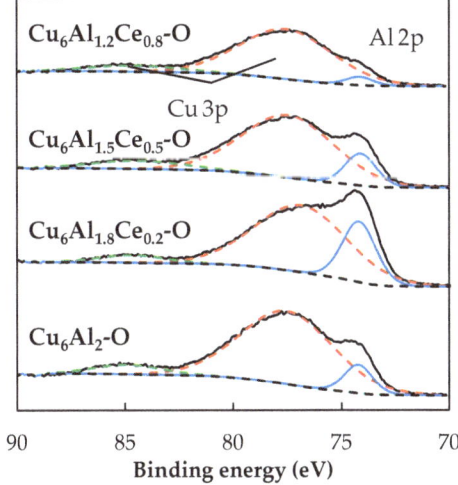

Figure 7. Al 2p photopeaks of $Cu_6Al_{2-x}Ce_x$-O samples.

The XPS spectra of O 1s are shown in Figure 8. Two photopeaks are distinct in the O 1s region at 530 (O-I) and 532 eV (O-II). O-II can be related to the oxygen of the surface hydroxyl or carbonate species [18,40,41], while O-I is related to the lattice oxygen of the metal oxide [19,40–42]. The binding energy and relative percentage of each photopeak are presented in Table 3. The lattice oxygen abundance increased progressively for the compounds with high Ce contents. This can be supported by the decline in the Ce^{III}/Ce^{IV} ratio presented previously and can also be attributed to the increase in the global valence of the metal oxide.

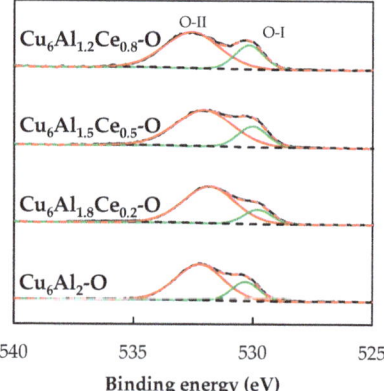

Figure 8. O 1s photopeaks of $Cu_6Al_{2-x}Ce_x$-O samples.

2.5. Catalytic Performance

Figure 9A shows the toluene conversion curves in the 150–400 °C range using the synthesized $Cu_6Al_{2-x}Ce_x$-O materials. The catalytic behavior has been compared to that of CuO and CeO_2 pure oxides in terms of T_{50}, which represents the temperature when 50% conversion is obtained. It can be seen that the toluene conversion into CO_2 increased with the rise in the ceria incorporation in catalytic materials. $Cu_6Al_{2-x}Ce_x$-O catalysts improved the reaction conversion in the following order: $Cu_6Al_{1.2}Ce_{0.8}$-O $\cong$ $Cu_6Al_{1.5}Ce_{0.5}$-O > CeO_2 > $Cu_6Al_{1.8}Ce_{0.2}$-O > Cu_6Al_2-O > CuO. The reaction pathway of toluene oxidation is mostly initiated by an attack on the methyl group, with subsequent oxidation steps. Thus, during toluene oxidation, formation of CO and/or benzene could be observed. However, with our catalysts, no formation of those by-products was detected, except in the case of the CeO_2 catalyst as mentioned in Figure S3.

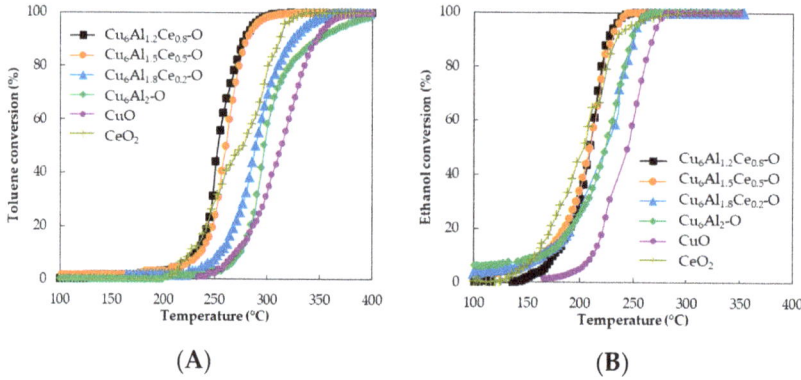

Figure 9. Light-off curves of toluene (**A**) and ethanol (**B**) total oxidation.

A similar trend is distinguished for the total oxidation of ethanol (Figure 9B). $Cu_6Al_{2-x}Ce_x$-O achieved complete ethanol conversion at lower temperatures than the binary catalyst Cu_6Al_2-O and the referenced pure oxides. The results on ceria-based catalysts indicate that the activity increases with an increase in Ce content, where $Cu_6Al_{1.2}Ce_{0.8}$-O and $Cu_6Al_{1.5}Ce_{0.5}$-O exhibit the highest catalytic activity (T_{50} = 220 °C). On the other hand, the formation of acetaldehyde as a major by-product was detected in the chromatographic analysis. This latter is, however, completely oxidized at a temperature

20–30 °C higher than the temperature for the 100% conversion of ethanol. After that, CO_2 and water are the main products that remain at the end of the reaction.

In order to compare the catalysts, the temperature that allows 50% of VOC conversion (T_{50}) is reported in Table 4. Moreover, the comparison of the $Cu_6Al_{2-x}Ce_x$-O catalysts is shown in Figure 10, as a function of the Ce^{III}/Ce^{IV} ratio (Figure 10A), percentage of O-I and O-II (Figure 10B) determined by XPS analysis, and the H_2 consumption of the α and β peaks (Figure 11) determined from H_2-TPR analysis.

Table 4. Summary of catalytic properties of $Cu_6Al_{2-x}Ce_x$-O.

Samples	T_{50} (°C)	
	Toluene	Ethanol
$Cu_6Al_{1.2}Ce_{0.8}$-O	254	210
$Cu_6Al_{1.5}Ce_{0.5}$-O	261	210
$Cu_6Al_{1.8}Ce_{0.2}$-O	286	230

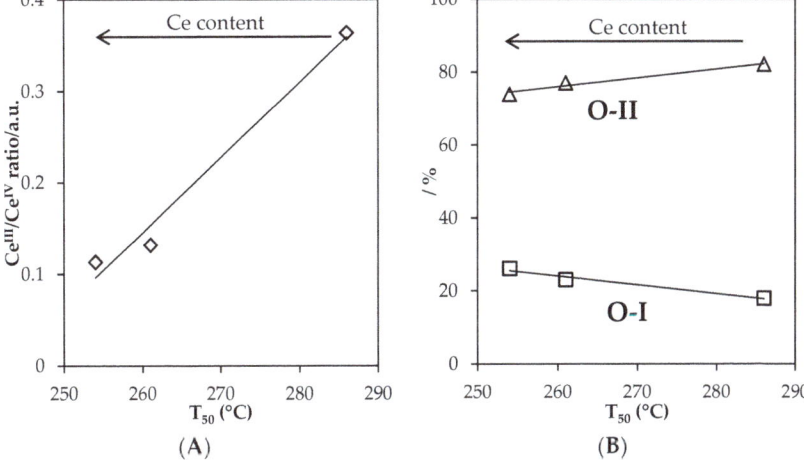

Figure 10. Evolution of XPS data ((**A**): Ce^{IV}/Ce^{III} ratio, (**B**): %O-I and %O-II) with T_{50} for toluene oxidation reaction.

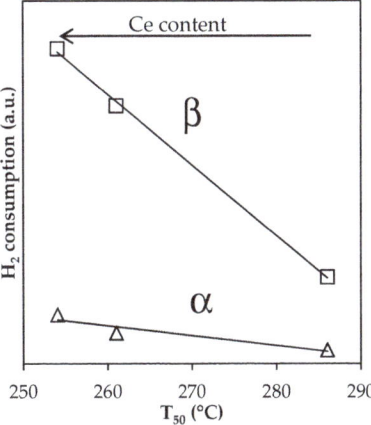

Figure 11. Evolution of H_2-TPR data with T_{50} for toluene oxidation reaction.

A correlation is noticed between the T_{50} of the catalysts in the total oxidation of toluene and the Ce^{III}/Ce^{IV} ratio obtained by XPS analysis (Figure 10A). However, no significant correlation is shown with nCu/nM ratio.

In order to determine which kind of oxygen is mainly related to the studied reactions, Figure 10B shows that an increase in the %O-I is well connected with a lower T_{50} in the toluene oxidation reaction. The %O-I increased linearly with the decrease in T_{50}, while the %O-II decreased. The reaction seems to be favored by a higher concentration of O-I at the surface. The lattice oxygen favors the reaction pathway via the Mars–Van Krevelen mechanism where it is directly involved as a reactive oxygen species. The lattice oxygen can thus positively contribute to and enhance the VOC oxidation. This finding is also related to the increase in the Ce-content that forces the lattice oxygen to migrate to the surface. Thus, despite the relatively low surface area of the synthesized catalysts, the catalytic activity will be enhanced. The correlation between the T_{50} and the H_2-consumption of the different peak observed in the H_2-TPR experiment is reported in Figure 11. As described above, in terms of redox properties, an increase in the consumption of H_2 of the α and β peaks (attributed to the reduction of smaller particles of CuO) is observed with increasing Ce content. The consumption of α and β peaks increased with a decrease in T_{50}, while the γ peak's consumption decreased (not presented here). The formation of a smaller cluster of CuO in interaction with ceria oxide and the diminution of larger particles of CuO seem to be the most advantageous for the activity of toluene oxidation. The presence of cerium in the copper catalyst can increase the dispersion of the copper species on the surface of the $Cu_6Al_{2-x}Ce_x$-O catalysts and therefore promote their reductions. It also assures a better availability and mobility of oxygen on the surface of the solid. This promotes the oxygen exchange with the well-dispersed copper species. In fact, this active phase could be reduced during the toluene oxidation and re-oxidized with oxygen released from the ceria. This synergistic effect between cerium and copper is a key factor for good catalytic activity in the toluene oxidation reaction.

Concerning ethanol oxidation, no correlation has been observed between T_{50} and the formation of CuO clusters. An optimum T_{50} in ethanol oxidation is obtained with $Cu_6Al_{1.5}Ce_{0.5}$-O catalysts (same T_{50} as $Cu_6Al_{1.2}Ce_{0.8}$-O with less Ce). It corresponds to the catalysts with the higher CuO crystallite size. It can be suggested that for the ethanol oxidation reaction, the active site is different from the toluene oxidation reaction site and present in bigger CuO crystallites.

The catalytic performances of $Cu_6Al_{1.2}Ce_{0.8}$-O materials are compared to the performances of supported noble metals and Cu-Ce catalysts in the literature in toluene and ethanol oxidation, based on T_{50} (Table 5). Concerning the toluene reaction, the T_{50} of our catalysts is higher than the one obtained with Pd/-Al_2O_3 [43]. That was expected for supported noble metals, but in the presence of those catalysts, the formation of CO and/or by-products such as benzene is possible [43]. However, using our catalyst, no formations of CO or benzene are detected. Compared to the CoAlCe-O [43] catalyst, a similar T_{50} is obtained by replacing Cu with Co. For Cu–Ce [23,44] catalysts, the T_{50} is lower than in this work. This could be explained by the higher proportion of Ce in those materials. Concerning the oxidation of ethanol, the T_{50} of our catalyst is 210 °C. The same T_{50} has been obtained with a higher content of Ce in those materials too [23]. Thus, our materials, which contain less cerium, are just as effective.

Table 5. Comparison of catalyst performances.

| Catalyst | T_{50} (°C) | | Ref. |
	Toluene	Ethanol	
Pd/Al_2O_3	218		[43]
CoAlCe-O	249		[43]
$Cu_{0.3}Ce_{0.7}$	187		[44]
$Cu_{0.15}Ce_{0.85}$	230	210	[23]
Ce/CuCoMgAl		262	[45]
$Cu_6Al_{1.2}Ce_{0.8}$-O	254	210	This work

3. Materials and Methods

3.1. Catalyst Synthesis

The copper-based hydrotalcite-like compounds, $Cu_6Al_{2-x}Ce_x$-HT, were synthesized by a coprecipitation method. The corresponding ratio of M^{2+}/M^{3+} used was equal to 3. A solution containing appropriate quantities of $Cu(NO_3)_2,3H_2O$ (PanReac Quimica, Barcelona, Spain), $Al(NO_3)_3,9H_2O$ (Chem Lab, Zedelgem, Belgium) and $Ce(NO_3)_3,6H_2O$ (Chem Lab, Zedelgem, Belgium) was added dropwise, under vigorous stirring, into NaOH (PanReac Quimica, Barcelona, Spain, 98%) (2 mol.L^{-1}) and Na_2CO_3 (Thermo Fisher Scientific, New Jersey, USA, 99%) (1 mol.L^{-1}) solution. The pH was maintained at 10.5 for 18 h at room temperature. The precipitate was then filtered, washed several times with hot deionized water (60 °C) and dried at 60 °C for 48 h. Three hydrotalcite-like compounds with different ratios were synthesized: $Cu_6Al_{2-x}Ce_x$ HT with x = 0, 0.2, 0.5 and 0.8. The $Cu_6Al_{2-x}Ce_x$-O were obtained after the thermal treatment of HT at 500 °C (1 °C.min^{-1}) for 4 h under an air flow of 33 mL.min^{-1}.

Copper oxides (99%) were purchased from Fluka Company. The coprecipitation of ($Ce(NO_3)_3$, $6H_2O$) in NaOH solution results in the formation of cerium hydroxide $Ce(OH)_4$. CuO and CeO_2 were used after calcination under air flow at 500 and 400 °C during 4 h, respectively.

3.2. Catalyst Characterization

The crystallinity of the hydrotalcite and calcined catalysts was analyzed at room temperature by X-ray diffraction analysis using a Bruker D8 Advance X-Ray Diffractometer (AXS) (Bruker, Champs-Sur–Marne, France) equipped with CuKα radiation (λ = 1.5418 Å) and a Lynx Eye Detector. The measurements were performed from 5° to 80° with a step size $\Delta(2\theta)$ = 0.02° and a counting time of 2 seconds per step.

The textural properties of all the calcined samples were determined using N_2 adsorption/desorption isotherms carried out using Micromeritics ASAP 2020 (Micromeritics, Norcross, USA). Prior to each analysis, the samples were degassed under vacuum for 2 h at 300 °C. The specific surface area was estimated using the Brunauer–Emmett–Teller (BET) equation, while the pore diameter and specific pore volume were obtained according to the Barrett–Joyner–Halenda (BJH) model.

H_2-temperature programmed reduction (H_2-TPR) studies were carried out using an AMI-200 instrument (Labor und Analysen Technik GmbH, Garbsen, Germany) equipped with a TCD detector. First, the calcined samples (30 mg) were treated in Ar flow (30 mL.min^{-1}) at 150 °C for 60 min. After cooling, the reduction of the solid was carried out in a 5 vol.% H_2/Ar gas mixture (20 mL.min^{-1}) while heating from 150 to 900 °C (ramping rate: 5 °C.min^{-1}).

X-ray Photoelectron Spectroscopy (XPS) (Kratos, Manchester, U.K.) spectra were recorded using monochromatic Al Kα (1486.6 eV) radiation at an operating source power of 13 kV and 10 mA. The hemispherical energy analyzer with a constant pass energy mode Ep = 50 eV was used in all measurements. The binding energy (BE) was calibrated based on the line position of C 1s (284.8 eV). The core-level spectra were decomposed into components with mixed Gaussian–Lorentzian lines using a subtraction of the Shirley-type background using the CasaXPS software (Casa Software Ltd. Teignmouth, UK).

The elemental compositions of the samples were analyzed using an Inductively Coupled Plasma-Optical Emission Spectrometer, ICP-OES (Thermo fisher, iCAP 6300 DUO) (Waltham, MA, USA), equipped with a charge injection device. Prior to the analysis, 50 mg of catalyst was dissolved into aqua regia (HNO_3/HCl 1:2) under microwaving for 30 min. The solution was then topped up to 50 mL with ultrapure water, diluted to 10% and filtered with a 0.45 µm cellulosic microfilter.

Scanning Electron Microscopy was performed with a JEOL JSM-711F (JEOL, Tokyo, Japan) apparatus equipped with an Energy-Dispersive X-ray Spectrometer (MEB-EDX). The calcined samples were adhered on aluminum stubs and then coated with a thin film of chromium.

3.3. Catalytic Activity Tests

Toluene and ethanol oxidations were carried out in a continuous-flow fixed-bed reactor loaded with 100 mg (35–45 mesh) of catalyst at atmospheric pressure. The gas mixture composed of 1000 ppm of VOC in air was passed through the reactor with a flow rate of 100 mL.min^{-1}, which corresponds to a gas hourly space velocity (GHSV) of about 30.000 h^{-1}. Before each test, the catalysts were preactivated at 300 °C for 1 h under flowing air (33 mL.min^{-1}).

For the toluene oxidation, the inlet and outlet gas stream concentrations were analyzed with a micro-gas chromatographer (Agilent 490MicroGC) coupled to an infrared analyzer (ADEV CO$_2$-CO Analyzer Model 4400IR) for CO$_2$ and CO analysis.

However, the reactants and the products of the oxidation of ethanol were analyzed using a micro-gas chromatographer (Varian CP- 4900) coupled to a Pfeiffer Vacuum OmniStar Quadrupole Mass Spectrometer (QMS-200).

4. Conclusions

A Cu$_6$Al$_{2-x}$Ce$_x$-O catalytic system with different Ce contents was successfully synthesized via a hydrotalcite route. The characterization of the products showed favorable physico-chemical and catalytic properties of the Cu$_6$Al$_{2-x}$Ce$_x$-O materials with Ce contents starting from x = 0.5. A homogeneous dispersion of Cu, Al and Ce species was distinguished for these materials. H$_2$-TPR suggests that combining Ce with the copper-based catalysts favors the dispersion of copper active species on the surface and promotes their reduction. XPS analysis confirmed the presence of Ce^{3+}, Ce^{4+} and Al^{3+} species on the surface of the Cu$_6$Al$_{2-x}$Ce$_x$-O samples, as well as the formation of copper Cu^{2+} species that would interact with cerium. The addition of Ce can favor a better dispersion of CuO particles on the surface of the materials. The catalyst with the best catalytic performance was found to be the one with the highest Ce content. The good catalytic activity for the total oxidation of toluene and ethanol of Cu$_6$Al$_{1.2}$Ce$_{0.8}$–O can be attributed to the high surface area, the good metal dispersion, the high proportion of Ce^{IV} and the high reducibility of the copper species on the surface. Thus, combining cerium with CuAl oxides using the LDH synthesis approach permits efficient catalysts to be obtained. A synergetic effect between copper and cerium species occurs and is favorable for the abatement of VOCs. The improvement of the textural, structural and redox properties of the oxides can lead to the enhancement of the catalytic performance for VOC oxidation.

Supplementary Materials: The following are available online at http://www.mdpi.com/2073-4344/10/8/870/s1. Figure S1: Cu 2p photopeaks of Cu$_6$Al$_{2-x}$Ce$_x$-O samples. Figure S2: Ce 3d photopeaks of Cu$_6$Al$_{2-x}$Ce$_x$-O samples.

Author Contributions: Conceptualization, R.C. and G.R.; methodology, R.C. and C.P.; validation, C.P. and S.S.; formal analysis, H.D. and R.E.K.; investigation, H.D., R.E.K. and G.R.; data curation, R.E.K. and G.R.; writing—original draft preparation, G.R.; writing—review and editing, R.C.; visualization, C.P.; supervision, R.C.; funding acquisition, R.C. and S.S.; All authors have read and agreed to the published version of the manuscript.

Funding: The authors gratefully acknowledge the financial support from the CPER-IRENE program ("COVO" project), the Hauts-de-France region and the European Community (Interreg V France-Wallonie-Vlaanderen project, "DepollutAir").

Conflicts of Interest: The authors declare no conflict of interest.

References

1. Huang, H.; Xu, Y.; Feng, Q.; Leung, D.Y.C. Low temperature catalytic oxidation of volatile organic compounds: A review. *Catal. Sci. Technol.* **2015**, *5*, 2649–2669. [CrossRef]
2. Torres, J.Q.; Royer, S.; Bellat, J.P.; Giraudon, J.M.; Lamonier, J.F. Formaldehyde: Catalytic oxidation as a promising soft way of elimination. *ChemSusChem* **2013**, *6*, 578–592. [CrossRef]
3. Brunet, J.; Genty, E.; Landkocz, Y.; Zallouha, M.A.; Billet, S.; Courcot, D.; Siffert, S.; Thomas, D.; De Weireld, G.; Cousin, R. Identification of by-products issued from the catalytic oxidation of toluene by chemical and biological methods. *C. R. Chim.* **2015**, *18*, 1084–1093. [CrossRef]
4. Santos, V.P.; Pereira, M.F.R.; Órfão, J.J.M.; Figueiredo, J.L. Mixture effects during the oxidation of toluene, ethyl acetate and ethanol over a cryptomelane catalyst. *J. Hazard. Mater.* **2011**, *185*, 1236–1240. [CrossRef]

5. Wu, H.; Wang, L.; Zhang, J.; Shen, Z.; Zhao, J. Catalytic oxidation of benzene, toluene and p-xylene over colloidal gold supported on zinc oxide catalyst. *Catal. Commun.* **2011**, *12*, 859–865. [CrossRef]
6. Burgos, N.; Paulis, M.; Mirari Antxustegi, M.; Montes, M. Deep oxidation of VOC mixtures with platinum supported on Al2O3/Al monoliths. *Appl. Catal. B Environ.* **2002**, *38*, 251–258. [CrossRef]
7. Liotta, L.F.; Ousmane, M.; Di Carlo, G.; Pantaleo, G.; Deganello, G.; Boreave, A.; Giroir-Fendler, A. Catalytic removal of toluene over Co3O4-CeO 2 mixed oxide catalysts: Comparison with Pt/Al2O 3. *Catal. Lett.* **2009**, *127*, 270–276. [CrossRef]
8. Genty, E.; Cousin, R.; Capelle, S.; Gennequin, C.; Siffert, S. Catalytic oxidation of toluene and CO over nanocatalysts derived from hydrotalcite-like compounds (X 62+Al 23+): Effect of the bivalent cation. *Eur. J. Inorg. Chem.* **2012**, 2802–2811. [CrossRef]
9. Busca, G.; Daturi, M.; Finocchio, E.; Lorenzelli, V.; Ramis, G.; Willey, R.J. Transition metal mixed oxides as combustion catalysts: Preparation, characterization and activity mechanisms. *Catal. Today* **1997**, *33*, 239–249. [CrossRef]
10. Sekine, Y.; Nishimura, A. Removal of formaldehyde from indoor air by passive type air-cleaning materials. *Atmos. Environ.* **2007**, *35*, 2001–2007. [CrossRef]
11. Trigueiro, F.E.; Ferreira, C.M.; Volta, J.C.; Gonzalez, W.A.; de Oliveria, P.G.P. Effect of niobium addition to Co/γ-Al2O3 catalyst on methane combustion. *Catal. Today* **2006**, *118*, 425–432. [CrossRef]
12. Garcia, T.; Agouram, S.; Sánchez-Royo, J.F.; Murillo, R.; Mastral, A.M.; Aranda, A.; Vázquez, I.; Dejoz, A.; Solsona, B. Deep oxidation of volatile organic compounds using ordered cobalt oxides prepared by a nanocasting route. *Appl. Catal. A Gen.* **2010**, *386*, 16–27. [CrossRef]
13. Papageorgiou, I.; Brown, C.; Schins, R.; Singh, S.; Newson, R.; Davis, S.; Fisher, J.; Ingham, E.; Case, C.P. The effect of nano- and micron-sized particles of cobalt-chromium alloy on human fibroblasts in vitro. *Biomaterials* **2007**, *28*, 2946–2958. [CrossRef] [PubMed]
14. Chattopadhyay, S.; Dash, S.K.; Tripathy, S.; Das, B.; Mandal, D.; Pramanik, P.; Roy, S. Toxicity of cobalt oxide nanoparticles to normal cells; An in vitro and in vivo study. *Chem. Biol. Interact.* **2015**, *226*, 58–71. [CrossRef] [PubMed]
15. Zhou, L.; He, J.; Zhang, J.; He, Z.; Hu, Y.; Zhang, C.; He, H. Facile in-situ synthesis of manganese dioxide nanosheets on cellulose fibers and their application in oxidative decomposition of formaldehyde. *J. Phys. Chem. C* **2011**, *115*, 16873–16878. [CrossRef]
16. Sekine, Y. Oxidative decomposition of formaldehyde by metal oxides at room temperature. *Atmos. Environ.* **2002**, *36*, 5543–5547. [CrossRef]
17. Kim, S.C. The catalytic oxidation of aromatic hydrocarbons over supported metal oxide. *J. Hazard. Mater.* **2002**, *91*, 285–299. [CrossRef]
18. Zedan, A.F.; Polychronopoulou, K.; Asif, A.; AlQaradawi, S.Y.; AlJaber, A.S. Cu-Ce-O catalyst revisited for exceptional activity at low temperature CO oxidation reaction. *Surf. Coat. Technol.* **2018**, *354*, 313–323. [CrossRef]
19. Zhang, X.; Zhang, X.; Song, L.; Hou, F.; Yang, Y.; Wang, Y.; Liu, N. Enhanced catalytic performance for CO oxidation and preferential CO oxidation over CuO/CeO2 catalysts synthesized from metal organic framework: Effects of preparation methods. *Int. J. Hydrogen Energy* **2018**, *43*, 18279–18288. [CrossRef]
20. Alejandre, A.; Medina, F.; Salagre, P.; Correig, X.; Sueiras, J.E. Preparation and study of Cu-Al mixed oxides via hydrotalcite-like precursors. *Chem. Mater.* **1999**, *11*, 939–948. [CrossRef]
21. Kannan, S.; Rives, V.; Knözinger, H. High-temperature transformations of Cu-rich hydrotalcites. *J. Solid State Chem.* **2004**, *177*, 319–331. [CrossRef]
22. Teixeira, C.D.O.P.; Montani, S.D.S.; Palacio, L.A.; Zotin, F.M.Z. The effect of preparation methods on the thermal and chemical reducibility of Cu in Cu-Al oxides. *Dalton Trans.* **2018**, *47*, 10989–11001. [CrossRef] [PubMed]
23. Delimaris, D.; Ioannides, T. VOC oxidation over CuO-CeO2 catalysts prepared by a combustion method. *Appl. Catal. B Environ.* **2009**, *89*, 295–302. [CrossRef]
24. Heynderickx, P.M.; Thybaut, J.W.; Poelman, H.; Poelman, D.; Marin, G.B. The total oxidation of propane over supported Cu and Ce oxides: A comparison of single and binary metal oxides. *J. Catal.* **2010**, *272*, 109–120. [CrossRef]
25. Qian, J.; Hou, X.; Wang, F.; Hu, Q.; Yuan, H.; Teng, L.; Li, R.; Tong, Z.; Dong, L.; Li, B. Catalytic reduction of NO by CO over promoted Cu3Ce0.2Al0.8 composite oxides derived from hydrotalcite-like Compounds. *J. Phys. Chem. C* **2018**, *122*, 2097–2106. [CrossRef]

26. Chen, C.; Wang, R.; Shen, P.; Zhao, D.; Zhang, N. Inverse CeO2/CuO catalysts prepared from heterobimetallic metal-organic framework precursor for preferential CO oxidation in H2-rich stream. *Int. J. Hydrogen Energy* **2015**, *40*, 4830–4839. [CrossRef]

27. Genty, E.; Dib, H.; Brunet, J.; Poupin, C.; Siffert, S.; Cousin, R. Effect of Ce Addition on mgal mixed oxides for the total oxidation of CO and toluene. *Top. Catal.* **2019**, *62*, 397–402. [CrossRef]

28. Sharma, M.K.; Melesse, S.F. Optimal block designs for CDC experiments method (2). *Metron* **2011**, *69*, 297–307. [CrossRef]

29. Alejandre, A.; Medina, F.; Rodriguez, X.; Salagre, P.; Sueiras, J.E. Preparation and activity of copper, nickel and copper-nickel-al mixed oxides via hydrotalcite-like precursors for the oxidation of phenol aqueous solutions. *Stud. Surf. Sci. Catal.* **2000**, *130 B*, 1763–1768. [CrossRef]

30. Dow, W.P.; Wang, Y.P.; Huang, T.J. TPR and XRD studies of yttria-doped ceria/γ-alumina-supported copper oxide catalyst. *Appl. Catal. A Gen.* **2000**, *190*, 25–34. [CrossRef]

31. Guo, X.; Zhou, R. A new insight into the morphology effect of ceria on CuO/CeO2 catalysts for CO selective oxidation in hydrogen-rich gas. *Catal. Sci. Technol.* **2016**, *6*, 3862–3871. [CrossRef]

32. Cecilia, J.A.; Arango-Díaz, A.; Marrero-Jerez, J.; Núñez, P.; Moretti, E.; Storaro, L.; Rodríguez-Castellón, E. Catalytic behaviour of CuO-CeO2 systems prepared by different synthetic methodologies in the CO-PROX reaction under CO2-H2O feed stream. *Catalysts* **2017**, *7*, 160. [CrossRef]

33. Qi, L.; Yu, Q.; Dai, Y.; Tang, C.; Liu, L.; Zhang, H.; Gao, F.; Dong, L.; Chen, Y. Influence of cerium precursors on the structure and reducibility of mesoporous CuO-CeO2 catalysts for CO oxidation. *Appl. Catal. B Environ.* **2012**, *119–120*, 308–320. [CrossRef]

34. Sun, S.; Mao, D.; Yu, J.; Yang, Z.; Lu, G.; Ma, Z. Lowerature CO oxidation on CuO/CeO2 catalysts: The significant effect of copper precursor and calcination temperature. *Catal. Sci. Technol.* **2015**, *5*, 3166–3181. [CrossRef]

35. Sumrunronnasak, S.; Chanlek, N.; Pimpha, N. Improved CeCuOx catalysts for toluene oxidation prepared by aqueous cationic surfactant precipitation method. *Mater. Chem. Phys.* **2018**, *216*, 143–152. [CrossRef]

36. Deng, C.; Li, B.; Dong, L.; Zhang, F.; Fan, M.; Jin, G.; Gao, J.; Gao, L.; Zhang, F.; Zhou, X. NO reduction by CO over CuO supported on CeO2-doped TiO2: The effect of the amount of a few CeO2. *Phys. Chem. Chem. Phys.* **2015**, *17*, 16092–16109. [CrossRef]

37. Biesinger, M.C. Advanced analysis of copper X-ray photoelectron spectra. *Surf. Interface Anal.* **2017**, *49*, 1325–1334. [CrossRef]

38. Du, L.; Wang, W.; Yan, H.; Wang, X.; Jin, Z.; Song, Q.; Si, R.; Jia, C. Copper-ceria sheets catalysts: Effect of copper species on catalytic activity in CO oxidation reaction. *J. Rare Earths* **2017**, *35*, 1186–1196. [CrossRef]

39. Khassin, A.A.; Yurieva, T.M.; Kaichev, V.V.; Bukhtiyarov, V.I.; Budneva, A.A.; Paukshtis, E.A.; Parmon, V.N. Metal-support interactions in cobalt-aluminum co-precipitated catalysts: XPS and CO adsorption studies. *J. Mol. Catal. A Chem.* **2001**, *175*, 189–204. [CrossRef]

40. Cheng, J.; Yu, J.; Wang, X.; Li, L.; Li, J.; Hao, Z. Novel CH4 combustion catalysts derived from Cu-Co/X-Al (X = Fe, Mn, La, Ce) hydrotalcite-like compounds. *Energy Fuels* **2008**, *22*, 2131–2137. [CrossRef]

41. Bai, Y.; Bian, X.; Wu, W. Catalytic properties of CuO/CeO2-Al2O3 catalysts for low concentration NO reduction with CO. *Appl. Surf. Sci.* **2019**, *463*, 435–444. [CrossRef]

42. Chang, Z.; Zhao, N.; Liu, J.; Li, F.; Evans, D.G.; Duan, X.; Forano, C.; De Roy, M. CuCeO mixed oxides from Ce-containing layered double hydroxide precursors: Controllable preparation and catalytic performance. *J. Solid State Chem.* **2011**, *184*, 3232–3239. [CrossRef]

43. Brunet, J.; Genty, E.; Barroo, C.; Cazier, F.; Poupin, C.; Siffert, S.; Thomas, D.; De Weireld, G.; de Bocarmé, T.V.; Cousin, R. The CoAlCeO mixed oxide: An alternative to palladium-based catalysts for total oxidation of industrial VOCs. *Catalysts* **2018**, *8*. [CrossRef]

44. He, C.; Yu, Y.; Yue, L.; Qiao, N.; Li, J.; Shen, Q.; Yu, W.; Chen, J.; Hao, Z. Low-temperature removal of toluene and propanal over highly active mesoporous CuCeOx catalysts synthesized via a simple self-precipitation protocol. *Appl. Catal. B Environ.* **2014**, *147*, 156–166. [CrossRef]

45. Pérez, A.; Montes, M.; Molina, R.; Moreno, S. Cooperative effect of Ce and Pr in the catalytic combustion of ethanol in mixed Cu/CoMgAl oxides obtained from hydrotalcites. *Appl. Catal. A Gen.* **2011**, *408*, 96–104. [CrossRef]

Article

Photoelectrocatalytic Degradation of Congo Red Dye with Activated Hydrotalcites and Copper Anode

Sara Argote-Fuentes [1], **Rossy Feria-Reyes** [2], **Esthela Ramos-Ramírez** [2,*], **Norma Gutiérrez-Ortega** [3,*] and **Gustavo Cruz-Jiménez** [4]

[1] Doctoral Program in Water Science and Technology, Engineering Division, Guanajuato Campus, University of Guanajuato, 77 Juarez St, Downtown, Guanajuato GTO 36000, Mexico; sg.argotefuentes@ugto.mx

[2] Chemistry Department, Natural and Exact Sciences Division, Guanajuato Campus, University of Guanajuato, w/n Noria Alta, Guanajuato GTO 36000, Mexico; rossyfr@gmail.com

[3] Department of Civil and Environmental Engineering, Engineering Division, Guanajuato Campus, University of Guanajuato, 77 Juarez St, Downtown, Guanajuato GTO 36000, Mexico

[4] Pharmacy Department, Natural and Exact Sciences Division, Guanajuato Campus, University of Guanajuato, w/n Noria Alta, Guanajuato GTO 36050, Mexico; cruzg@ugto.mx

* Correspondence: ramosre@ugto.mx (E.R.-R.); normagut@ugto.mx (N.G.-O.); Tel.: +52-4737320006 (ext. 1457) (E.R.-R.); +52-4737320006 (ext. 2227) (N.G.-O.)

Citation: Argote-Fuentes, S.; Feria-Reyes, R.; Ramos-Ramírez, E.; Gutiérrez-Ortega, N.; Cruz-Jiménez, G. Photoelectrocatalytic Degradation of Congo Red Dye with Activated Hydrotalcites and Copper Anode. *Catalysts* 2021, *11*, 211. https://doi.org/10.3390/catal11020211

Academic Editors: Ioan-Cezar Marcu and Octavian Dumitru Pavel

Received: 1 January 2021

Accepted: 3 February 2021

Published: 5 February 2021

Publisher's Note: MDPI stays neutral with regard to jurisdictional claims in published maps and institutional affiliations.

Abstract: Photoelectrocatalysis is a novel technique that combines heterogeneous photocatalysis with the application of an electric field to the system through electrodes for the degradation of organic contaminants in aqueous systems, mainly of toxic dyes. The efficiency of these combined processes depends on the semiconductor properties of the catalysts, as well as on the anodic capacity of the electrode. In this study, we propose the use of active hydrotalcites in the degradation of Congo red dye through processes assisted by ultraviolet (UV) irradiation and electric current. Our research focused on evaluating the degradation capacity of Congo red by means of photolysis, catalysis, photocatalysis, electrocatalysis, and photoelectrocatalysis, as well as identifying the effect of the properties of the active hydrotalcites in these processes. The results show that a maximum degradation was reached with the photoelectrocatalysis process with active hydrotalcites and a copper anode at 6 h with 95% in a half-life of 0.36 h. The degradation is favored by the attack of the OH$^\bullet$ radicals under double bonds in the diazo groups where the electrode produces Cu^{2+} ions, and with the photogenerated electrons, the recombination speed of the electron–hole in the hydrotalcite catalyst is reduced until the complete degradation.

Keywords: photocatalysts; Cu electrodes; diazo dyes; electrocatalysts; layer double hydroxides; photoelectrochemical degradation

1. Introduction

Wastewater, mainly from the textile industry, contains large amounts of azo dyes, as well as inorganic salts such as NaCl and Na$_2$SO$_4$ [1]. These azo dyes are highly toxic, carcinogenic, and mutagenic in nature and can even bioaccumulate in the food chain [2,3]; therefore, we must develop effective methods for the treatment of industrial wastewater that can degrade the synthetic dyes contained in it [4]. Azo dyes are the most consumed in industry, mainly in textiles, representing up to 35% of dye consumption. They are characterized by a structure that contains at least two aromatic compounds linked together by azo type chromophore group (-N=N-) [4]. Various methods, such as adsorption [5], biodegradation [6], chemical oxidation [7], and microbial or enzymatic treatment [8], have been implemented for the elimination of textile dyes from water; however, because these methods are not destructive, a transfer of the contaminant from one phase to another occurs, which requires additional treatments, such as advanced oxidation processes, which, in recent years, have been extensively investigated.

Heterogeneous photocatalysis is considered an emerging destructive technology that leads to the total mineralization of diverse organic contaminants [9–11], in which the degradation of the contaminating compound dissolved in water occurs by the action of semiconductor materials irradiated by light, mainly ultraviolet. Some of the most used photocatalyst metallic semiconductor materials are TiO_2, ZnO, SnO_2, ZrO_2, V_2O_5, WO_3, CeO_2, and g-C_3N_4, as well as several mixtures of these have been studied [12–16]. In the case of TiO_2, despite being the most used photocatalyst, it presents some disadvantages, the main disadvantage is that it presents a relatively high value of the forbidden band energy (approximately, 3.2 eV for the anatase phase and 3.0 eV for the rutile phase), which limits its absorption in the spectrum of the ultraviolet-visible (UV-VIS) region, while also presenting a high rate of recombination of photogenerated electron/hole (e^-/h^+) pairs, which results in a decrease of its photocatalytic activity. That is why much research is focused on improving the application of TiO_2, with the intent to reduce the band gap due to structural modifications [17,18]. Most studies focused on doping TiO_2 with metallic ions such as nickel, manganese, cobalt, titanium, chrome, iron, vanadium, zinc, and copper, as well as with nonmetallic elements or even through the formation of composites with MnO_2, In_2O_3, CeO_2, and MoS_2 [19–21]. There is another alternative currently being studied to improve the photocatalytic activity of TiO_2 in photoactivation using an electric current, which suggests that if a better photoactivation of TiO_2 corresponds to a higher degeneration rate of reactive species, it would also correspond to a higher overall degradation performance [22,23].

Within the alternative material options to TiO_2 are the layered double hydroxides or hydrotalcites, which are compounds used as precursors of oxides, which have shown to be a viable option as photocatalytically active materials [24–27]. Hydrotalcites are composed of inorganic layers with a laminar structure and are chemically characterized by the general formula $[M_{1-x}^{2+}M_x^{3+}(OH)_2]x+(A^{n-})_{x/n} \cdot mH_2O$, where M^{2+} ($M^{2+} = Mg^{2+}$, Ni^{2+}, Co^{2+}, Cu^{2+}, Zn^{2+}) is a divalent cation and M^{3+} ($M^{3+} = Al^{3+}$, Fe^{3+}, Cr^{3+}, etc.) is a trivalent cation, where both cations' octahedral networks with positive residual charge are structured. A^{n-} are anions that lodge in the interlaminar space to compensate the residual charge of the sheets ($A^{n-} = NO_3^-$, CO_3^{2-}, SO_4^{2-}, Cl^-, Br^-, I^-, etc.) and m is the amount of water molecules [28,29]. These materials can be easily and economically synthesized at laboratory level. Figure 1 shows the laminar structure of a hydrotalcite Mg/Al with carbonate ions and water molecules in the interlaminar space.

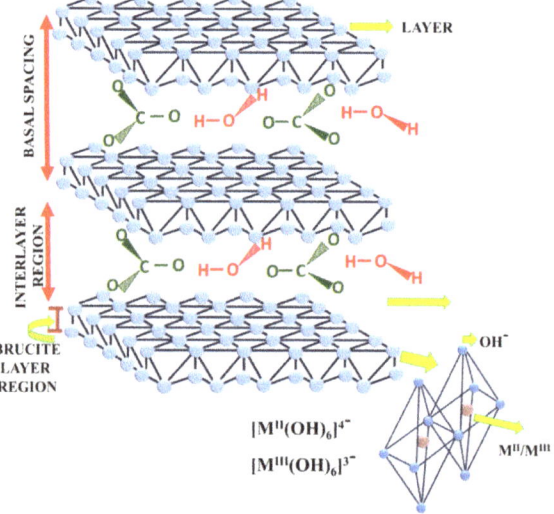

Figure 1. The laminar structure of Mg/Al-CO_3^{2-} hydrotalcite.

Due to its anion exchange capacity in the interlaminar space, hydrotalcites have been used to remove from water mono-, di-, and polyvalent anions, organochlorine compounds [30,31], heavy metals [32,33], and synthetic dyes [5,34–36]. In addition, the chemical composition of hydrotalcites, as well as their physicochemical properties, have allowed their application in various areas such as catalysis [37], electrocatalysis [38,39], energy storage [40,41], and photocatalysis [5,29,31].

Photoelectrocatalysis or photoelectrochemical degradation is a novel form of electrochemistry that combines the techniques of heterogeneous photocatalysis with electrochemical oxidation techniques for the improved removal of organic contaminants in aqueous systems. Its general, rationale is that metal oxides act as semiconductors, and like that in conventional photocatalysis, they produce pairs of electrons–holes that are photogenerated when the anodic material absorbs photons from light irradiation [42,43]. In this research work, the influence of the structural, textural, and thermal properties of activated hydrotalcites on the degradation capacity of synthetic Congo red dye in aqueous solutions was studied using a combined system of photoelectrocatalytic operating parameters.

2. Results and Discussion

2.1. Characterization of Catalytic Precursors of the Hydrotalcites

The X-ray diffractogram obtained from the synthesized solid with a ratio of Mg/Al = 1, hydrotalcite (HT-1) is shown in Figure 2a. The figure shows that the structure is crystalline, corresponding to a hydrotalcite-type material, by showing characteristic peaks with Miller's indexes (003), (006), (009), (012), (015), (018), (110), and (113) (JCPDS 22 0700 Card), with rhombohedral symmetry. In addition, traces of the crude crystalline phase (**001) (Joint Committee on Powder Diffraction Standards-JCPDS card 7-239) and phases of boehmite or hydrated aluminum oxide compound (*, *120, *022) can be identified (JCPDS card 1-1283). This synthesized solid contains the precursor crystalline phases to be activated by heat treatment at 400 °C.

For the HT-2 sample, corresponding to the hydrotalcite of ratio Mg/Al = 1 calcined at 400 °C, the X-ray diffractogram is shown in Figure 2b. It shows that the synthesized hydrotalcite has structurally evolved when treated thermally at 400 °C, and a mixture of crystalline and amorphous phases can be identified. The crystalline phase of the periclase type has characteristic peaks of MgO (JCPDS 4-0829) with the index (111) at 35.48°, the index (200) at 44.40°, and the index (220) at, approximately, 63.04°, which are wider than those of the initial solid, indicating that it has lost crystallinity due to the thermal treatment. Additionally, a dispersed phase of low crystallinity with an amorphous predominance of the magnesium–aluminum spinel-type $MgAl_2O_4$ (***111, ***311, ***440) (JCPDS magnesium–aluminum spinel 21-1152) is observed, which would imply that the solid starts to rehydrate thanks to its memory effect. The oxides obtained by calcination at 400 °C acquire surface properties necessary for the semiconductor activity in the catalytic photoelectrodegradation of the Congo red dye.

Figure 2c shows the X-ray diffraction pattern of the solid calcined at 400 °C after the photodegradation process of the Congo red dye (HT-RC). The material calcined at 400 °C made up of a mixture of mixed and simple Mg-Al oxides regenerated its original hydrotalcite-type structure due to the capacity of these materials to recover their structure through a process called the memory effect. This memory effect is carried out on the structure of the solid after calcination when these come into contact with solutions containing anions or other dissolved substances, as is the case with the Congo red dye, so that the solid uses the carbonate ions present in the solution to reconstruct the original double-laminar structure. The pattern of the solid after photodegradation is the same as that of the fresh hydrotalcite, even with a higher purity since no segregated phases of $Mg(OH)_2$ or $Al(OH)_3$ are observed, and since no displacement is observed in the angle 2θ at the maximum reflection peak, it is confirmed that the interlaminar space of the solid is not increased by the presence of the Congo red dye.

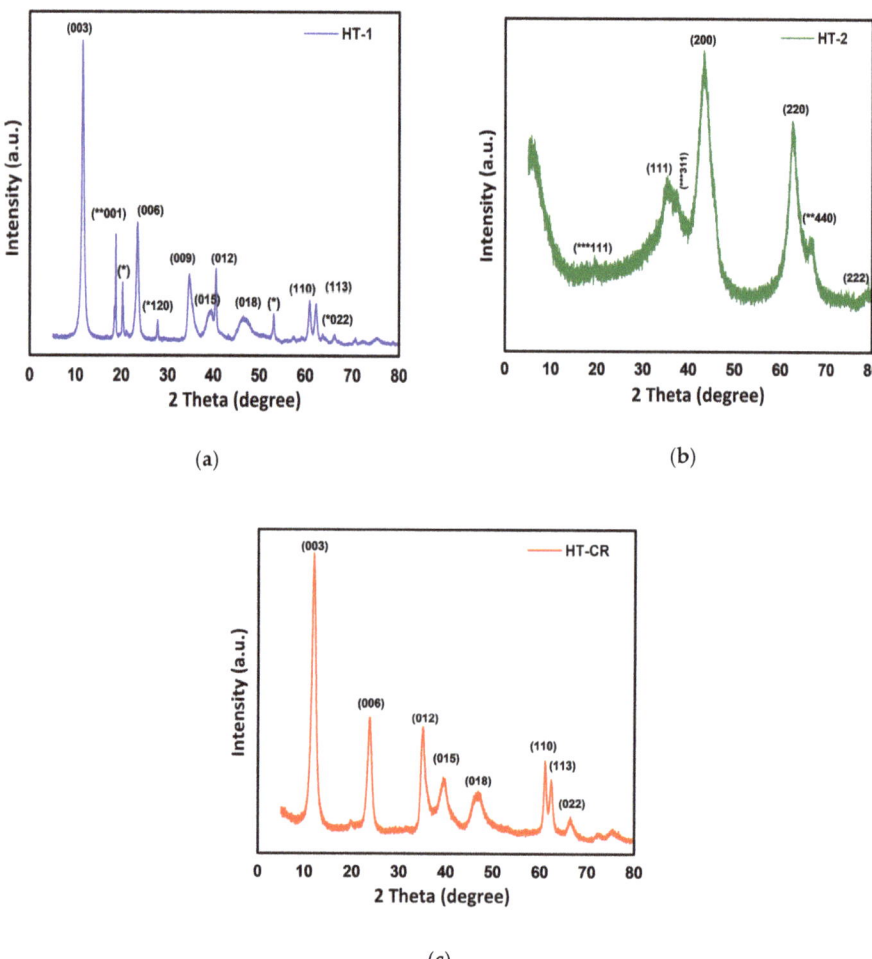

Figure 2. X-ray diffractogram of: (**a**) fresh synthesized hydrotalcite (HT-1) with ratio Mg/Al = 1, (**b**) hydrotalcite calcined at 400 °C HT-2, and (**c**) hydrotalcite after photodegradation of the Congo red.

The thermogravimetric (TGA) and differential thermal (DTA) analysis of the fresh synthesized solid (HT-1) is shown in Figure 3. In this study, the main changes that the synthesized solid exhibits because of the temperature increase associated with the thermal evolution of the solid and its thermal decomposition can be observed. This analysis allows us to identify the appropriate temperature to activate the hydrotalcites to be used as semiconductors in photoelectrodegradation processes. The temperature increases at 198 °C correspond to the first signal of the DTA and are attributed to the endothermic reaction generated by the elimination of the water occluded in the interlaminate of the hydrotalcite's crystalline structure, which represents a loss of 29.7% of the initial mass of the solid in the TGA graph. The second endothermic reaction centered at 355 °C is attributed to the decomposition of the carbonate anions (CO_3^{2-}) eliminated in the form of carbon dioxide and the dehydroxylation reaction of the hydrotalcite sheets that corresponds to the TGA curve in the range of 240–400 °C, presenting a mass loss of 24.5%. The solids are stable up to 500 °C, where finally, their laminar structure collapses and the spinel of the mixed oxide Mg/Al and periclase is formed. The total mass loss was approximately 53%.

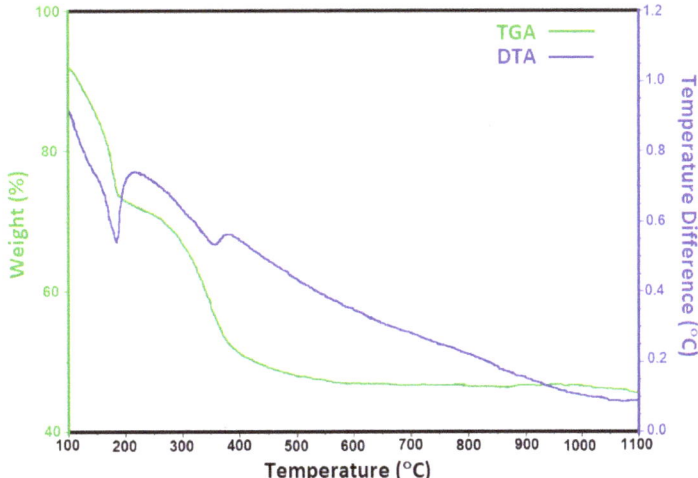

Figure 3. Thermogravimetric analysis (TGA) and differential thermal analysis (DTA) of freshly synthesized hydrotalcite (HT-1).

Figure 4 shows the infrared spectra of the HT-1, HT-2, and HT-RC solids. In all the solids, the characteristic signal of the stretching of the O-H groups of the brucite-type layer with a wide band ranging from 3750 to 3000 cm^{-1} was observed, where for the HT-1 solid, the connection of several signals of greater intensity can be seen, corresponding to the vibration of the OHs of the various crystalline species identified by X-ray diffraction (XRD), as well as the vibration with the interlaminar carbonates and water molecules. This signal becomes less intense and more continuous for the HT-2 solid because the structure has been partially dehydrated. For the HT-RC solid, the intensity increases again due to the reconstruction of the hydrotalcite structure without the presence of other crystalline phases. The signal around 1626 cm^{-1} corresponds to the H-OH vibration of the water molecules, which was observed in the HT-1 solid, disappears in the HT-2 solid that has been calcined, and reappears in the HT-RC solid by the effect of reconstruction of the lamellar structure. Regarding the band present in 1380 cm^{-1}, a fine and intense signal was observed for the HT-1 solid that corresponds to the vibration of the interlaminar CO_3^{2-}, which decreases in intensity for the HT-2 solid due to partial decarbonization with the appearance of a band at 1529 cm^{-1} of the free CO_3^{2-} to be adsorbed on the solid, and finally, in the HT-RC solid, the band intensifies again around 1274 cm^{-1} of the carbonates again in the interlaminar space.

The bands below 1000 cm^{-1} were assigned to the stretching of the metallic oxides Al-O, Mg-O, Al-O-Al, and Al-O-Mg. Specifically, the bands at 669 and 551 cm^{-1} disappear for the HT-2 solid due to the segregation of the crystalline phases of the simple oxides; they are formed again in the HT-RC solid due to the lamellar reconstruction of the hydrotalcite. Additionally, in the HT-RC solid, besides the abovementioned signals of the hydrotalcites, the spectrum is characterized by the presence of a small amount of the Congo red dye adsorbed in the solid, given the signals of the absorption bands at 1725 cm^{-1} that correspond to the presence of double bonds in the diazo groups -N=N, at 1047 cm^{-1} that is attributed to the vibration frequency of the C-N bond, and at 793 cm^{-1}, for the elongation vibration of the SO_3^- group.

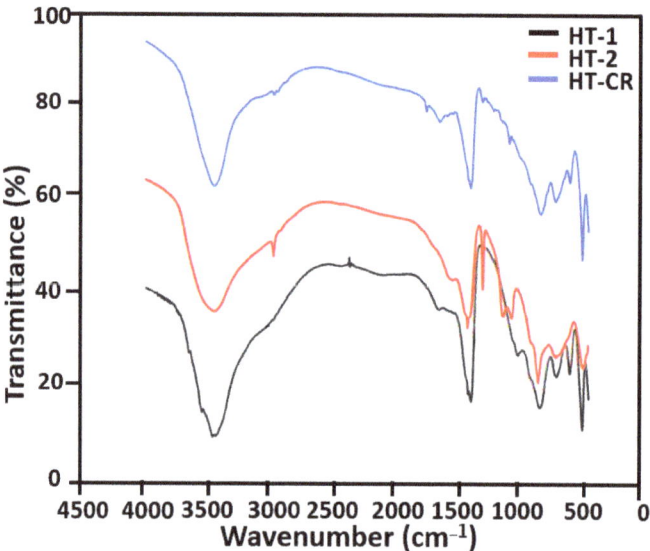

Figure 4. Fourier-transform infrared (FTIR) spectra of HT-1, HT-2, and Congo red dye (HT-CR) samples.

The N_2 adsorption–desorption isotherms of the HT-1 and HT-2 solids are presented in Figure 5. The fresh hydrotalcite HT-1 and the one calcined at 400 °C, HT-2, present type IV isotherms according to the IUPAQ classification with a hysteresis loop of H3 type, which confirms the existence of laminar particle aggregates that, in turn, originate pores in the form of cracks or fissures. This behavior is associated with the presence of mesopores formed by the accommodation of crystalline structures. Regarding the values of the specific area, it is observed that the HT-2 solid has a larger specific area than the HT-1 sample, which is attributed to the thermal decomposition that produces the decarbonation and partial decarboxylation that increases the size and volume of pores.

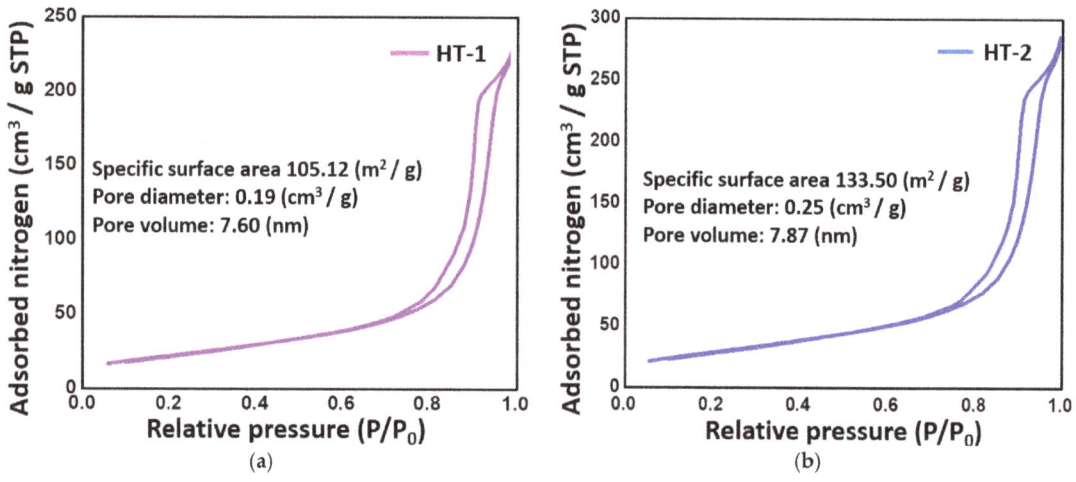

Figure 5. Isotherms of N_2 adsorption-desorption of solids and textural property values: (**a**) HT-1 and (**b**) HT-2.

Figure 6a shows the scanning electron microscopy images of the HT-1 solid corresponding to the synthesized HT-1 hydrotalcite, in which particles formed by lamellar aggregates can be observed as scales. Figure 6b shows the HT-2 solid of calcined hydrotalcite, which also shows particles with crystalline aggregates larger than the fresh solid and with smaller particles attached. Finally, Figure 6c shows the hydrotalcite reconstructed after the photodegradation process where particles formed by thin sheet aggregates like the original structure are observed.

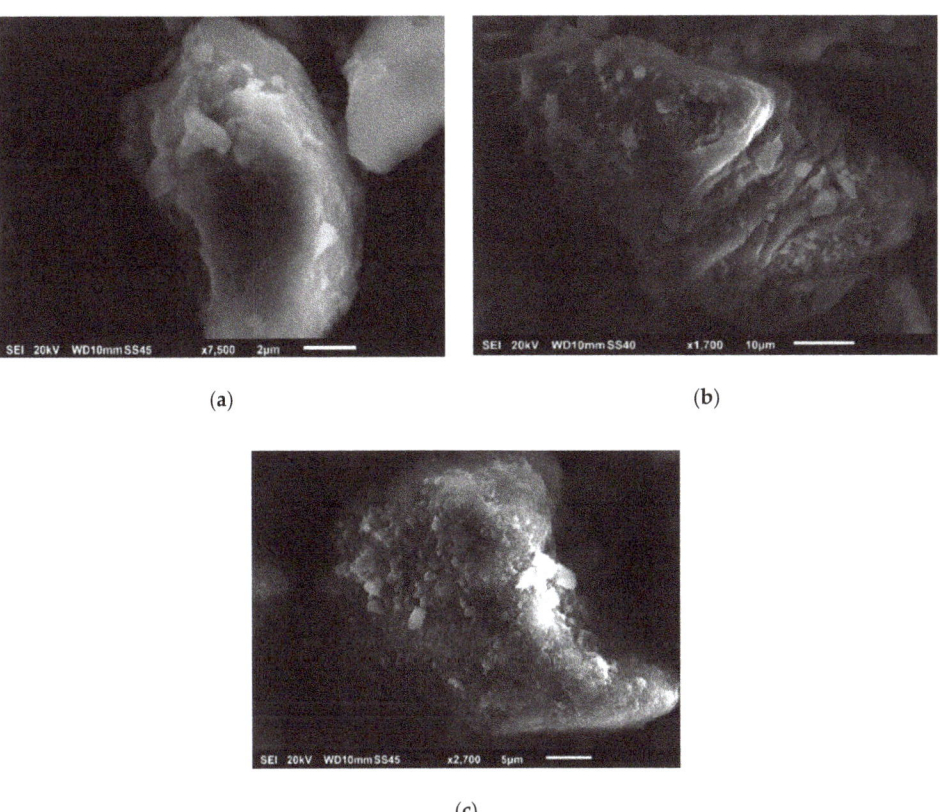

(a)   (b)

(c)

Figure 6. Scanning electron microscopy (SEM) images of: (**a**) fresh synthesized hydrotalcite HT-1 with ratio Mg/Al = 1, (**b**) hydrotalcite calcined at 400 °C HT-2, and (**c**) hydrotalcite after photodegradation of the Congo red.

Figure 7a shows the UV-VIS diffuse reflectance spectrum of the HT-2 photocatalyst used to degrade Congo red. For calculating the optical band gap energy (Eg), Tauc's graphical method was used (Figure 7b). The result for the catalyst treated at 400 °C before photodegradation was 3.28 eV. According to this value, this photocatalyst has a potential application in the degradation of Congo red, so it could favor the photodegradation capacity, considering that it has similar values compared to TiO_2, whose Eg values are reported to be between 3.0 and 3.2 eV [23].

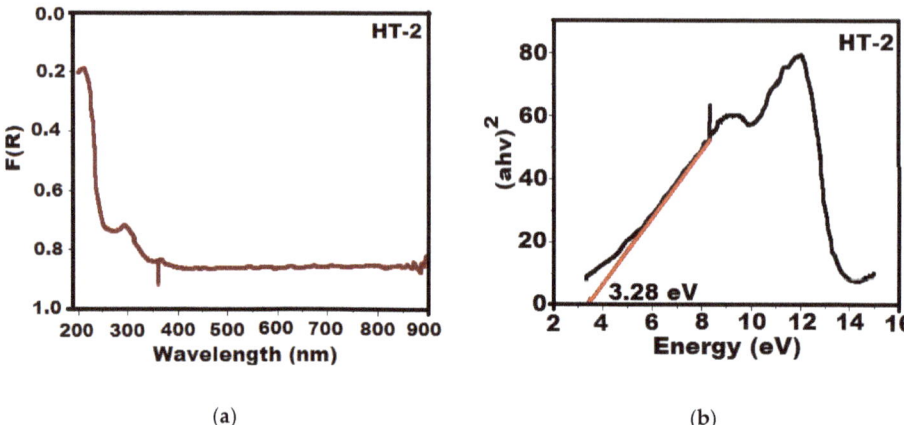

Figure 7. Determination of the energy of the forbidden optical band: (**a**) ultraviolet-visible (UV-VIS) diffuse reflectance spectrum of the HT-2 photocatalyst and (**b**) graph of the calculation of the optical band gap energy (Eg) value by Tauc's method for the catalyst at 400 °C before photodegradation.

2.2. Evaluation of the Degradation Capacity of the Congo Red Dye by Photolysis, Catalysis, Photocatalysis, Electrocatalysis, and Photoelectrocatalysis

The standardization of the analytical method for the quantification method of the dye was done using high-performance liquid chromatography (HPLC). The calibration curve for the industrial-grade synthetic dye Congo red in a concentration range of 0.5–40 mg/L, as shown in Figure S1, showed a linear correlation of 0.9991, which allowed us to validate the method for use. With the HPLC-optimized method, samples of each of the degradation processes of the Congo red dye were analyzed.

Figure 8 shows the degradation kinetics of a Congo red dye solution from an initial concentration of 400 mg/L using different systems: photolysis, catalysis, photocatalysis, electrocatalysis, and photoelectrocatalysis in a maximum time of 6 h. The aim of the study was to identify the effect of each of the elements involved in the system of photoelectro-catalysis and its degree of participation in the catalytic activity of 200 mg of catalyst. In the figure, we can see that the effect of photolysis results in a gradual decrease in the concentration of the dye by 8.75% at 4.5 h, remaining constant after that time. In the case of catalysis, that result implies the contact of the catalyst with the dye producing a decrease in the concentration until reaching 46% at 5.5 h. In photocatalysis, the effect of light on the catalyst accelerates the speed of degradation, reaching, in the first hour, more than 50% of the degradation up to 75% at 6 h. Concerning electrocatalysis, the effect of current is greater and more effective on the catalyst, reaching 62% degradation in 1 h and a maximum capacity of 85% at 3.5 h. Finally, the combined effect of UV light irradiation with an electric current on the catalyst reaches a degradation of 87% in 0.5 h, 91% in 1 h, and 96% at 6 h.

With the data from the kinetic studies, the models for each process of degradation of the Congo red dye for the Langmuir–Hinshelwood isotherm for the pseudo-first order were applied as shown in Figure 9a and for the pseudo-second order Ho model as shown in Figure 9b.

Figure 9, coupled with the data in Table 1 shows that only the photolysis and catalysis systems fit the pseudo-first-order kinetic model, which is attributed to the low concentration of hydroxyl radical species formed in the process. In the case of photocatalysis, electrocatalysis, and photoelectrocatalysis processes, it fits better to the pseudo-second-order kinetic model of Ho, which is attributed to the fact that the degradation process requires different stages that imply the formation of different species, mainly the hydroxyl radicals that can be identified by high-resolution liquid chromatography. The pseudo-first-order model assumes that the speed of the degradation reaction is linearly proportional to the concentration. In contrast, the pseudo-second-order model assumes that the degradation process is controlled by the reactions that occur in the system due to chemisorption in the catalyst, as well as the process of sharing or exchanging electrons between the catalyst, the electrode, and the Congo red. In recent years, degradation processes have been reported that present this type of kinetic behavior, mainly when composites or combined systems are used [44–47].

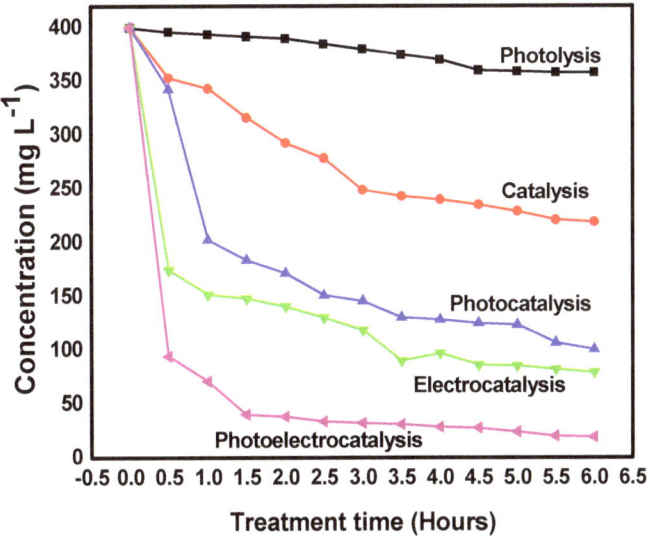

Figure 8. Degradation of Congo red dye by photolysis, catalysis, photocatalysis, electrocatalysis, and photoelectrocatalysis.

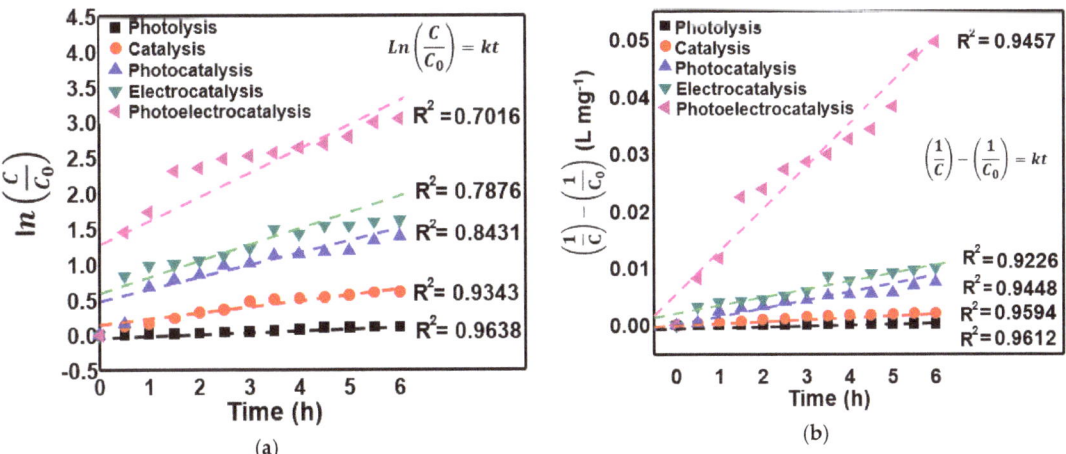

Figure 9. Kinetic models of the different degradation processes of the Congo red dye: (**a**) pseudo-first-order reaction and (**b**) pseudo-second-order reaction.

Table 1. Order of reaction, speed constant, and half-life of the degradation of the Congo red dye for each of the processes.

Process	Pseudo-First Order			Pseudo-Second Order		
	K (h^{-1})	$t_{1/2}$ (h)	R^2	K $(L \cdot mg^{-1} \cdot h^{-1})$	$t_{1/2}$ (h)	R^2
Photolysis	0.0212	32.69	0.9638	0.00006	41.67	0.9612
Catalysis	0.0990	7.00	0.9343	0.0004	6.25	0.9594
Photocatalysis	0.1975	3.51	0.8421	0.0011	2.27	0.9448
Electrocatalysis	0.2027	3.42	0.7876	0.0015	1.67	0.9226
Photoelectrocatalysis	0.3515	1.97	0.7016	0.0072	0.35	0.9457

Figure 10 shows the chromatograms corresponding to the degradation kinetics of 400 mg/L and the degradation by products of Congo red when using hydrotalcites activated at 400 °C as catalysts. The signal corresponding to the dye appears in an elution time of 1.6 min, and it can be observed that the concentration of the dye gradually decreases with respect to the treatment time in each process analyzed. Thus, in Figure 10a, the chromatograms correspond to the catalysis process of hydrotalcite with the Congo red dye, and at 1.2 min, a signal appears that is attributed to the formation of reactive oxygen species (ROS) such as $^\bullet O_2$, $OH^\bullet$, H_2O_2, hydroperoxyl ($ROOH^\bullet$), and peroxynitrile ($ONOO^\bullet$)—these are some of the reactive species formed during the processes studied and that can be detected in concentrations of picomoles, employing chromatography techniques [48–50]. It should be emphasized that the established technique was used only to detect the signal corresponding to the Congo red color and not the reactive oxygen species. Thus, during the catalytic process, we observed that the maximum concentration of the reactive species occurs during the first hour of the process with a value of 1.5×10^{-3} absorbance units (A.U.), and that the generation of the reactive oxygen species decreases with the treatment time; this can be attributed to the fact that the MII/MIII pairs decrease their redox pairs. In Figure 10b, the degradation of the dye is favored by ultraviolet radiation, which causes an increase in reactive species of up to 3.5×10^{-3} A.U. in 4 h of treatment and in which it can be observed that no secondary products are generated during its degradation process. In this case, the degradation is greater because the hydrotalcite is photoactivated by the presence of ultraviolet light, where the photoactivity is related to the forbidden band energy of the crystalline phases and its textural properties [51,52].

For the degradation process using the photocatalyst of hydrotalcite calcined at 400 °C confirmed mainly by the amorphous spine phase $MgAl_2O_4$ and the crystalline periclase phase, it can be considered that the catalyst absorbs a photon, which causes an electron to be transferred from the valence band to the conduction band, leaving a hole in the valence band [53], as shown in Equation (1).

$$MgAl_2O_4 + MgO + hv \rightarrow h^+{}_{VB} + e^-{}_{CB} + MgAl_2O_4 + MgO \qquad (1)$$

This contributes to the dissociation of water molecules and to the formation of free radicals due to the presence of ultraviolet light, according to Equations (2)–(6) [51].

$$H_2O \longleftrightarrow H^+ + OH^- \qquad (2)$$

$$h^+{}_{VB} + H_2O \rightarrow H^+ + {}^\bullet OH \qquad (3)$$

$$e^-{}_{CB} + O_2 \rightarrow O^{\bullet-}{}_2 \qquad (4)$$

$$O^{\bullet-}{}_2 + 2H_2O \rightarrow 2^\bullet OH + 2OH^- + O_2 \qquad (5)$$

$$O^{\bullet-}{}_2/^\bullet OH + \text{organic compounds} \rightarrow \text{degradation} \qquad (6)$$

For the electrocatalysis process in Figure 10c, the application of the electric field increases the generation of the reactive oxygen species from 2.0×10^{-3} to 4.0×10^{-3} A.U.

in 3 h of treatment. This is because the Cu^0/Cu^{2+} ions that are generated on the surface of the electrode and the free electrons of the MII/MIII pair of hydrotalcites that accelerate the degradation of the dye due to the application of the electric field are present in the aqueous medium. In the photoelectrocatalytic process shown in Figure 10d, the reactive oxygen species reach a value of 6.0×10^{-3} A.U. in the first hour of treatment due to the simultaneous application of the electric field and the ultraviolet radiation, which favors the formation of a higher concentration of $^\bullet OH$ radicals in the cathode and accelerates the degradation of the Congo red dye. The electrolysis reactions in the aqueous medium are described in Equations (7)–(9) and occur simultaneously with Equations (1)–(6) described above due to the application of the electric field [54]:

$$\text{Anode } H_2O \rightarrow 2H^+ + \frac{1}{2} O_{2(g)} + 2e^- \tag{7}$$

$$\text{Cathode } 2H_2O + 2e^- \rightarrow 2OH^- + H_{2(g)} \tag{8}$$

The above reaction, coupled with the reaction of copper oxidation in an aqueous medium by the application of an electrical potential, produces the reaction shown in Equation (9).

$$Cu^0 \rightarrow Cu^{2+} + 2e^- \tag{9}$$

Under these conditions in the photochemical reactor and in the presence of ultraviolet light, copper used in a wide range of accessible oxidation states, such as Cu0, Cu^+, Cu^{2+}, and Cu^{2+}, generated by the electrolysis reaction could react with the photographed electrons according to the reaction shown in Equation (10) [55].

$$Cu^{2+} + e^-_{CB} \rightarrow Cu^+ \tag{10}$$

In the presence of an electric field, the OH^- and Cu^{2+} ions are generated on the surface of the cathode and the anode simultaneously, and the Cu^{2+} and OH^- ions react in the solution to produce CuO and CuO_2.

$$2Cu^+ + 2OH^- \rightarrow CuO + Cu_2O \tag{11}$$

$$Cu^+ + (O_2, H_2O_2, O_2, \text{other oxidants}) \rightarrow Cu^{2+} + e^- \tag{12}$$

In the case of photoelectrocatalysis, the presence of the catalyst formed by the structure of the activated hydrotalcites (HT-2) containing Mg^{2+} and Al^{3+} favors an increase in the catalytic activity by slowing down the recombination of the electron–hole pairs for the reaction system shown in Equation (9), which considers the formation of Cu^{2+} ions and the generation of e^-. The electrons in the valence band of the activated hydrotalcite are transferred directly to copper, which produces oxidation to Cu^{2+} and Cu^+ species, and the voids created in the hydrotalcite valence band contribute to the degradation of the organic compounds. Figure 11 graphically depicts the potentials for interfacial load transfer and the potentials for recombination of each of the reactive species present in the heterogeneous photoelectrocatalytic system.

The excess of electrons contributed by the copper electrode and the recombination of the same one on the hollow electron pairs of the hydrotalcites accelerate the supply of electrons so that the degradation of the red dye occurs in the photoelectrocatalytic system. For this degradation process to occur, the number of electrons involved in the total combustion of the Congo red dye was set at 178, assuming that the main process is the formation of nitrate and sulfate ions, as shown in Equation (13) [12,56–58].

$$C_{32}H_{22}S_{22}- + 84H_2O \rightarrow 32CO_2 + 6NO_3^- + 2SO_4^{2-} + 19H^+ + 178e^- \tag{13}$$

The Cu^{2+} and Cu^+ ions with the photogenerated electrons decrease the recombination speed of the electron–hole pair, generating a higher catalytic activity in the degradation of the Congo red dye, which visually produces the loss of color. For the photoelectrocatalysis process with which different active chemical species for the degradation are generated, the elimination of 95% of the Congo red dye is achieved.

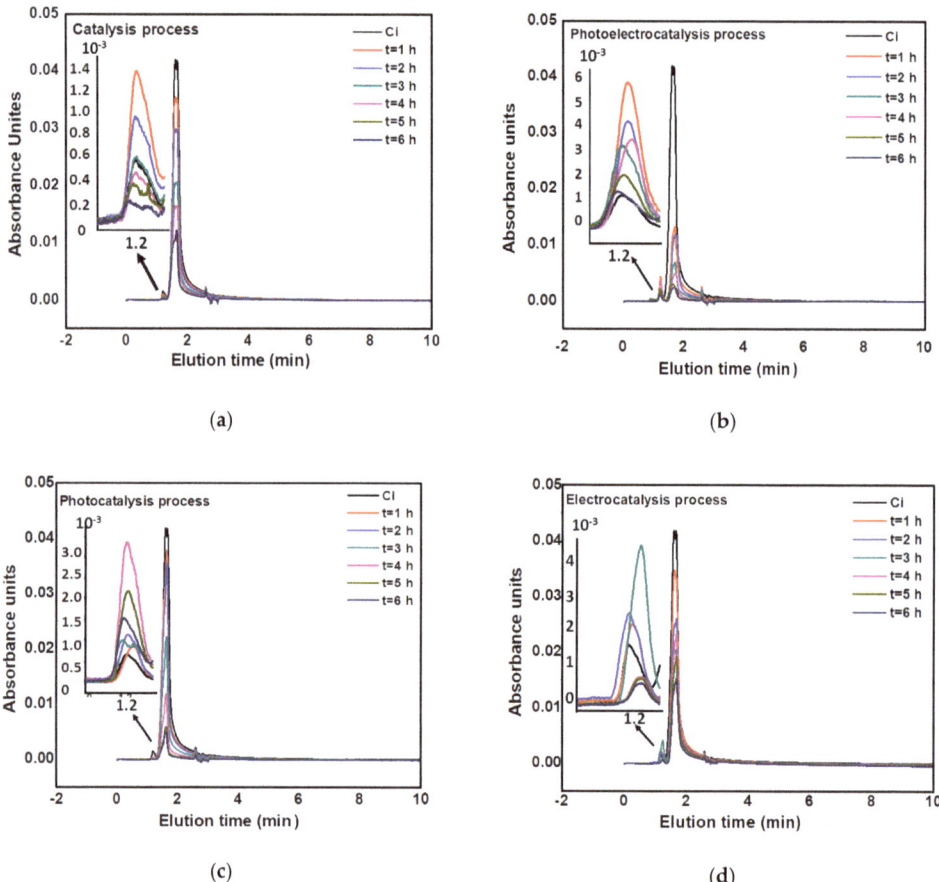

Figure 10. Chromatograms of the degradation of 400 mg/L Congo red by the different processes of degradation: (**a**) catalysis, (**b**) photocatalysis, (**c**) electrocatalysis, and (**d**) photoelectrocatalysis. Isocratic flow 1.0 mL/min = 480 nm using a Thermo chromatographic column Scientific™ Hypersil™ BDS C18 reverse-phase (250 mm × 4.6 mm) and 5 μm particle size. pH 6.7; methanol/water mobile phase 60:40 (*v*/*v*).

Degradation products formed at the end of each process were analyzed by liquid chromatography–mass spectrometry (LC–MS) and identified by interpreting their mass spectra data presenting their molecule ion peaks with respect to m/z (where m is the molecular weight of the intermediates in the mass spectra). The main species detected in the solutions are presented in Figure S2, including LC–MS spectra for the aqueous solution of 10.0 mg/L Congo red.

Figure S2a shows the chromatogram of the 10 mgL^{-1} standards of Congo red. Congo red is a sulfonated compound with two sulfonic acid groups, and degradation can occur in various stages by an electrophilic attack on the benzene rings or a rupture of the bond -C-S-between the aromatic ring and the sulfonate groups. In each one of the chromatograms of the different processes, the intermediate products can be identified; at 633 and 550 m/z, two pseudomolecular ions are identified, where the 633 m/z ions were due to the "M − 2Na + H" [59]. The elimination of the two sulfonate groups and a benzene ring generates an intermediate with 468.8 and 386.8 m/z; the intermediate by direct fragmentation is 2-nitrosonaphthalene (m/z = 157). The additional oxidation by OH$^-$ radicals is produced by the presence of 3-carboxy propanoate (m/z = 117), 4-carboxybutanoate (m/z = 131), acetic acid (m/z = 60), achieving, in this way, the degradation of the red Congo in the medium. According to the results, the Congo red degradation route as indicated in Figure S2f was used, showing the species detected for the Congo red aqueous solution, and showing possible species formed after the irradiation of the Congo red/HT-2 solution.

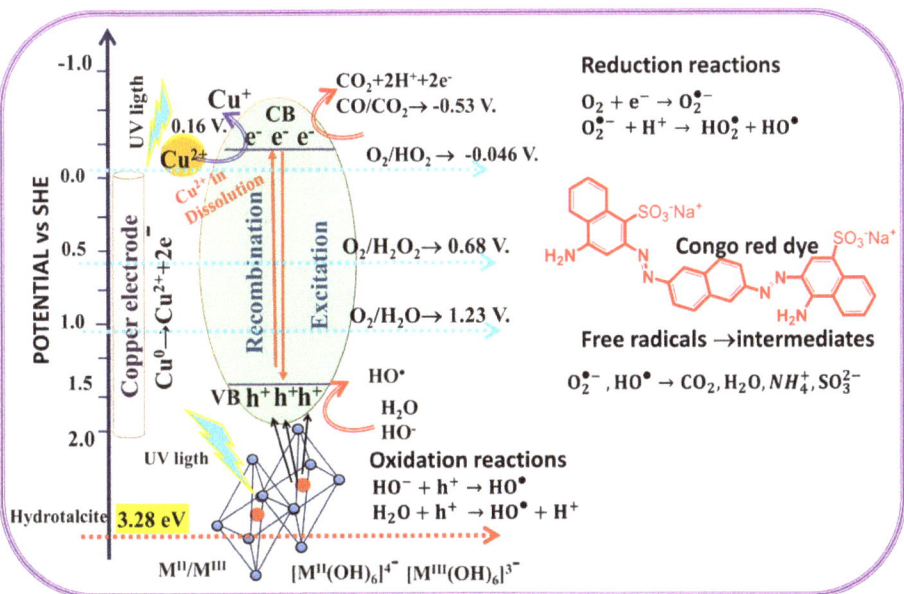

Figure 11. The mechanism for generating photoelectrocatalytic activity of Cu(II)/HT-2 under ultraviolet (UV) light irradiation. The UV light irradiation induces interfacial charge transfer (IFCT) from the hydrotalcite (HT) valence band (VB) to the Cu^{2+} ion. [47]. Copyright American Chemical Society, 2009.

The degradation of the Congo red molecule is favored due to the presence of double bonds in the diazo groups (-N=N-), which is the preferred site for the attack of the •OH radicals. Such an attack can occur in the proximal or distal nitrogen atom of the azo group leading to a variety of cleavage pathways resulting in several identifiable products. The complete destruction of the chromophore leads to the discoloration of the solution with subsequent degradation of the rest of the dye's organic structure.

Figure S3 shows how the Congo red dye is significantly degraded by the photocatalytic process, as shown in Figure S3b, and how the degradation of the dye is complete using the photoelectrocatalytic process according to Figure S3c.

Table 2 shows research that has used TiO$_2$ reference photocatalysts for Congo red degradation, comparing the photocatalyst used, the concentration of the dye, the degradation time, and the source of irradiation.

Table 2. Comparative study of the photodegradation of the Congo red dye with different TiO$_2$ materials.

Photocatalyst	Congo Red Concentration (mg/L)	Catalyst Concentration (g/L)	Time (h)	Removal (%)	Irradiation Source	Reference
Nanocrystalline TiO$_2$ Degussa P-25	20	0.25	0.5	98	Suntest 675 Wm^{-2}	[10]
Mg-TiO$_2$-P25 NPs	7	0.5	7	80	400 nm vis lamp	[60]
TiO-W	30	0.15	0.83	96	365 nm vis lamp	[61]
TiO$_2$	55	1.0	8	20	254 nm UV lamp	[62]
Photoelectrocatalysis	400	0.66	1	91	254 nm UV lamp	This work

TiO$_2$ nanoparticles under the same UV irradiation conditions can only degrade a low percentage despite having low concentrations and a higher amount of catalyst. TiO$_2$ improves its photodegradation capacity but requires combining with other metals and can only reach high removal percentages if the concentration of Congo red is low and/or with long irradiation times. The results of this research suggest that degradation of Congo red dye using combined photoelectrolytic systems with activated hydrotalcites and Cu anode has improved efficiency at high concentrations of Congo red, making it a promising system for application in effective degradation of the contaminant in aqueous systems. Among the main advantages offered by the use of activated catalysts as photo and electrostimulated catalysts in a degradation system, are the ability to generate pairs of electron holes, as well as the possibility of recovering the catalyst for other recycles thanks to the effect memory.

3. Materials and Methods

3.1. Obtaining the Catalytic Precursors of the Hydrotalcites

Hydrotalcites were synthesized by the coprecipitation method from magnesium nitrate hexahydrate [Mg(NO$_3$)$_2$·6H$_2$O], aluminum nitrate nonahydrate [Al(NO$_3$)$_3$·9H$_2$O], and sodium carbonate (Na$_2$CO$_3$) in a basic medium at pH 11.5, from stoichiometric amounts of Mg/Al = 1. The synthesized hydrotalcite was dried in an oven at 80 °C for 24 h, and then, a portion was calcined in a muffle at 400 °C for 4 h at a heating rate of 10 °C/min. The fresh solid and solid calcined at 400 °C were ground in an agate mortar until a fine particle size was obtained and were labeled as HT-1, to the fresh hydrotalcite, and HT-2, to the hydrotalcite calcined.

3.2. Structural, Thermal, and Textural Characterization of Hydrotalcite-type Catalytic Precursors

For the characterization and identification of the crystalline structures present, an Inel Equinox Powder Diffractometer coupled to a copper anode and using a monochromatic radiation source from CuKα, whose wavelength (λ) was 1.5418 Å, in a range of 5–80 theta degrees with a step size of 0.02° and a scanning speed of 2°/min was used.

The thermal, differential, and thermogravimetric evolution profile (DTA and TGA) was carried out using Thermo-analyzer TA instruments (New Castle, DE, USA), at a heating rate of 10 °C/min in an air atmosphere at a speed of 100 mL/min, and using α-alumina as a reference standard.

For the identification of the functional groups, including the different types of vibration in their links present in the materials that could suggest a structural composition, as well as for the identification of the presence of anions in the interlaminar space, the infrared spectra with Fourier transforms (FTIR) were obtained using a Perkin Elmer Spectrum 100 FT-IR Spectrometer equipment in (Waltham, MA, USA) a wavelength range between 500 and 4500 cm^{-1}; the samples were prepared by mixing the solid with KBr (the target) in a 1:100 ratio.

The textural characterization was carried out using the technique of nitrogen (N$_2$) physisorption in an automated Micromeritics TriStar ll Plus equipment. Before being analyzed, the samples were degassed by heating to 110 °C for 3 h in an atmosphere of helium, and then, they were left to cool down to room temperature; the distribution of the size of the pore was calculated based on the curves of desorption of the isotherms of Barrett–Joyner–Halenda (BJH). Using the Brunauer, Emmett, and Teller (BET) theory, the textural properties of the solids, such as specific area, pore size, and pore diameter, were determined, as well as the nitrogen adsorption–desorption isotherms.

The morphology was analyzed in a JSM 6010/ LV field-emission scanning electron microscope at an acceleration voltage of 20 Kv.

The determination of the band gap energy of the catalyst was carried out by the diffuse reflectance technique in a Shimadzu UV-Vis model 2401-PC equipment with integration sphere.

3.3. Optimization of the Analytical Method for the Quantification of the Degradation of the Congo Red Dye

A chromatography analysis of each of the processes was made in triplicate and aliquots of 2 mL were taken every 30 min and centrifuged at 14,000 rpm for 15 min. The samples were then passed through 0.2 μm Iso-Disc™ filters N-25-2, Nylon 25 μm × 0.2 μm Supelco, to separate the suspended particles from the supernatant. From the previously filtered solutions, 100 μL of the samples are taken and diluted with 900 μL milli-Q water. For the detection of high-resolution liquid chromatography, 20 μL of the solution was injected through a Thermo Chromatographic Column Scientific™ Hypersil™ BDS C18 reverse-phase (250 mm × 4.6 mm and particle size of 5 μm) at a wavelength of 480 nm, 2150 psi, and 26 °C using a Perkin Elmer 1000 chromatograph with a diode array detector. Analysis of degradation products of Congo red was done by liquid chromatography–mass spectrometry (LC–MS) measurements were performed using Agilent 1100 series (Waldbronn, BW, Germany) with electrospray ionization mass spectrometry. All the reagents used for the analysis were of analytical grade; acetonitrile (J.T. Baker 99%) and methanol (J.T. Baker 99%), and water was from Milli-Q of 18.0 μS.

3.4. Evaluation of the Degradation Capacity of Congo Red Dye by Photolysis, Catalysis, Photocatalysis, Electrocatalysis, and Photoelectrocatalysis

All these degradation tests of the Congo red dye were performed in triplicate using a photochemical reactor RFQ-500 in a closed system with constant agitation and room temperature, taking aliquots of 2.0 mL every 30 min during the experiment. For the catalytic degradation tests, 300 mL of Congo red dye solution at a concentration of 400 mg/L were used in contact with 200 mg of HT-2 catalyst (hydrotalcite calcined at 400 °C). For the photocatalytic degradation tests, a 120 V, 9-watt UV lamp of radiation of 254 nm was used. The initial reaction temperature was 25 °C with a maximum of 29 °C in all tests. In the electrocatalysis and photoelectrocatalysis tests, a potentiostat (BioLogic Science Instrument, EC-Lab Version 10.12 software, France) was used, and a potential of 25.0 millivolts was applied through a copper electrode with an area of 3.0 cm^2 as the anode and L302 stainless steel as the cathode, as shown in Figure S4.

4. Conclusions

In this study, Mg/Al hydrotalcite type materials were prepared by the coprecipitation method to be used as heterogeneous catalysts in the degradation of Congo red dye under different degradation systems. The hydrotalcite type catalysts activated at 400 °C showed a combination of crystalline and amorphous phases of $MgAl_2O_4$ and MgO that show semiconductor surface properties associated with a forbidden band energy of 3.28 eV, confirming particles with slit type mesopores with specific areas above 133 m^2/g that provide them with favorable textural properties for dye access for its degradation assisted by ultraviolet (UV) irradiation and/or electric current. The results show that the solid, at 2 h of contact, can catalyze the degradation of 29%, improving its degradation capacity up to 60% when the process is assisted by UV irradiation, 65% when assisted by electric current, and 90% when assisted by UV irradiation and electric current. The maximum degradation was reached with the photoelectrocatalysis process at 6 h with a degradation of 95% and a half-life of 0.36 h. The efficiency of the catalytic process assisted by irradiation and electric

Catalysts **2021**, *11*, 211

current with a Cu electrode is because the Cu^{2+} ions and with the photogenerated electrons decrease the speed of recombination of the electron–hole in the catalyst, and this generates a greater catalytic activity in the degradation of the Congo red dye. The degradation of the dye is favored by the presence of double bonds in the diazo groups (-N=N-) of the Congo red dye that become the preferred site for the attack of the OH radicals—the proximal or distal nitrogen atom of the azo group that leads to a variety of cleavage pathways resulting in several intermediary products that are completely degraded. At the conclusion of the degradation process, it is observed that the hydrotalcite is completely reconstructed with a high purity of crystalline phases, which can be used for further studies in the degradation of the azo type dye.

Supplementary Materials: The following are available online at https://www.mdpi.com/2073-4344/11/2/211/s1. Figure S1. (a) Linear correlation of the calibration curve by high-performance liquid chromatography (HPLC) of the industrial-grade synthetic dye Congo red and (b) solution of the calibration curve. Figure S2. Chromatograms of the liquid chromatography–mass spectrometry (LC–MS) analysis of Congo red after degradation: (a) Congo red standard, (b) catalyst, (c) photocatalyst, (d) electrocatalyst, (e) photoelectrocatalyst, and (f) possible degradation products of Congo red in the presence of hydrotalcites. Figure S3. (a) Solutions of dye at different concentrations for degradation, (b) solutions after degradation by the process of photocatalysis, and (c) solutions after degradation by photoelectrocatalysis. Figure S4. Diagram of the photoelectrocatalysis process.

Author Contributions: Conceptualization, E.R.-R., N.G.-O. and R.F.-R.; methodology, E.R.-R. and N.G.-O.; software R.F.-R. and S.A.-F., G.C.-J.; formal analysis, E.R.-R., N.G.-O., R.F.-R. and S.A.-F.; investigation, R.F.-R. and S.A.-F.; resources, E.R.-R. and G.C.-J.; data curation, S.A.-F. and G.C.-J.; writing—original draft preparation, E.R.-R. and R.F.-R.; writing—review and editing, N.G.-O.; E.R.-R.; funding acquisition, E.R.-R. and N.G.-O. All authors have read and agreed to the published version of the manuscript.

Funding: This research received no external funding. Funding was received from the resources of the University of Guanajuato.

Institutional Review Board Statement: Not applicable.

Informed Consent Statement: Not applicable.

Data Availability Statement: Not applicable.

Acknowledgments: We would like to especially thank the Directorate for Research and Postgraduate Support (DAIP) at the University of Guanajuato for their support in developing this project. We also thank the University of Guanajuato-CONACyT National Laboratories (National Laboratory for Characterization of Physicochemical Properties and Molecular Structure UG-UAA-CONACyT and Laboratory for Research and Technological Development in Advanced Coatings UG-CINVESTAV-IPN-CONACyT). Additionally, we thank the CONACyT for the graduate scholarship and the financing of the infrastructure project 255270.

Conflicts of Interest: The authors declare no conflict of interest.

References

1. Robinson, T.; McMullan, G.; Marchant, R.; Nigam, P. Remediation of dyes in textile effluent: A critical review on current treatment technologies with a proposed alternative. *Bioresour. Technol.* **2001**, *77*, 247–255. [CrossRef]
2. El Gaini, L.; Lakraimi, M.; Sebbar, E.; Meghea, A.; Bakasse, M. Removal of indigo carmine dye from water to Mg–Al–CO_3-calcined layered double hydroxides. *J. Hazard. Mater.* **2009**, *161*, 627–632. [CrossRef]
3. Chen, F.; Wu, X.; Bu, R.; Yang, F. Co–Fe hydrotalcites for efficient removal of dye pollutants via synergistic adsorption and degradation. *RSC Adv.* **2017**, *7*, 41945–41954. [CrossRef]
4. Van Der Zee, F.P.; Font, J.; Fortuny, A.; Fabregat, A. Towards advanced aqueous dye removal processes: A short review on the versatile role of activated carbon. *J. Environ. Manag.* **2012**, *102*, 148–160.
5. Guzmán-Vargas, A.; Limb, E.; Uriostegui-Ortega, G.A.; Oliver-Tolentino, M.A.; Rodríguez, E.E. Adsorption and subsequent partial photodegradation of methyl violet 2B on Cu/Al layered double hydroxides. *Appl. Surf. Sci.* **2016**, *363*, 372–380. [CrossRef]
6. Bilińska, L.; Gmurek, M.; Ledakowicz, S. Comparison between industrial and simulated textile wastewater treatment by AOPs—Biodegradability, toxicity and cost assessment. *Chem. Eng. J.* **2016**, *306*, 550–559. [CrossRef]

7. Ersoz, G.; Napoleoni, A.; Atalay, S. Comparative study using chemical wetoxidation for removal of reactive black 5 in the presence of activated carbon. *J. Environ. Eng.* **2013**, *139*, 1462–1469. [CrossRef]

8. Li, X.L.; Zhang, J.; Jiang, Y.C.; Hu, M.C.; Li, S.N.; Zhai, Q.G. Highly efficient biodecolorization/degradation of Congo red and alizarin yellow R bychloroperoxidase from Caldariomyces fumago: Catalytic mechanism anddegradation pathway. *Ind. Eng. Chem. Res.* **2013**, *52*, 13572–13579. [CrossRef]

9. Wang, Y.S.; Shen, J.H.; Horng, J.J. Chromate enhanced visible light driven TiO$_2$ photocatalytic mechanism on acid orange 7 photodegradation. *J. Hazard. Mater.* **2014**, *274*, 420–427. [CrossRef] [PubMed]

10. Erdemoglu, S.; Aksu, S.K.; Sayilkan, F.; Izgi, B.; Asilturk, M.; Sayılkan, H.; Frimmel, F.; Gucer, S. Photocatalytic degradation of Congo red by hydrothermally synthesized nanocrystalline TiO$_2$ and identification of degradation products by LC–MS. *J. Hazard. Mater.* **2008**, *155*, 469–476. [CrossRef] [PubMed]

11. Starukh, G. Photocatalytically Enhanced Cationic Dye Removal with Zn-Al Layered Double Hydroxides. *Nanoscale Res. Lett.* **2017**, *12*, 391. [CrossRef]

12. Janczarek, M.; Kowalska, E. On the Origin of Enhanced Photocatalytic Activity of Copper-Modified Titania in the Oxidative Reaction Systems. *Catalysts* **2017**, *7*, 317. [CrossRef]

13. Dorian, A.H.H.; Sorrell, C.C. Sand supported mixed-phase TiO$_2$ photocatalysts for water decontamination applications. *Adv. Eng. Mater.* **2014**, *16*, 248–254.

14. Pattnaik, S.P.; Behera, A.; Acharya, R.; Parida, K. Green exfoliation of graphitic carbon nitride towards decolourization of Congo-Red under solar irradiation. *J. Environ. Chem. Eng.* **2019**, *7*, 103456. [CrossRef]

15. Kumru, B.; Antonietti, M. Colloidal properties of the metal-free semiconductor graphitic carbon nitride. *Adv. Colloid Interfac. Sci.* **2020**, *283*, 102229. [CrossRef] [PubMed]

16. Zhao, Z.; Sun, Y.; Dong, F. Graphitic carbon nitridebased nanocomposites: A review. *Nanoscale* **2015**, *7*, 15–37. [CrossRef]

17. Hieu-Nguyen, C.; Chun-Chieh, F.; Ruey-Shin, J. Degradation of methylene blue and methyl orange by palladium-doped TiO$_2$ photocatalysis for water reuse: Efficiency and degradation pathways. *J. Clean. Prod.* **2018**, *202*, 413–427. [CrossRef]

18. Chong, M.N.; Jin, B.; Chow, C.W.K.; Saint, C. Recent developments in photocatalytic water treatment technology: A review. *Water Res.* **2010**, *44*, 2997–3027. [CrossRef]

19. Kushwaha, H.S.; Vaish, R. Enhanced visible light photocatalytic activity of curcumin-sensitized perovskite Bi$_{0.5}$Na$_{0.5}$TiO$_3$ for rhodamine 6G degradation. *Int. J. Appl. Ceram. Technol.* **2016**, *13*, 333–339. [CrossRef]

20. Subhan, M.A.; Saha, P.C.; Uddin, N.; Sarker, P. Synthesis, structure, spectroscopy and photocatalytic studies of nano multimetal oxide MgO·Al$_2$O$_3$ ·ZnO and MgO·Al$_2$O$_3$ ·ZnO curcumin composite. *Int. J. NanoSci. NanoTechnol.* **2017**, *13*, 69–82.

21. Li, Y.; Cui, W.; Liu, L.; Zong, R.; Yao, W.; Liang, Y.; Zhu, Y. Removal of Cr(VI) by 3D TiO$_2$ -graphene hydrogel via adsorption enriched with photocatalytic reduction. *Appl. Catal. B Environ.* **2016**, *199*, 412–423. [CrossRef]

22. Arotiba, O.A.; Orimolade, B.O.; Koiki, B.A. Visible light–driven photoelectrocatalytic semiconductor heterojunction anodes for water treatment applications. *Curr. Opin. Electrochem.* **2020**, *22*, 25–34. [CrossRef]

23. Turolla, A.; Bestetti, M.; Antonelli, M. Optimization of heterogeneous photoelectrocatalysis on nanotubular TiO2 electrodes: Reactor configuration and kinetic modelling. *Chem. Eng. Sci.* **2018**, *182*, 171–179. [CrossRef]

24. Figueras, F. Base Catalysis in the Synthesis of Fine Chemicals. *Top. Catal.* **2004**, *29*, 189–196. [CrossRef]

25. Cavani, F.; Trifirò, F.; Vaccari, A. Hydrotalcite-type anionic clays: Preparation, properties and applications. *Catal. Today* **1991**, *11*, 173–302. [CrossRef]

26. Rives, V. *Layered Double Hydroxides: Present and Future*, 1st ed.; Nova Science Publishers, Inc.: New York, NY, USA, 2001; pp. 9–321.

27. Timar, V.; Varga, G.; Murath, S. Synthesis, characterization and photocatalytic activity of crystalline Mn(II) Cr(III)-layered double hydroxide. *Catal. Today* **2017**, *284*, 195–201. [CrossRef]

28. Ballarin, B.; Mignani, A.; Scavetta, E.; Giorgetti, M.; Tonelli, D.; Boanini, E.; Mousty, C. Ruta de síntesis para hidróxidos dobles en capas de nanopartículas de oro soportadas como catalizadores eficaces en la electro oxidación del metanol. *Langmuir* **2012**, *28*, 15065–15074. [CrossRef] [PubMed]

29. Zhang, H.; Chen, H.; Azat, S.; Mansurov, Z.; Liu, X.; Wang, J.; Su, X.; Wu, R. Super adsorption capability of rhombic dodecahedral Ca-Al layered double oxides for Congo red removal. *J. Alloy. Comp.* **2018**, *768*, 572–581. [CrossRef]

30. Ramos-Ramírez, E.; Tzompantzi-Morales, F.; Gutiérrez-Ortega, N.; Mojica-Calvillo, H.G.; Castillo-Rodríguez, J. Photocatalytic Degradation of 2,4,6-Trichlorophenol by MgO–MgFe$_2$O$_4$ Derived from Layered Double Hydroxide Structures. *Catalysts* **2019**, *9*, 454. [CrossRef]

31. Ramos-Ramírez, E.; Gutiérrez-Ortega, N.L.; Tzompantzi-Morales, F.; Barrera-Rodríguez, A.; Castillo-Rodríguez, J.C.; Tzompantzi-Flores, C.; Guevara-Hornedo, M.P. Photocatalytic Degradation of 2,4-Dichlorophenol on NiAl-Mixed Oxides Derivates of Activated Layered Double Hydroxides. *Top. Catal.* **2020**, *63*, 546–563. [CrossRef]

32. Jawad, A.; Peng, L.; Liao, Z.; Zhou, Z.; Shahzad, A.; Ifthikar, J.; Chen, Z. Selective removal of heavy metals by hydrotalcites as adsorbents in diverse wastewater: Different intercalated anions with different mechanisms. *J. Clean. Prod.* **2018**, *211*, 1112–1126. [CrossRef]

33. Setshedi, K.; Ren, J.; Aoyi, O.; Onyango, M.S. Removal of Pb(II) from aqueous solution using hydrotalcite-like nanostructured material. *Int. J. Phys. Sci.* **2012**, *7*, 63–72.

34. Dias, A.C.; Fontes, M.P.F. Arsenic (V) removal from water using hydrotalcites as adsorbents: A critical review. *Appl. Clay. Sci.* **2020**, *191*, 105615. [CrossRef]

35. Nong, L.; Xiao, C.; Jiang, W. Azo dye removal from aqueous solution by organic-inorganic hybrid dodecanoic acid modified layered Mg-Al hydrotalcite. *Korean J. Chem. Eng.* **2011**, *28*, 933–938. [CrossRef]

36. Mustapha, M.; Derriche, Z.; Denoyel, R.; Prevot, V.; Forano, C. Thermodynamical and structural insights of orange II adsorption by MgRAlNO3 layered double hydroxides. *J. Solid State Chem.* **2011**, *184*, 1016–1024. [CrossRef]

37. Sels, B.F.; de Vos, D.E.; Jacobs, P.A. Hydrotalcite-like anionic clays in catalytic organic reactions. *Catal. Rev.* **2001**, *43*, 443–488. [CrossRef]

38. Shakeel, M.; Arif, M.; Yasin, G.; Li, B.; Khan, H.D. Layered by Layered Ni-Mn-LDH/g-C3N4 Nanohybrid for Multi-Purpose Photo/electrocatalysis: Morphology Controlled Strategy for Effective Charge Carriers Separation. *Appl. Catal. B* **2018**, *242*, 485–498. [CrossRef]

39. Chen, X.; Yang, Z.; Wang, L.; Qin, H.; Tong, M. Synthesis of Hollow Spherical Zinc-Aluminum Hydrotalcite and Its Application as Zinc Anode Material. *J. Electrochem. Soc.* **2019**, *166*, A2589–A2596. [CrossRef]

40. Pan, L.; Huang, H.; Niederberger, M. Layered cobalt hydrotalcite as an advanced lithium-ion anode material with high capacity and rate capability. *J. Mater. Chem. A* **2019**, *7*, 21264–21269. [CrossRef]

41. Zhou, F.; Pan, N.; Chen, H.; Xu, X.; Wang, C.; Du, Y.; Li, L. Hydrogen production through steam reforming of toluene over Ce, Zr or Fe promoted Ni-Mg-Al hydrotalcite-derived catalysts at low temperature. *Energy Convers. Manag.* **2019**, *196*, 677–687. [CrossRef]

42. Garcia-Segura, S.; Brillas, E. Applied photoelectrocatalysis on the degradation of organicpollutants in wastewaters. *J. PhotoChem. Photobiol. C* **2017**, *31*, 1–35. [CrossRef]

43. Zhang, Y.; Cui, W.; An, W.; Liu, L.; Liang, Y.; Zhu, Y. Combination of photoelectrocatalysis and adsorption for removal of bisphenol A over TiO2-graphene hydrogel with 3D network structure. *Appl. Catal. B* **2018**, *221*, 36–46. [CrossRef]

44. Ghenaatgar, A.; Tehrani, R.M.A.; Khadir, A. Photocatalytic degradation and mineralization of dexamethasone using WO3 and ZrO2 nanoparticles: Optimization of operational parameters and kinetic studies. *J. Water Process Eng.* **2019**, *32*, 100969. [CrossRef]

45. Tavakoli-Azar, T.; Reza-Mahjoub, A.; Seyed-Sadjadi, M.; Farhadyar, N.; Hossaini-Sadr, M. Improving the photocatalytic performance of a perovskite ZnTiO3 through ZnTiO3@S nanocomposites for degradation of Crystal violet and Rhodamine B pollutants under sunlight. *Inorg. Chem. Commun.* **2020**, *119*, 108091. [CrossRef]

46. Kamarudin, N.S.; Jusoh, R.; Jalil, A.A.; Setiabudi, H.D.; Sukora, N.F. Synthesis of silver nanoparticles in green binarysolvent for degradation of 2,4-D herbicide:Optimization and kinetic studies Faculty. *Chem. Eng. Res. Des.* **2020**, *159*, 300–314. [CrossRef]

47. Zhang, S.; Lin, T.; Chen, W.; Xu, H.; Tao, H. Degradation kinetics, byproducts formation and estimated toxicity of metronidazole (MNZ) during chlor(am)ination. *Chemosphere* **2019**, *235*, 21–31. [CrossRef]

48. Tantawi, O.; Baalbaki, A.; El Asmar, R.; Ghauch, A. A rapid and economical method for the quantification of hydrogen peroxide (H2O2) using a modified HPLC apparatus. *Sci. Total Environ.* **2019**, *654*, 107–117. [CrossRef]

49. Pinkernell, U.; Effkemann, S.; Karst, U. Simultaneous HPLC Determination of Peroxyacetic Acid and Hydrogen Peroxide. *Anal. Chem.* **1997**, *69*, 3623–3627. [CrossRef] [PubMed]

50. Huang, Y.; Wang, L.; Chen, B.; Zhang, Q.; Zhu, R. Detecting hydrogen peroxide reliably in water by ion chromatography: A method evaluation update and comparison in the presence of interfering components. *Environ. Sci. Water Res. Technol.* **2020**, *6*, 2396–2404. [CrossRef]

51. Nakata, K.; Fujishima, A. TiO2 photocatalysis: Design and applications. *J. Photochem. Photobiol. C* **2012**, *13*, 169–189. [CrossRef]

52. Li, F.; Zhao, Y.; Liu, Y.; Hao, Y.; Liu, R.; Zhao, D. Solution combustion synthesis and visible light-induced photocatalytic activity of mixed amorphous and crystalline MgAl2O4 nanopowders. *Chem. Eng. J.* **2011**, *173*, 750–759. [CrossRef]

53. Sánchez-Cantú, M.; Galicia-Aguilar, J.A.; Santamaría-Juárez, D.; Hernández-Moreno, L.E. Evaluation of the mixed oxides produced from hydrotalcite-like compound's thermal treatment in arsenic uptake. *Appl. Clay. Sci.* **2016**, *121–122*, 146–153.

54. Nassar, M.Y.; Ahmed, I.S.; Samir, I. A novel synthetic route for magnesium aluminate (MgAl2O4) nanoparticles using sol–gel auto combustion method and their photocatalytic properties. *Spectrochim. Acta A* **2014**, *131*, 329–334. [CrossRef]

55. Bard, A.J.; Parsons, R.; Jordan, J. *Standard Potentials in Aqueous Solutions*; Marcel Dekker: New York, NY, USA, 1985.

56. Araña, J.; Peña Alonso, A.; Doña Rodríguez, J.M.; Herrera Melián, J.A.; González Díaz, O.; Pérez Peña, J. Comparative study of MTBE photocatalytic degradation with TiO2 and Cu-TiO2. *Appl. Catal. B* **2008**, *78*, 355–363. [CrossRef]

57. Irie, H.; Kamiya, K.; Shibanuma, T.; Miura, S.; Tryk, D.A.; Yokoyama, T.; Hashimoto, K. Visible light-sensitive Cu(II)-grafted TiO2 photocatalysts: Activities and X-ray absorption fine structure analyses. *J. Phys. Chem. C* **2009**, *113*, 10761–10766. [CrossRef]

58. Jalife-Jacobo, H.; Feria-Reyes, R.; Serrano-Torres, O.; Gutiérrez-Granados, S.; Peralta-Hernández, J.M. Diazo dye Congo red degradation using a Boron-doped diamond anode: An experimental study on the effect of supporting electrolytes. *J. Hazard. Mater.* **2016**, *319*, 78–83. [CrossRef] [PubMed]

59. Thomas, M.; Naikoo, G.A.; Sheikh, M.U.D.; Bano, M.; Khan, F. Effective photocatalytic degradation of Congo red dye using alginate/carboxymethyl cellulose/TiO2 nanocomposite hydrogel under direct sunlight irradiation. *J. Photochem. Photobiol. A* **2016**, *327*, 33–43. [CrossRef]

60. Bhagwat, U.O.; Wu, J.J.; Asiri, A.M.; Anandan, S. Sonochemical Synthesis of Mg-TiO2 nanoparticles for persistent Congo red dye degradation. *J. PhotoChem. Photobiol. A* **2017**, *346*, 559–569. [CrossRef]

61. Ullah, I.; Haider, A.; Khalid, N.; Ali, S.; Ahmed, S.; Khan, Y.; Ahmed, N.; Zubair, M. Tuning the band gap of TiO$_2$ by tungsten doping for efficient UV and visible photodegradation of Congo red dye. *Spectrochim. Acta A* **2018**, *204*, 150–157. [CrossRef]
62. Curkovic, L.; Ljubas, D.; Juretic, H. Photocatalytic decolorization kinetics of diazo dye Congo red aqueous solution by UV/TiO$_2$ nanoparticles. *React. Kinet. Mech. Catal.* **2010**, *99*, 201–208.

Article

MgCr-LDH Nanoplatelets as Effective Oxidation Catalysts for Visible Light-Triggered Rhodamine B Degradation

Susanginee Nayak * and Kulamani Parida *

Centre for Nano Science and Nano Technology, Institute of Technical Education and Research (ITER), Siksha 'O' Anusandhan Deemed to be University, Bhubaneswar 751030, Odisha, India
* Correspondence: susanginee@gmail.com (S.N.); kulamaniparida@soauniversity.ac.in or paridakulamani@yahoo.com (K.P.); Tel.: +91-674-2351777 (K.P.)

Abstract: In this work, we successfully exfoliated MgCr-(NO_3^-) LDH with large purity by a simple formamide method followed by post-hydrothermal treatment and characterized by different physicochemical techniques. The UV-DRS study persuades the red-shifted absorption band and suitable band gap of MgCr-(NO_3^-) LDH for optimum light harvestation ability related to the optical properties. Alternatively, the production of elevated photocurrent density of MgCr-(NO_3^-) LDH (3:1, 80 °C) in the anodic direction was verified by the LSV study, which further revealed their effective charge separation efficacy. These MgCr-LDH nanosheets (3:1, 80 °C) displayed the superior Rhodamine B (RhB) degradation efficiency of 95.0% at 0.80 kW/m^2 solar light intensity in 2 h. The tremendous catalytic performances of MgCr-LDH (3:1, 80 °C) were typically linked with the formation of surface-active sites for the charge trapping process due to the presence of uncoordinated metallocenters during the exfoliation process. Furthermore, the maximum amount of the active free atoms at the edges of the hexagonal platelet of MgCr-LDH causes severance of the nanosheets, which generates house of platelets of particle size ~20–50 nm for light harvestation, promoting easy charge separation and catalytic efficiency. In addition, radical quenching tests revealed that h^+ and •OH play as major active species responsible for the RhB degradation.

Keywords: LDH; exfoliation; nanosheets; oxidation; photocatalysts

Citation: Nayak, S.; Parida, K. MgCr-LDH Nanoplatelets as Effective Oxidation Catalysts for Visible Light-Triggered Rhodamine B Degradation. *Catalysts* **2021**, *11*, 1072. https://doi.org/10.3390/catal11091072

Academic Editors: Ioan-Cezar Marcu and Octavian Dumitru Pavel

Received: 12 July 2021
Accepted: 31 August 2021
Published: 3 September 2021

1. Introduction

Layered structure material represents an emerging class of two-dimensional (2D) materials that acquire sheet-like morphology with the thickness of single or few-layered atoms [1–3]. The importance of layered materials is credited due to their rich interlayer chemistry, such as intercalation and ion exchange properties, which modified their electronic and optical properties. There are lot of many-layered solid materials with few-layer, single-layer, or stacked-layered structures that have been identified as such layered double hydroxide (LDH) [3–9], layered metal hydroxides [10], layered graphene oxide [4,9], and layered graphitic carbon nitride (g-C_3N_4) [11–14], in photocatalytic dye degradation and energy conversion reactions. Amongst these, LDHs have been considered as one of the most efficient photocatalysts in producing clean H_2 energy along with environmental abatement following a green technological aspect. Although the direct semiconducting capability of LDH is very much restricted, its lamellar structure smooths the photoinduced electron transfer from bulk to the surface, which is useful in water splitting, and the photogenerated holes are thereof involved in the pollutant degradation [15,16]. Normally, LDHs represents a group of anionic layered materials consisting of positively charged layers with interlayer anions and H_2O molecules for charge recompense, which is widely used as catalysts, catalyst support, ion exchangers, and electrocatalytic and photocatalytic materials [5]. The generic formula of LDHs is $[M(II)_{1-x}M(III)_x(OH)_2]^{x+}[A^{n-}{}_{x/n} \cdot mH_2O]^{x-}$, where M(II) and M(III) specify the metal atoms and A^{n-} represents the anions, n is the charge upon interlayer anion, m is amount of H_2O, and x = $M^{III}/(M^{II} + M^{III})$. The atomic arrangement of LDHs is correlated to that of brucite-like layers of $Mg(OH)_2$, and the development of excess charge owing to the inclusion of M(III) cations is balanced by the intercalation of A^{n-} together with the presence of H_2O in the interlayer gallery. Stable LDHs requisites M(II)/M(III) ratio among 2:1, 3:1, and 4:1, where x lies in the range of 0.20 < x < 0.33. However, the bulk LDHs is restricted to certain applications owing to the detachment of the inner portion of the layer host structure. 2D LDH usually crystallizes into small hexagonal platelets or house-of-card morphology owing to their 2D layered structure possessing large surface area and could be simply prepared with NO_3^- anions by co-precipitation method. An intriguing feature of 2D material is their potential to exfoliate/delaminate into resultant uni/multi-lamellar crystallites or nanosheets (NS) having unusual structural aspects, which are of critical 2D anisotropy [17]. According to this reason, LDHs could be exfoliated into single/multiple 2D NS of thickness ~1 nm, and that may be used as building block for various functional nanoheterostructures materials with positively charged NS framework in comparison to positively charged montmorillonite framework [18].

In order to optimize the formamide exfoliation procedure of strongly bound positively charged NS of LDHs, several factors have been verified in the recent days; e.g., chemical nature and composition of the 2D-brucite-like layers [19,20], organic solvents systems (acrylates, butanol, toluene, etc.) [21,22], charge of anions [22], drying procedure [23], water content [24], hydrothermal treatment [25], and morphological variation [20]. A challenge for the thorough investigations is associated with the sufficient control on the interlayer interaction with brucite-like layers, stacking fault, and inclusion of polytype pattern. The substitution of M(II) by M(III) cation in binary LDHs induces excess layer charge and affects the interaction with the polar solvent formamide molecules. For illustration, $Mg_{1-x}Al_x(OH)_2(glycinate)_x \cdot mH_2O$, x = 0.25 exfoliates better in formamide [19]. During the exfoliation process, most specifically, the intercalated or surface bound solvent molecules connected via H-bonding with the interlayer anions and the –OH group of the metal hydroxide layers and electrostatically interacted with the positively charged 2D LDH NS. The interlayer gallery expansion is dependent upon the layer charge density and at low layer charge density; the basal spacing is tightly bound with horizontal orientation of NO_3^- ions. Alternatively, the NO_3^- anions and H_2O molecules are oriented out of the horizontal plane at elevated layer charge density, enhancing gallery height extension. However, exfoliation is triggered by hydration process but much improved by tuning hydrothermal temperature, time, and at fixed compositions, e.g., x = 0.25.

Herein, the present work is related to reference [7], and our aim was to promote the exfoliation of MgCr-LDH (Mg(II): Cr(III) = 2:1, 3:1, and 4:1) using the formamide method followed by post-hydrothermal treatment (70, 80, and 90 °C). By adopting this robot synthetic methodology, the MgCr-(NO_3^-) LDH (3:1, 80 °C) preserved house-of-cards or nano-platelet-like morphology, which clearly disclosed a high degree of exfoliation of the materials. Consequently, MgCr-LDH (3:1, 80 °C) displayed 95.0% RhB degradation in 2 h under visible light exposure.

2. Results and Discussion

2.1. XRD Characterization

PXRD proceeded to recognize the crystal structure, crystallinity, and phase purity of the materials [26]. In the process of hydrolysis of Mg/Cr nitrate solution by aqueous NaOH solution and formamide, the crystal clear MgCr-LDH dispersion displayed the occurrence of colloidal MgCr-LDH. Then, centrifuging the colloidal MgCr-LDH dispersion yielded an MgCr-LDH gel containing freely and arbitrarily stacked LDH NS. Later, post-hydrothermal treatment of these MgCr-LDH gels produced uni/multilayer NS. Figure 1a (i–iii), displays the variation of characteristic diffraction peaks of MgCr-LDH (2:1), MgCr-LDH (3:1), and MgCr-LDH (4:1) without hydrothermal treatment along with hydrothermal-treated samples at different temperatures (70, 80, and 90 °C), which clearly resembled exfoliated LDH NS with a small enough crystalline nature (Figure 1a (iv–vi)). The XRD patterns of the as-synthesized MgCr-LDH (3:1) samples without hydrothermal treatment exhibit broadening of peak at $2\theta = 30°$ to $60.4°$ owing to the presence of formamide and corresponds to the (012) and (110) planes, respectively (Figure 1a (ii)). The absence of (003) and (006) characteristic planes in MgCr (3:1) material reveals the exfoliation of bulk MgCr-LDH. Similarly, MgCr-LDH (2:1) and MgCr-LDH (4:1) follow the similar XRD pattern as that of MgCr-LDH (3:1) with a slight appearance of the (006) plane in MgCr-LDH (4:1). The (006) planes in the PXRD pattern of MgCr-LDH (3:1, 80 °C) at lower $2\theta \sim 11$–23° revealed broad and symmetrical basal reflections, while the spiky and asymmetrical reflections of (012) and (110) were expressed at higher $2\theta \sim 34$–66°, respectively. The broad (006) and spike (110) planes of MgCr-LDH (3:1, 80 °C) could be correlated with the interlayer height differences and stacking disorder of distinct NS in materials [27,28], whereas the rest of crystalline plane was absence in the PXRD pattern of hydrothermally-treated MgCr-LDH (Figure 1b (v)). The interlayer spacings (d) were deliberate via Braggs law, $n\lambda = 2d \sin(\theta)$, in which $n = 1$, λ is the wavelength of the target and θ is the angle of incidence. The d value of MgCr-LDH (3:1, 80 °C) was found to be 0.46 nm corresponding to d (006) plane and NO_3^- as interlayer anion. Alternatively, the d value of MgCr-LDH (3:1) without hydrothermal was found to be 0.16 nm corresponding to d (110) plane as there were absence of (003) and (006) plane in the PXRD pattern of the material. Hence, hydrothermal temperature played a major role and dramatically altered the crystal growth of the exfoliated MgCr-LDH NS, and the utmost crystallinity was noted at 80 °C for MgCr-LDH (3:1) [29].

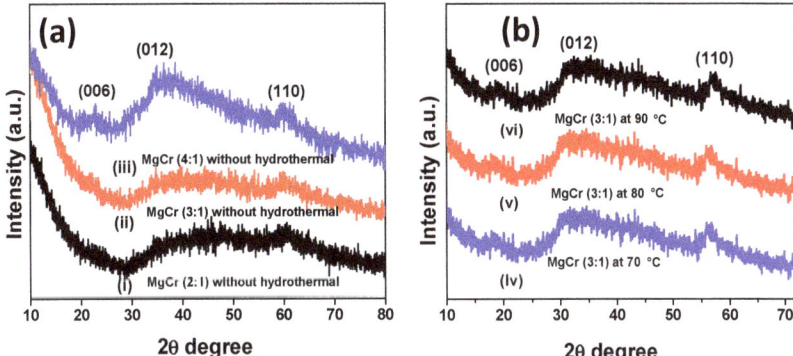

Figure 1. XRD spectra of exfoliated MgCr-LDH (2:1), MgCr-LDH (3:1), and MgCr-LDH (4:1) without hydrothermal ((**a**) (i–iii)); exfoliated MgCr-LDH (3:1) at different hydrothermal temperature of 70, 80, and 90 °C ((**b**) (iv–vi)).

2.2. Morphological Analysis

TEM images were acquired to trace out the arrangement of the LDH uni/multi-lamellar NS (Figure 2) [30]. In the moderately dissipate area, the MgCr-LDH lamellae (3:1, 80 °C), possess platelet like shape oriented at random with house-of-card morphology (Figure 2a,b). Contrary to the typical hexagon, at mild hydrothermal temperature of 80 °C, the elevated amount of dynamic in free atoms at the edges of the hexagonal platelet causes intersection of the NS, which generates house of platelets type morphology of MgCr-LDH as shown in Figure 2b. Figure 2c reveals distinct lattice distance ~0.38 nm in MgCr-LDH (3:1, 80 °C), which is approximately related to the (006) plane of the 2D MgCr-LDH nanocrystals. The particle diameter of MgCr-LDH (3:1, 80 °C) NS is expected to be average size of 20–50 nm. Furthermore, the morphology of MgCr-LDH (3:1, 80 °C) NS was verified with FESEM analysis. In Figure 2c,d, the morphology of MgCr-LDH (3:1, 80 °C) was composed of irregular matrix and aggregated into larger nanoparticles with rough and porous surface during mild hydrothermal treatment at 80 °C. This might be due to the hydrothermal treatment effect on MgCr-LDH (3:1) materials. Nevertheless, the TEM image of MgCr-LDH (3:1, 80 °C) easily revealed their exact sheet-like morphology of hexagonal nanoplatelets, which are of distinctive features of exfoliated LDH material, and approximately matching with the XRD outcome. Additionally, the EDX analysis of MgCr-LDH (3:1, 80 °C) as shown in Figure 2c confirms that the system contains all the elements like Mg, Cr, and O without any impurity, which proves its compositional purity.

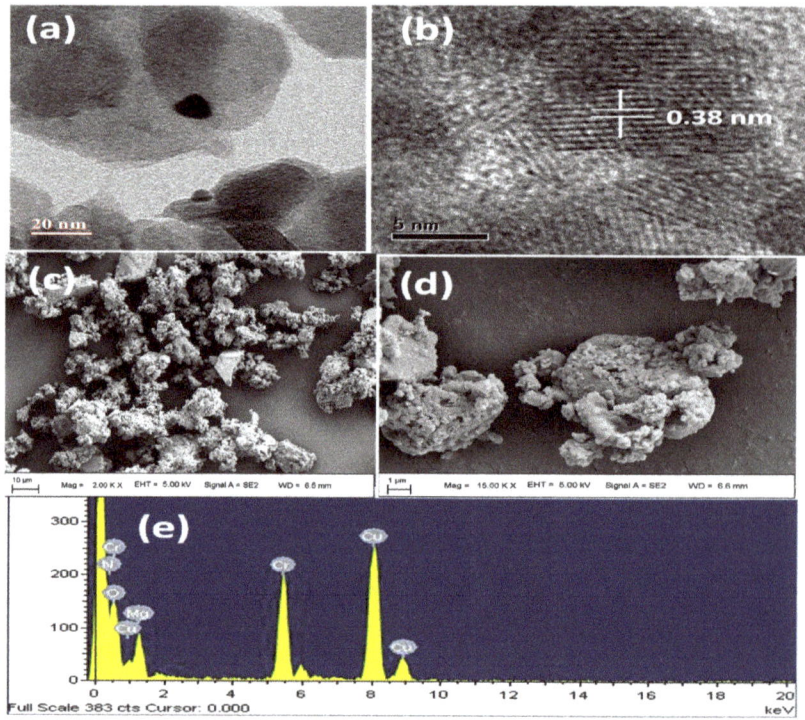

Figure 2. (**a**) TEM morphology of exfoliated MgCr-LDH (3:1, 80 °C), (**b**) lattice fringe of MgCr-LDH (3:1, 80 °C) as detected from HR-TEM image analysis, (**c,d**) FESEM morphology of MgCr-LDH (3:1, 80 °C), and (**e**) EDX spectrum of MgCr-LDH (3:1, 80 °C).

2.3. Optical Study

The electronic and optical characteristic properties of the exfoliated MgCr-LDH (2:1, 3:1, 4:1) NS without hydrothermal treatment and MgCr-LDH (3:1) NS with hydrothermal treatment (70, 80, and 90 °C) were determined by UV-Vis DRS spectroscopy technique (Figure 3a). The Figure 3a clearly reveals that each MgCr-LDH photocatalyst possess sufficient potential to absorb visible light and displayed remarkable increase in the absorption intensity, which indicated the enhanced excitonic charge pairs partition efficiency of the resultant materials. In concise, MgCr-LDH (3:1) without hydrothermal displayed broad absorption band within 300–500 nm owing to the ligand to metal charge transfer (LMCT) as $O2p \rightarrow Cr\text{-}3d\text{-}t_{2g}$ and absorption band at 500–700 nm is assigned to the 2Eg (D) $\rightarrow$ 2T$_{2g}$ spin allowed transition of Cr^{3+} in the tetrahedral coordination sites [9,26]. Specifically, the gradual increases in the red shift of the absorption band in MgCr-LDH (3:1) are owing to the increase in Cr^{3+} content in the brucite-like host layers. The exfoliation of MgCr-LDH during mild hydrothermal treatment causes severance and folding of NS with formation of tunnels of hexagonal plates, which acts as light harvestation antenna and triggers electronic transition in the respective orbital of the metallocenters in the catalysts. Importantly, the absorption edge of MgCr-LDH (3:1, 80 °C) moved towards a longer wavelength with broad and intense absorption from 400 to 800 nm, which is due to the reduced thickness of the exposed atomic sites of the nanolayers that minimized the charge transfer distance and endorsed for effective compilation of charge concentrated over the conductive NS. Furthermore, the inimitable structure in MgCr-LDH (3:1, 80 °C) endorsed light to scatter frequently inside the structure to increase the optical distance and enhance light absorption capacity. In addition, the atomically reduced thickness of the exposed atomic sites of the uni/multi-nanolayers of MgCr-LDH (3:1, 80 °C) NS is another factor responsible for the broad and intense visible light absorption capability within 400–800 nm.

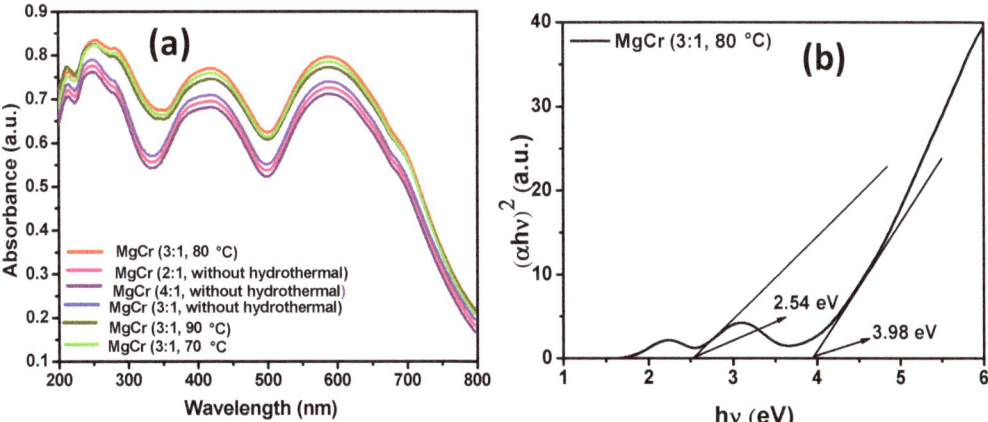

Figure 3. (**a**) UV-Vis DRS spectra of the series of exfoliated MgCr-LDH (2:1, 3:1, 4:1) without hydrothermal and MgCr-LDH (3:1) with hydrothermal treatment of 70, 80, and 90 °C. (**b**) Band gap energy of MgCr-LDH (3:1, 80 °C) as derived from Kubelka–Munk equation through Tauc plot.

The band-gap energy of MgCr-LDH (3:1, 80 °C) can be deliberate by Tauc plot as reported earlier following the Kubelka–Munk Equation (1) [5–13].

$$(\alpha h\nu)^{1/n} = A(h\nu - E_g) \tag{1}$$

where ν and α are the light frequency and adsorption coefficient, respectively

Figure 3b shows two types of band-gap energy resides in MgCr-LDH (3:1, 80 °C), which is of 2.54 (E_g1) and 3.98 eV (E_g2) due to the presence of LDH phase, and falls under directly allowed transition. In addition, the E_g1 can also be assigned to the direct electronic transition from O2p to Crnd levels of MgCr-LDH [12]. Furthermore, the E_g2 of MgCr-LDH samples can be assigned to the existence of electronic transition from O2p to Mgns/np [12].

2.4. FTIR Study

FTIR plot of MgCr-LDH (3:1, 80 °C) is represented in Figure 4 [31]. The sample displayed a broad absorption band at around 3374 cm^{-1}, which signifies the occurrence of -OH group of H_2O molecules [32]. Similarly, the two distinct bands at 1639 and 1440 cm^{-1}, corresponded to the stretching mode of vibration in Mg-O and bending mode of vibration in adsorbed H_2O over MgCr-LDH surface [32]. The absorption band at lower frequency level of 1000 cm^{-1} corresponds to M–O (Mg–O, Cr–O) and M–O–M (Mg–O–Cr) vibrations, respectively [33]. This functional band gives strong evidence of the formation of MgCr-LDH samples.

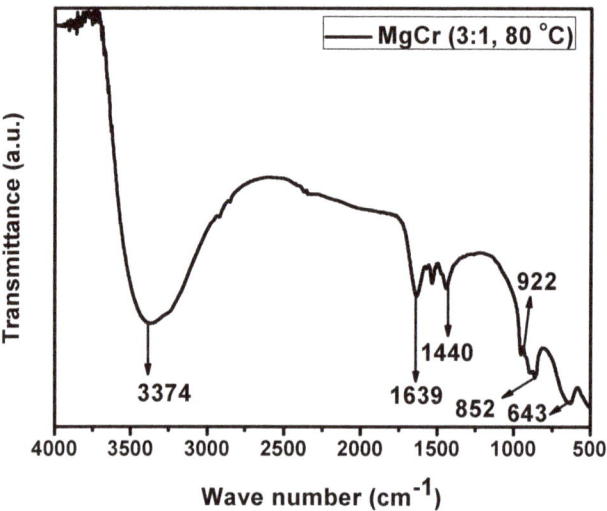

Figure 4. FTIR spectrum of exfoliated MgCr−LDH (3:1, 80 °C).

2.5. Electrochemical Study

LSV study of MgCr-LDH (3:1, 80 °C) samples was conducted in a potential panel of −1.0 to 1.2 V, using 0.1 M Na_2SO_4, and scan rate of 10 mV·s^{-1} to reveal the photocurrent retaliation of the as-synthesized catalysts. Figure 5 exposes the photocurrent capacity measures under dark and light environments. As shown in Figure 5, MgCr-LDH (3:1, 80 °C) could be able to produce current density of 1.20 µA/cm^{-2} under light exposure. The progress of the photocurrent in the anodic direction shows that all the MgCr-LDH samples filling the properties of n-type semiconductor [11]. The oxidation peak intensity decreases gradually at optimal loading density of Cr^{3+} in MgCr-LDH (3:1) and further mild hydrothermal treatment at 80 °C provides compact NS structure of MgCr-LDH, where the oxidation peak of $Cr \rightarrow Cr^{3+}$ decreases gradually and disappear, suggesting the enhance stability of MgCr-LDH (3:1, 80 °C) in the nanostructure. The enhanced stability of MgCr-LDH (3:1, 80 °C) causes superior visible light driven photocatalytic activity. The flat band potential of the entire MgCr-LDH sample was detected at −0.60 V vs. Ag/AgCl, pH = 6.5. The flat band potential is directly correlated to baseline of the conduction band (CB) of an n-type semiconductor [11]. As 2.54 eV is the primary E_g of MgCr-LDH, the corresponding valence band maximum (VB) was +1.94 V. Particularly, electrode potential was transformed to NHE by the subsequent Equation (2) [34]:

$$E(NHE) = E(Ag/AgCl) + E^0(Ag/AgCl) + 0.059pH \tag{2}$$

Hence, CB and VB of MgCr-LDH in NHE scale were determined to be −0.01 and +2.53 V. Similarly, the dark current measurement of MgCr-LDH (3:1, 80 °C) showed a slight incremental current density as compared to light current density.

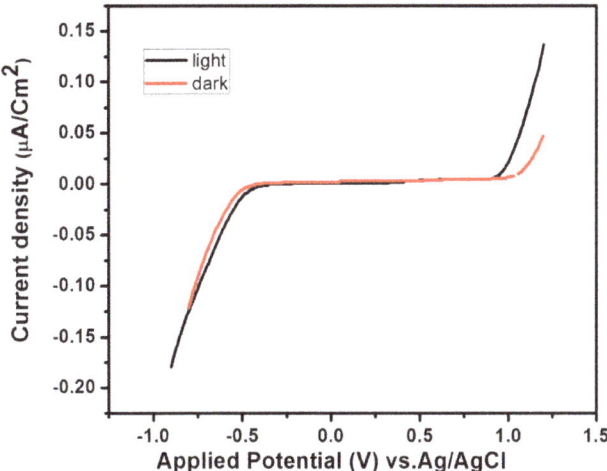

Figure 5. Light and dark current density of MgCr−LDH (3:1, 80 °C).

2.6. Photocatalytic RhB Degradation Activity

The RhB degradation activity of MgCr-LDH samples was studied under solar light exposure. Self-degradation study of RhB was performed by exposing under the solar energy for 30 min, and the results showed that the RhB self-degradation was roughly insignificant. Furthermore, the adsorption study of the catalyst was performed under dark condition for 30 min. The RhB degradation study was activated by dispersion of 0.03 g of the catalyst in 20 ppm of RhB (20 mL) under solar energy for 120 min. The exfoliation of MgCr-LDH under mild hydrothermal condition generates uncoordinated metallocenters and dense amount of free atoms at the edges of hexagonal platelet responsible for oxygen related vacancies and causes intersection of the NS for enhancing light harvestation ability of the materials and corresponding exciton separation efficiency directly or indirectly responsible for the photooxidation of RhB to non-toxic products. The RhB photodegradation activities of all of the as-synthesized MgCr-LDH samples were measured in accordance with the following Equation (3):

$$\text{Photodegradation rate (\%)} = (C_0 - C/C_0) \times 100 \qquad (3)$$

The order of intensification of RhB degradation for series of MgCr-LDH was 75% (MgCr 4:1), 85% (MgCr 2:1), 90% (MgCr 3:1), 93% (MgCr 3:1, 70 °C), 95% (MgCr 3:1, 80 °C), and 91% (MgCr 3:1, 90 °C), respectively (Figure 6a). These outcomes evidently show that (MgCr 3:1, 80 °C) exhibits enhanced RhB degradation. In addition, an excess substitution of Cr^{3+} to Mg^{2+} (MgCr 4:1) results decreases in activity because of the blocking of the reactive phases of MgCr-LDH. The rate of RhB degradation and related spectral changes of absorbance are depicted in Figure 6b. MgCr (3:1, 80 °C) shows excessive potential for maintaining higher stability approximately to the extent of 3 cycles (Figure 6c) than other as-prepared materials as discussed so far. The degradation activity results are in fine matching with the characterization results.

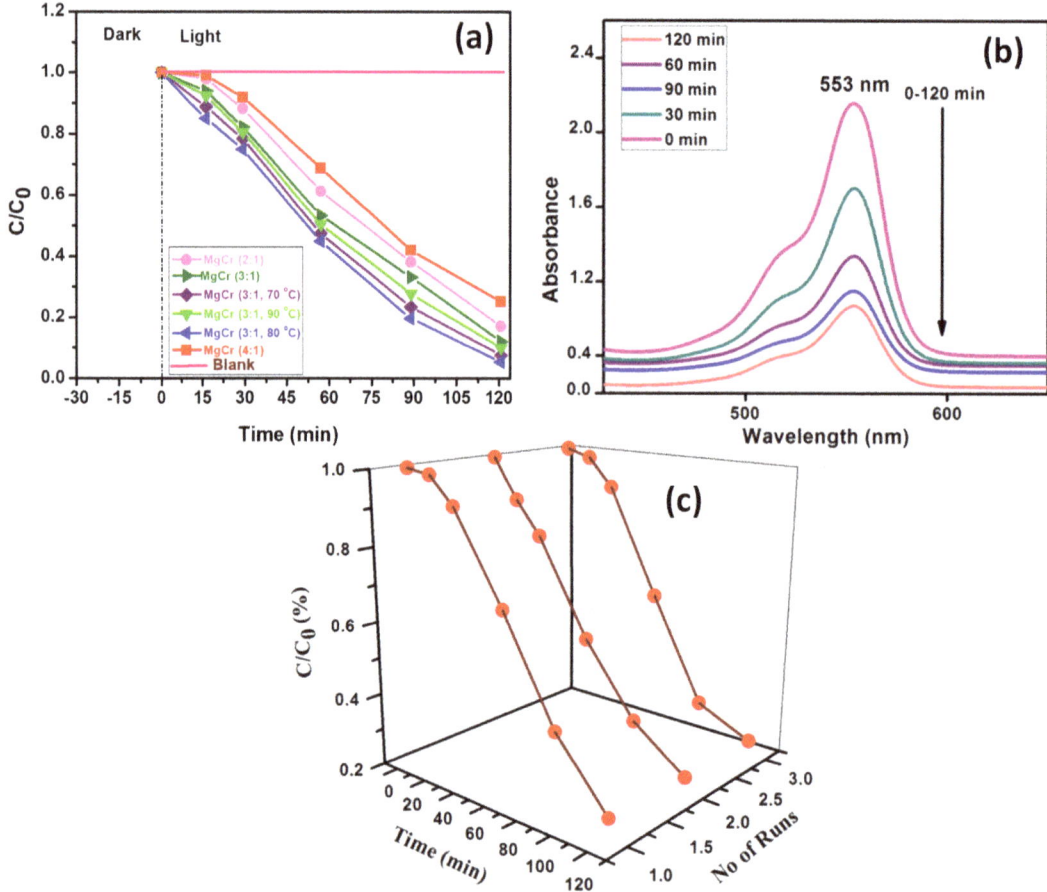

Figure 6. (**a**) Photocatalytic degradation rate of RhB over variant of MgCr−LDH samples. (**b**) Spectral alteration of RhB absorption in different time interval by MgCr 3:1, 80 °C. (**c**) Stability test of MgCr 3:1, 80 °C for three different cycles.

2.7. Kinetics of the RhB Degradation

The kinetics of RhB degradation follows pseudo-first-order and the experimental data were fitted to the Langmuir−Hinshelwood kinetic model using the following Equations (4) and plotted in Figure 7a:

$$\ln C_0/C = kt \tag{4}$$

The k denoted as the apparent rate constant. The rate constant (k) of the RhB degradation reaction for different MgCr-LDH samples was calculated by linear fitting of the ln (C_0/C) vs. time plot (Figure 7a) and consequently, the measured slope of ln (C_0/C) vs. time plot provides the value of k. The k value depicted in Table 1 clearly shows that RhB degradation for MgCr-LDH based samples follows pseudo-first order kinetics in Langmuir−Hinshelwood model. The fitted line parameters related to the regression coefficient (R^2) are also given in Table 1.

The superior rate constant value of MgCr 3:1, 80 °C (0.02361 min^{-1}), reveals its great potential as unilamellar NS towards environmental oxidative reactions.

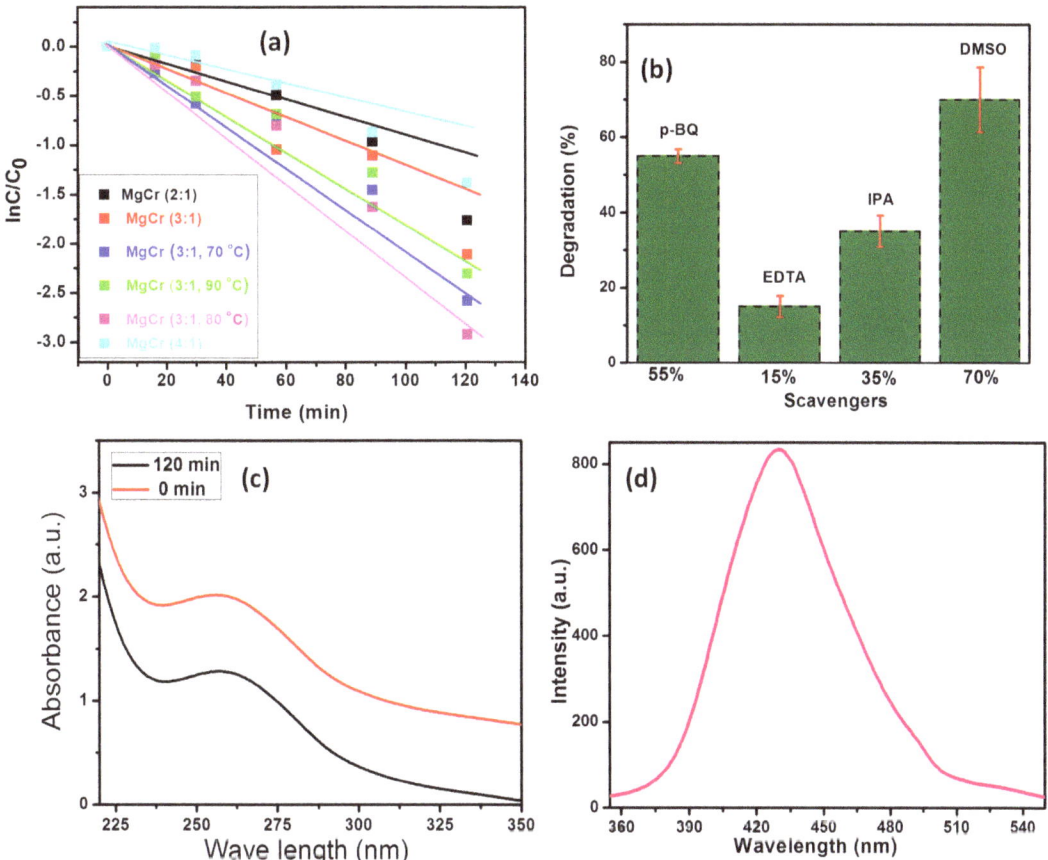

Figure 7. (**a**) Kinetics of RhB degradation by linear fitting of C_0/C vs. time using a series of exfoliated MgCr−LDH samples. (**b**) Histogram showing % of degradation using different scavenging agents. (**c**) Absorbance plot of NBT by MgCr−LDH (3:1, 80 °C). (**d**) PL spectra of MgCr−LDH (3:1, 80 °C) by TA.

Table 1. Rate constant and regression co-efficient value of exfoliated MgCr-LDH NS.

Catalyst	Rate Constant (min^{-1})	R^2
MgCr (2:1)	0.01467	0.93
MgCr (3:1)	0.01739	0.92
MgCr (3:1, 70 °C)	0.01998	0.93
MgCr (3:1, 80 °C)	0.02361	0.93
MgCr (3:1, 90 °C)	0.01836	0.94
MgCr (4:1)	0.01193	0.94

2.8. Scavenger Study for the Radicals

The scavengers taken active part in the RhB degradation study was examined by distinct trapping reagents as para-benzoquinone (p-BQ), dimethyl sulfoxide (DMSO), iso-propanol (IPA), and ethylenediaminetetraacetic acid (EDTA) for scavenging superoxide ($\bullet O_2^-$), electron (e^-), hydroxyl ($\bullet OH$), and hole (h^+) radicals, sequentially. In the experiment process, 5 mM of each trapping agent was incorporated to 20 mL of 20 ppm RhB added with 0.03 g of catalyst and set for degradation reactions. The scavenger test results shows an increased rate of RhB degradation (15, 35, 55, 70%) with addition of EDTA, IPA, p-BQ, and DMSO scavenging reagents, as depicted in Figure 7b. This result shows the participation of hole and hydroxyl radicals as primary and superoxide as secondary active species for RhB degradation process.

2.9. Confirmatory Test for •O_2^- Radicals

The •O_2^- radical was confirmed via a nitroblue tetrazolium (NBT) test [35]. Then, NBT was used to determine dense of •O_2^- and disclose the photodegradation efficiency of MgCr-LDH. Figure 7c shows no such extent of variation in the NBT concentration by MgCr-LDH, before and after the RhB photodegradation reactions, which reveals that CB potential of MgCr-LDH is not sufficient to directly produce •O_2^- radicals.

2.10. Confirmatory Test for •OH Radicals

Terephthalic acid (TA) PL probe was utilized for the detection of •OH radical [36]. TA directly reacts with the •OH radical, producing 2-hydroxyterephthalic acid (TAOH) by emission band at 426 nm for excitation at 315 nm. Then, a ~0.02-g catalyst was supplied to 0.004 M of NaOH solution consisting of 20 mL of TA, and subsequently the suspension was manifest to solar energy for 2 h. Afterward, intense PL peak of TAOH solution quantify the •OH radical formed through photooxidation reactions. The maximum intensity of TAOH peak extends the highest percentage of •OH formation. Figure 7d shows the formation of primary •OH radicals in RhB degradation catalyzed by MgCr-LDH.

2.11. Mechanism of RhB Degradation by MgCr-LDH

The mechanistic path of RhB degradation by using MgCr-LDH catalyst (Figure 8), could be clearly elucidated in terms of structural and morphological features, together with band gap sequence, and active sites, which correlated to the formation of exfoliated NS with photoinduced carrier charge separation for enhanced activity. The apparent enhancement in the photocatalytic RhB degradation over MgCr-LDH is owing to the formation of constant layer charge density, morphological aspects by mild hydrothermal temperature, and surface-active sites during the exfoliation process. Normally, the exfoliation rate decreases at a lower x = 0.2, due to tight interlayer gallery for which the polar interactions towards formamide are reasonable owing to low layer charge density. Though Columbic/electrostatic interactions are considered as an extraneous parameter in LDH containing divalent anion that limits the rate of intercalation, but we noticed the well exfoliation of MgCr-(NO_3^-) LDH at x = 0.3, and provided maximum space for formamide intercalation and consequently increases the basal spacings. It is further noted that high layer charge density could certainly result in strong interaction with formamide, and in this energy balance process, the electrostatic interactions may stabilize the LDH compound. Further MgCr-(NO_3^-) LDH at x = 0.3 provided platelets such as morphology as revealed from TEM image. It is also expected that the degree of exfoliation of $Mg^{II}_{0.9}Cr^{III}_{0.3}$ $(OH)_2(NO_3)_{0.3}$. nH_2O reached to approximately complete exfoliation by post-hydrothermal treatment [13]. Further XRD also revealed a clear node of (006) plane considerably on hydrothermal post-treatment, in context to Cr(III) proportion. This shows a homogeneous allocation of H_2O molecules and out of plane NO_3^- ion orientation and causes interlace crystallites. The LSV analysis also confirmed that MgCr-LDH (3:1, 80 °C) possess an intrinsic n-type characteristic properties of materials. By utilizing the Tauc plot (Figure 3b) derived from the Kubelka–Munk equation and LSV plot (Figure 5), the CB and VB position of MgCr-LDH (3:1, 80 °C) was found to be −0.01 and +2.53 V, corresponding to E_g of 2.54 eV, respectively. Further, MgCr-LDH (3:1, 80 °C)-acquired nano-platelets slanted at random (TEM image) and these morphological features of MgCr-LDH harvests light energy with reflection at the core of tunneling of intersected NS for superior activities. Under visible light exposure, photoexcited charge pairs were generated over the surface of MgCr-LDH. In this way, the migration and separation of photoinduced carriers charge pairs is due to the presence of surface-active sites at nanometric region for electron and hole hopping process. However, the CB position of MgCr-LDH (−0.01 V vs. NHE) is insufficient for the production of $\bullet O_2^-$ and $\bullet HO_2$ ($E^{\ominus}(O_2/\bullet O_2^-) = -0.33$ eV vs. NHE) and $E^{\ominus}(O_2/\bullet HO_2, -0.05$ eV vs. NHE) [15,37]. Alternatively, the depth VB position of MgCr-LDH (+2.53 V vs. NHE) was relatively sufficient for the direct oxidation reaction of holes with adsorbed H_2O molecules to form $\bullet OH$ radicals $E^{\ominus}(\bullet OH/OH^- = +1.99$ eV vs. NHE) [38]. These concepts were verified by the scavenger test, where h^+ were of the primary active-species in MgCr-LDH, which were responsible for the photodegradation of RhB. Consequently, these $\bullet OH$ radicals reacted with RhB over the surface of MgCr-LDH (3:1, 80 °C) to produce non-toxic products.

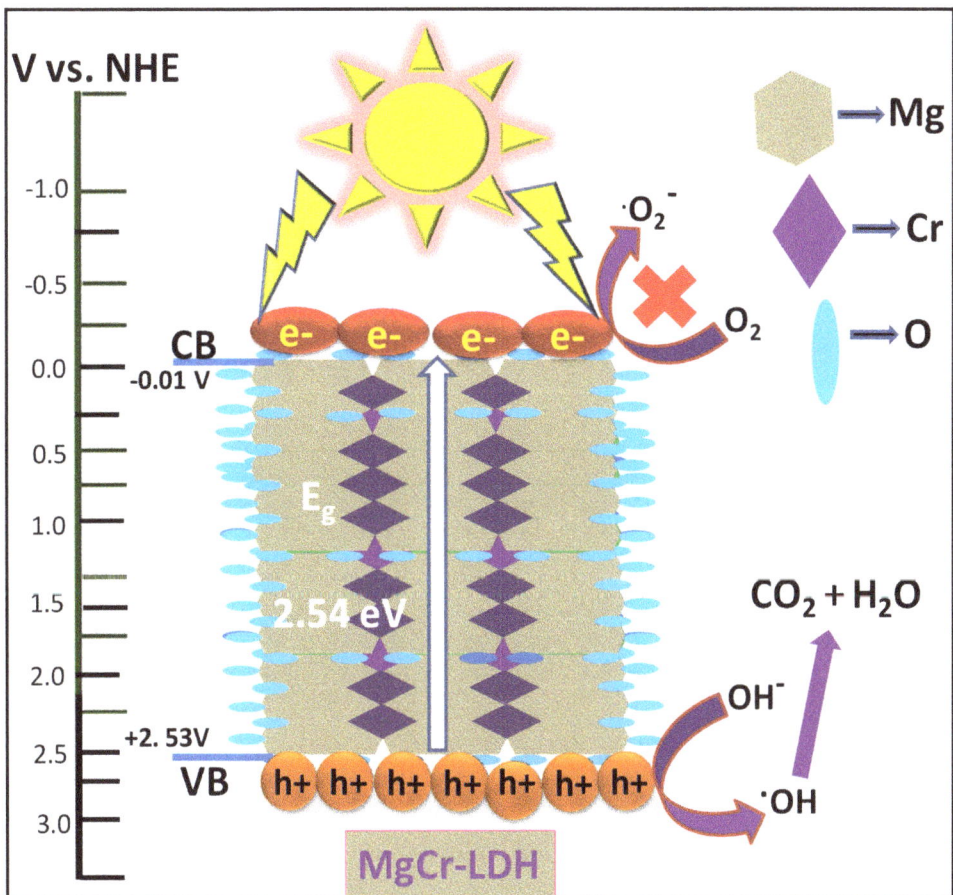

Figure 8. Plausible mechanism of charge separation by MgCr−LDH (3:1, 80 °C) for photocatalytic RhB degradation.

Based on the activities in scavengers test along with band edge position, the mechanism of charge separation in MgCr-LDH (3:1, 80 °C) for enhanced RhB degradation could be explained by following Equations (5)–(8):

$$MgCr + h\nu \rightarrow MgCr\,(h^+) + MgCr\,(e^-) \tag{5}$$

$$OH^- + MgCr\,(h^+)\,_{VB} \rightarrow \bullet OH \tag{6}$$

$$MgCr\,(h^+) + H_2O \rightarrow \bullet OH + H^+ \tag{7}$$

$$RhB + \bullet OH \rightarrow \text{several steps} \rightarrow \text{non-toxic products} \tag{8}$$

3. Experimental Section

3.1. Chemicals

Mg(NO$_3$)$_2$·6H$_2$O (98%, Sigma–Aldrich, India), and Cr(NO$_3$)$_3$·9H$_2$O (98%, Sigma–Aldrich), 23 vol% formamide (Sigma–Aldrich), anhydrous NaOH (98%, Sigma–Aldrich), were directly used for reaction. The mixed metal salt solutions were prepared by using deionized water.

3.2. Synthesis of Exfoliated MgCr-LDH NS by Formamide Method (2:1, 3:1, and 4:1)

Firstly, exfoliated MgCr-LDH gel was synthesized by co-precipitation method. In a distinct synthetic process, 20 mL of solution of mixed metal nitrate containing Mg (NO$_3$)$_2$·6H$_2$O (0.020 M) with Cr (NO$_3$)$_3$·9H$_2$O (0.010 M), Mg (NO$_3$)$_2$·6H$_2$O (0.030 M) with Cr (NO$_3$)$_3$·9H$_2$O (0.010 M), and Mg (NO$_3$)$_2$·6H$_2$O (0.040 M) with Cr (NO$_3$)$_3$·9H$_2$O (0.010 M) were drop-wise added to the specific quantity of 23 vol% formamide. The pH adjustment of the mixed metal solution was varied to pH 7 by slow addition of aqueous 1-M NaOH solution till the saturation of the precipitate of MgCr-LDH. The MgCr-LDH gel precipitate was again diffused in formamide solution and subsequently ultra-sonicated for 30 min. Afterwards, the MgCr-LDH (3:1) gel suspension was transferred into 100 mL Teflon lined autoclave reactor and treated at 70, 80, and 90 °C for 24 h. Then the product was centrifuged and washed 3–4 times with distilled H$_2$O and vacuum dried at 40 °C. The coded name of the dried MgCr-LDHs samples at different atomic ratio are MgCr (2:1), MgCr (3:1), and MgCr (4:1), respectively. In addition, MgCr-LDH (2:1, 3:1, 4:1) without hydrothermal treatment was prepared for comparison.

3.3. Photocatalytic RhB Degradation Activity

(Conditions: exfoliated MgCr-LDH = 0.03 g, [RhB] = 20 ppm, exposer time = 120 min).

The photodegradation of RhB was executed in batch mode using a 20-ppm-concentrated RhB aqueous solution and 0.03 g of the catalyst. The suspension was uncovered to sunlight (~solar intensity = 0.80 kW/m^2) in closed Pyrex conical flasks with steady stirring during hot summer days. Prior to the solar experiments, dark reactions were carried out for comparison. The RhB degradation was analyzed by the spectrophotometric technique at 554 nm. After 2 h of solar irradiation, the conversion was reached up to 95.0%. Furthermore, the stability test of the catalyst for RhB degradation was repeated for three cycles. The stability test of the photocatalyst was executed in each run by simply washing with ethanol and deionized water followed by oven-drying at 80 °C for next use in second cycle.

3.4. Materials Characterization

Powder X-ray diffraction (PXRD) was performed by a Rigaku Miniflex powder diffraction meter, using Cu Kα source (λ = 1.54 Å, 30 kV, 50 mA). The bending and stretching modes of vibration of the materials were carried out by JASCO Fourier transform infrared (FTIR)-4600, using the KBr reference. The ultraviolet-visible diffuse reflectance spectra (UV−Vis DRS) were produced by a JASCO-V-750 UV−Vis spectrophotometer using BaSO$_4$ as a reference. Photoluminescence (PL) was analyzed using an FP-8300 JASCO spectrofluorometer. A high-resolution transmission electron microscopy (HR-TEM) and energy dispersive X-ray (EDX) study was carried by JEM-2100F at an accelerating voltage of 200 kV. The field emission scanning electron microscopy (FESEM) micrograph was acquired by a HITACHI 3400N microscope. The entire photoelectrochemical (PEC) measurements were carried out by potentiostat−galvanostat (IVIUM n STAT multi-channel electrochemical analyzer), with accessories of a 300 W Xenon lamp for visible light supply, a three-electrode system containing Pt, Ag/AgCl, and fluorine-doped tin oxide (FTO), as a counter, reference, and working electrode, respectively. The working electrode was made by an electrophoretic deposition process by our earlier reported method [15]. The electrolyte contained 0.1 M of Na$_2$SO$_4$ solution. The linear sweep voltammetry (LSV) study was completed by applied bias within −1.0 to +1.2 V at scanning rate of 5 mV s^{-1} in visible light exposure.

4. Conclusions

In summary, our thorough investigations on MgCr-(NO_3^-) LDH disclosed the exfoliation capability of MgCr-(NO_3^-) LDH in formamide into uni/multi-layer NS, which strongly depended on layer charge (Mg^{2+}/Cr^{3+} = 2:1, 3:1, and 4:1) and hydrothermal temperature (70, 80, and 90 °C). With an optimum metallic (Mg^{2+}/Cr^{3+}) ratio of 3:1, and hydrothermal treatment of 80 °C for 24 h, the exfoliated MgCr-(NO_3^-) LDH, displayed superior photocatalytic RhB degradation (95.0%) under solar light exposure for 2 h. This synthetic process is simple, cost-effective and thus potential strategy for the production of stable and exfoliated MgCr-LDH into uni/multi-layer NS. Finally, the morphology is a vital aspect in determining the nanosheet structure and by applying an optimized synthetic protocol; we tried to maintain house-of-nano-platelet morphology in MgCr-(NO_3^-) LDH (3:1, 80 °C) for light absorption, fast charge separation, and transfer for superior catalytic activities. Consequently, post-hydrothermal treatment might be suitable to augment productivity in exfoliation procedure. This work is an effectual approach to optimize novel catalytic system with high efficiency and stability, but initiate new ground for the vast application of MgCr-LDH in environmental remediation as well as energy production.

Author Contributions: S.N. and K.P. conceptualized and designed the experiments. S.N. synthesized the materials, executed the experiments, wrote, and edited the manuscript. All authors have read and agreed to the published version of the manuscript.

Funding: This research received no external funding.

Institutional Review Board Statement: Ethical review and approval were waived for this study.

Informed Consent Statement: Not applicable.

Data Availability Statement: Data sharing is not applicable to this article.

Acknowledgments: The authors gratefully acknowledge the support of the management of Siksha 'O' Anusandhan Deemed to be University to carry out this research work. The author S. Nayak is gratefully acknowledged to the CSIR-India for awarding CSIR-RA fellowship vide file no. 09/969 (0011)/2020 EMR-1 dated 13/10/2020.

Conflicts of Interest: The authors declare no conflict of interest.

References

1. Brisebois, P.P.; Siaj, M. Harvesting graphene oxide–years 1859 to 2019: A review of its structure, synthesis, properties and exfoliation. *J. Mater. Chem. C* **2020**, *8*, 1517–1547. [CrossRef]
2. Sultana, S.; Mansingh, S.; Parida, K.M. Facile synthesis of CeO_2 nanosheets decorated upon BiOI microplate: A surface oxygen vacancy promoted Z-scheme-based 2D-2D nanocomposite photocatalyst with enhanced photocatalytic activity. *J. Phys. Chem. C* **2018**, *122*, 808–819. [CrossRef]
3. Sakita, A.M.P.; Vallés, E.; Della Noce, R.; Benedetti, A.V. Novel NiFe/NiFe-LDH composites as competitive catalysts for clean energy purposes. *Appl. Surf. Sci.* **2018**, *447*, 107–116. [CrossRef]
4. Liang, D.; Yue, W.; Sun, G.; Zheng, D.; Ooi, K.; Yang, X. Direct synthesis of unilamellar MgAl-LDH nanosheets and stacking in aqueous solution. *Langmuir* **2015**, *31*, 12464–12471. [CrossRef]
5. Nayak, S.; Parida, K.M. Superactive NiFe-LDH/graphene nanocomposites as competent catalysts for water splitting reactions. *Inorg. Chem. Front.* **2020**, *7*, 3805–3836. [CrossRef]
6. Nayak, S.; Parida, K.M. Nanostructured CeO_2/MgAl-LDH composite for visible light induced water reduction reaction. *Int. J. Hydrog. Energy* **2016**, *41*, 21166–21180. [CrossRef]
7. Nayak, S.; Pradhan, A.C.; Parida, K.M. Topotactic transformation of solvated MgCr-LDH nanosheets to highly efficient porous $MgO/MgCr_2O_4$ nanocomposite for photocatalytic H2 evolution. *Inorg. Chem.* **2018**, *57*, 8646–8661. [CrossRef]
8. Nayak, S.; Parida, K. Comparison of NiFe-LDH based heterostructure material towards photocatalytic rhodamine B and phenol degradation with water splitting reactions. *Mater. Today Proc.* **2021**, *35*, 43–246. [CrossRef]
9. Nayak, S.; Parida, K. Recent progress in LDH@graphene and analogous heterostructure for highly active and stable photocatalytic and photoelectrochemical water splitting. *Chem. Asian J.* **2021**, *16*, 2211–2248. [CrossRef]
10. Sahoo, D.P.; Nayak, S.; Reddy, K.H.; Martha, S.; Parida, K. Fabrication of a $Co(OH)_2$/ZnCr LDH "p–n" heterojunction photocatalyst with enhanced separation of charge carriers for efficient visible-light-driven H_2 and O_2 evolution. *Inorg. Chem.* **2018**, *57*, 3840–3854. [CrossRef]

11. Nayak, S.; Mohapatra, L.; Parida, K. Visible light-driven novel g-C$_3$N$_4$/NiFe-LDH composite photocatalyst with enhanced photocatalytic activity towards water oxidation and reduction reaction. *J. Mater. Chem. A* **2015**, *3*, 18622–18635. [CrossRef]
12. Nayak, S.; Parida, K.M. Dynamics of charge-transfer behavior in a plasmon-induced quasi-type-II p–n/n–n dual heterojunction in Ag@ Ag$_3$PO$_4$/g-C$_3$N$_4$/NiFe LDH nanocomposites for photocatalytic Cr(VI) reduction and phenol oxidation. *ACS Omega* **2018**, *3*, 7324–7343. [CrossRef] [PubMed]
13. Nayak, S.; Parida, K.M. Deciphering Z-scheme charge transfer dynamics in heterostructure NiFe-LDH/N-rGO/g-C$_3$N$_4$ nanocomposite for photocatalytic pollutant removal and water splitting reactions. *Sci. Rep.* **2019**, *9*, 2458–2481. [CrossRef]
14. Biswal, L.; Nayak, S.; Parida, K. Recent progress on strategies for the preparation of 2D/2D MXene/g-C$_3$N$_4$ nanocomposites for photocatalytic energy and environmental applications. *Catal. Sci. Technol.* **2021**, *11*, 1222–1248. [CrossRef]
15. Nayak, S.; Swain, G.; Parida, K. Enhanced photocatalytic activities of RhB degradation and H2 evolution from in situ formation of the electrostatic heterostructure MoS$_2$/NiFe LDH nanocomposite through the Z-scheme mechanism via p–n heterojunctions. *ACS Appl. Mater. Interfaces* **2019**, *11*, 20923–20942. [CrossRef]
16. Gholami, P.; Khataee, A.; Soltani, R.D.C.; Dinpazhoh, L.; Bhatnagar, A. Photocatalytic degradation of gemifloxacin antibiotic using Zn-Co-LDH@ biochar nanocomposite. *J. Hazard. Mater.* **2020**, *382*, 121070–121081. [CrossRef] [PubMed]
17. Wang, Q.; O'Hare, D. Recent advances in the synthesis and application of layered double hydroxide (LDH) nanosheets. *Chem. Rev.* **2012**, *112*, 4124–4155. [CrossRef]
18. Ma, R.; Liu, Z.; Li, L.; Iyi, N.; Sasaki, T. Exfoliating layered double hydroxides in formamide: A method to obtain positively charged nanosheets. *J. Mater. Chem. A* **2006**, *16*, 3809–3813. [CrossRef]
19. Hibino, T.; Jones, W. New approach to the delamination of layered double hydroxides. *J. Mater. Chem. A* **2001**, *11*, 1321–1323. [CrossRef]
20. Karthikeyan, J.; Fjellvåg, H.; Knudsen, K.; Vistad, Ø.B.; Sjåstad, A.O. Quantification and key factors in delamination of (Mg$_{1-y}$Ni$_y$)$_{1-x}$Al$_x$(OH)$_2$(NO$_3$)$_x$·mH$_2$O. *Appl. Clay Sci.* **2016**, *124–125*, 102–110. [CrossRef]
21. Li, L.; Ma, R.; Ebina, Y.; Iyi, N.; Sasaki, T. Positively charged nanosheets derived via total delamination of layered double hydroxides. *Chem. Mater.* **2005**, *17*, 4386–4391. [CrossRef]
22. Wu, Q.; Olafsen, A.; Vistad, Ø.B.; Roots, J.; Norby, P. Delamination and restacking of a layered double hydroxide with nitrate as counter anion. *J. Mater. Chem. A* **2005**, *15*, 4695–4700. [CrossRef]
23. Adachi-Pagano, M.; Forano, C.; Besse, J.-P. Delamination of layered double hydroxides by use of surfactants. *Chem. Commun.* **2000**, *1*, 91–92. [CrossRef]
24. Hibino, T. Delamination of layered double hydroxides containing amino acids. *Chem. Mater.* **2004**, *16*, 5482–5488. [CrossRef]
25. Liang, H.; Meng, F.; Cabán-Acevedo, M.; Li, L.; Forticaux, A.; Xiu, L.; Wang, Z.; Jin, S. Hydrothermal continuous flow synthesis and exfoliation of NiCo layered double hydroxide nanosheets for enhanced oxygen evolution catalysis. *Nano Lett.* **2015**, *15*, 1421–1427. [CrossRef] [PubMed]
26. Acharya, L.; Nayak, S.; Pattnaik, S.P.; Acharya, R.; Parida, K. Resurrection of boron nitride in pn type-II boron nitride/B-doped-g-C$_3$N$_4$ nanocomposite during solid-state Z-scheme charge transfer path for the degradation of tetracycline hydrochloride. *J. Colloid Interface Sci.* **2020**, *566*, 211–223. [CrossRef]
27. Yu, J.; Liu, J.; Clearfield, A.; Sims, J.E.; Speiegle, M.T.; Suib, S.L.; Sun, L. Synthesis of layered double hydroxide single-layer nanosheets in formamide. *Inorg. Chem.* **2016**, *55*, 12036–12041. [CrossRef]
28. Yu, J.; Martin, B.R.; Clearfield, A.; Luo, Z.; Sun, L. One-step direct synthesis of layered double hydroxide single-layer nanosheets. *Nanoscale* **2015**, *7*, 9448–9451. [CrossRef]
29. Sharma, S.K.; Kushwaha, P.K.; Srivastava, V.K.; Bhatt, S.D.; Jasra, R.V. Effect of hydrothermal conditions on structural and textural properties of synthetic hydrotalcites of varying Mg/Al ratio. *Ind. Eng. Chem. Res.* **2007**, *46*, 4856–4865. [CrossRef]
30. Yang, X.; Makita, Y.; Liu, Z.-h.; Sakane, K.; Ooi, K. Structural characterization of self-assembled MnO$_2$ nanosheets from birnessite manganese oxide single crystals. *Chem. Mater.* **2004**, *16*, 5581–5588. [CrossRef]
31. Cavani, F.; Trifiro, F.; Vaccari, A. Hydrotalcite-type anionic clays: Preparation, properties and applications. *Catal. Today* **1991**, *11*, 173–301. [CrossRef]
32. Kustrowski, P.; Sulkowska, D.; Chmielarz, L.; Rafalska-Lasocha, A.; Dudek, B.; Dziembaj, R. Influence of thermal treatment conditions on the activity of hydrotalcite-derived Mg-Al oxides in the aldol condensation of acetone. *Microporous Mesoporous Mater.* **2005**, *78*, 1–22. [CrossRef]
33. Abello, S.; Medina, F.; Tichit, D.; Ramirez, J.P.; Groen, J.C.; Sueiras, J.E.; Salagre, P.; Cesteros, Y. Aldol condensations over reconstructed Mg-Al hydrotalcites: Structure-activity relationships related to the rehydration method. *Chem. A Eur. J.* **2005**, *11*, 728–739. [CrossRef] [PubMed]
34. Yin, W.; Bai, L.; Zhu, Y.; Zhong, S.; Zhao, L.; Li, Z.; Bai, S. Embedding metal in the interface of a p-n heterojunction with a stack design for superior Z-scheme photocatalytic hydrogen evolution. *ACS Appl. Mater. Interfaces* **2016**, *8*, 23133–23142. [CrossRef] [PubMed]
35. Ye, L.; Deng, K.; Xu, F.; Tian, L.; Peng, T.; Zan, L. Increasing visible-light absorption for photocatalysis with black BiOCl. *Phys. Chem. Chem. Phys.* **2012**, *14*, 82–85. [CrossRef] [PubMed]
36. Aguirre, M.E.; Zhou, R.; Eugene, A.J.; Guzman, M.I.; Grel, M.A. Cu$_2$O/TiO$_2$ heterostructures for CO$_2$ reduction through a direct Z-scheme: Protecting Cu$_2$O from photocorrosion. *Appl. Catal. B* **2017**, *217*, 485–493. [CrossRef]

37. Zeng, H.; Zhang, W.; Deng, L.; Luo, J.; Zhou, S.; Liu, X.; Pei, Y.; Shi, Z.; Crittenden, J. Degradation of dyes by peroxymonosulfate activated by ternary CoFeNi-layered double hydroxide: Catalytic performance, mechanism and kinetic modeling. *J. Colloid Interface Sci.* **2018**, *515*, 92–100. [CrossRef]

38. Zhao, X.; Niu, C.; Zhang, L.; Guo, H.; Wen, X.; Liang, C.; Zeng, G. Co-Mn layered double hydroxide as an effective heterogeneous catalyst for degradation of organic dyes by activation of peroxymonosulfate. *Chemosphere* **2018**, *204*, 11–21. [CrossRef] [PubMed]

Article

Low-Temperature Oxidation Removal of Formaldehyde Catalyzed by Mn-Containing Mixed-Oxide-Supported Bismuth Oxychloride in Air

Xiaoli Wang [1,*], Gongde Wu [2,*] and Yanwen Ma [3,*]

1   School of Environment and Technology, Nanjing Institute of Technology, Nanjing 211167, China
2   Energy Research Institute, Nanjing Institute of Technology, Nanjing 211167, China
3   Key Laboratory for Organic Electronics and Information Displays, Institute of Advanced Materials, Nanjing University of Posts and Telecommunications, Nanjing 210023, China
*   Correspondence: wangxiaoli212@njit.edu.cn (X.W.); wugongde@njit.edu.cn (G.W.); amywma@njupt.edu.cn (Y.M.)

Abstract: The Mn-containing mixed-oxide-supported bismuth oxychloride (BiOCl) catalysts were prepared by calcining their corresponding parent hydrotalcite supported BiOCl. The crystal structure of BiOCl was found to be intact during calcination, while significant differences appeared in the chemical state of Mn and the redox capacities of the catalysts before and after calcination. Compared to the hydrotalcite-supported catalysts, the mixed-oxide-supported BiOCl showed much higher catalytic performance in the oxidation removal of formaldehyde due to the synergetic catalysis of more surface oxygen vacancies and higher surface basicity. The complete removal of formaldehyde could be achieved at 70 °C, and the removal efficiency was maintained more than 90% for 21 h. A possible reaction mechanism was also proposed.

Keywords: formaldehyde; oxidation removal; BiOCl; mixed oxides; manganese

Citation: Wang, X.; Wu, G.; Ma, Y. Low-Temperature Oxidation Removal of Formaldehyde Catalyzed by Mn-Containing Mixed-Oxide-Supported Bismuth Oxychloride in Air. *Catalysts* **2022**, *12*, 262. https://doi.org/10.3390/catal12030262

Academic Editors: Ioan-Cezar Marcu and Octavian Dumitru Pavel

Received: 15 January 2022
Accepted: 22 February 2022
Published: 25 February 2022

Publisher's Note: MDPI stays neutral with regard to jurisdictional claims in published maps and institutional affiliations.

1. Introduction

Among volatile organic pollutants (VOCs), formaldehyde is one of the most harmful gases with the widest sources. As early as 2004, formaldehyde was listed as the first carcinogen by the international agency for research on cancer (IARC) [1–3]. In order to eliminate the harm of formaldehyde to human health, researchers from all over the world have carried out a lot of work, and many formaldehyde removal technologies have been developed, such as adsorption, photocatalysis, low-temperature plasma and catalytic oxidation [4–7]. Among them, catalytic oxidation can completely decompose formaldehyde into harmless CO_2 and H_2O without secondary pollution, and thus is considered the most effective method [4–7]. Especially the effective removal of formaldehyde by catalytic oxidation under the condition of natural air atmosphere and low temperature shows more practical significance and application prospect [8]. Additionally, the key of this technology is to design efficient catalysts and catalytic systems.

When it comes to the reported catalysts for the oxidation removal of formaldehyde, noble metal catalysts were represented by Pt or Au catalysts [9–15], while transition metal catalysts were mainly Mn or Co catalysts [16–19]. Owing to the low cost and rich resources, the transition metal catalysts, especially transition metal oxides, had attracted much attention in academia and industry. Among them, manganese oxides usually exhibited excellent catalytic performance in the removal reaction of formaldehyde than the other transition metal oxides, such as CoO, TiO_2 and CeO_2 [20–24]. This might be because manganese oxides could mobilize electrons and generate the mobile-electron environment, which is exactly needed by oxidation removal reaction of formaldehyde due to the variable oxidation states of Mn from −3 to +7. A series of effective manganese oxides with different valences, structures, morphology and matrixes had been prepared and widely used as catalysts. Rong et al. impregnated PET into ultrathin MnO_2 nanosheets, and the obtained MnO_2/PET catalyst could remove formaldehyde by oxidation with a removal efficiency of 81% at room temperature. The impregnation of PET was thought to benefit the formaldehyde adsorption and removal [25]. Bai et al. prepared a three-dimensional (3D) ordered mesoporous MnO_2 using KIT-6 mesoporous molecular sieves as a hard template. Complete conversion of formaldehyde was achieved at 130 °C, which was attributed to the specific mesoporous structure, high specific surface area and a large number of surface Mn^{4+} ions [26]. Ag/MnO_2 was also further synthesized for catalytic removal of formaldehyde, and complete removal efficiency was achieved at 100 °C due to the synergetic catalysis of Ag nanoparticles and three-dimensional ordered mesoporous MnO_2 supports [27]. Clearly, high removal efficiency and low reaction temperature were a research direction of formaldehyde removal reaction. Besides single manganese oxide catalysts, several binary or ternary Mn-containing mixed oxides were found to be more active in the oxidation removal of formaldehyde. Zhu et al. found that MnO_xCeO_2 composite catalyst with the nominal Ce/Mn ratio of 1:10 exhibited the best activity and achieved complete formaldehyde conversion at 100 °C [28]. Huang et al. discovered that $Co_xMn_{3-x}O_4$ nanosheet catalysts showed substantially higher catalytic activity for formaldehyde oxidation, and complete conversion of formaldehyde also appeared at 100 °C [29]. O'Shea reported that PdMnO/Al_2O_3 could completely catalytically oxidize the mixture of formaldehyde and methanol at 70 °C [30]. It was generally accepted that the introduction of the second or third metal oxides could influence the crystallization process of mixed oxides, which would induce a decrease in the particle sizes and an increase in the surface area. This might afford the enhanced catalytic performance of mixed oxides in the removal reaction of formaldehyde in comparison with the single manganese oxide. The previous studies proved that the Mn-containing mixed oxides were promising catalysts or catalyst supports for the removal of formaldehyde.

On the other hand, BiOX (X = Cl, Br, I) is a kind of layered semiconductor with high anisotropy, and its light absorption capacity can be regulated by changing the type and content of halogen elements. In particular, BiOCl with high thermal stability had been used wildly in photocatalytic degradation of organic pollutants to CO_2, H_2O and other inorganic small molecules [31,32]. Our consistent interests in the design of heterogeneous catalysts and the removal of organic pollutants pushed us to combine the catalytic characteristics of Mn-containing mixed oxides and BiOCl.

In our previous report, we had found that CuNiAl-HT supported BiOCl exhibited effective catalytic performance in the selective oxidation of glycerol by 3% H_2O_2 [33]. Inspired by this work, in view of characteristics and catalytic requirements of the formaldehyde removal reaction, we tried to load active BiOCl on the Mn-containing mixed oxides that were usually thought to be active for the complete oxidation reaction. The catalytic performance and stability of the obtained catalysts in the oxidation removal of formaldehyde were investigated, and the structure–activity relationship of catalysts and reaction mechanisms were also discussed.

2. Results

2.1. Characterization of Catalysts

Figure 1 illustrates the powder XRD patterns of samples. The xBiOCl/MnMgAl-HT showed typical hydrotalcite-like reflections of (003), (006) and (110) planes and BiOCl reflections of (101), (110) and (102) planes [31–35] (see Figure 1A). Moreover, with the increase in the loading amount of BiOCl, the intensity of the characteristic reflections assigned to BiOCl strengthened, while no significant change was found in the hydrotalcite-like reflections. As expected, there was no significant difference in the XRD patterns of 0.01BiOCl/MnMgAl-HT and 0.012BiOCl/MnMgAl-HT due to the small difference in BiOCl content. Thus, for the purpose of simplification, 0.012BiOCl/MnMgAl-HT did not appear in the following characterization figures. After calcination, it was found that the characteristic peaks of hydrotalcite-like almost disappeared, while the BiOCl reflections were still present clearly (see Figure 1B), suggesting that the crystal structure of BiOCl was intact during calcination. Simultaneously, two new peaks at about 43° and 63° associated with the MgAl mixed oxides were also detected. This indicated that the layered structure of the hydrotalcite-like matrix was destroyed, and the new phases of mixed oxides appeared during calcination. No significant manganese oxides crystalline phase was present, probably because of their amorphous state. In addition, for 0.015BiOCl/MnMgAlO, some weak reflections of Bi_2O_3 were also detected, which were absent in the other samples, might originate from the decomposition of excessive BiOCl during calcination.

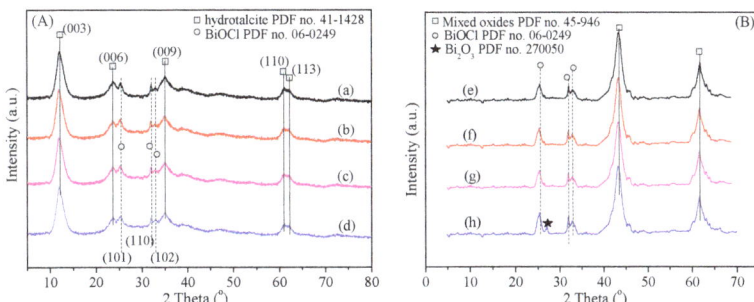

Figure 1. XRD patterns of (**A**) xBiOCl/MnMgAl-HT and (**B**) xBiOCl/MnMgAlO (a) 0.005BiOCl/MnMgAl-HT, (b) 0.01BiOCl/MnMgAl-HT, (c) 0.012BiOCl/MnMgAl-HT, (d) 0.015BiOCl/MnMgAl-HT, (e) 0.005BiOCl/MnMgAlO, (f) 0.01BiOCl/MnMgAlO and (g) 0.012BiOCl/MnMgAlO, (h) 0.015BiOCl/MnMgAlO.

The N_2 adsorption/desorption isotherms of xBiOCl/MnMgAl-HT and xBiOCl/MnMgAlO in Figure 2 all showed type IV isotherms, indicating that mesoporous structure had been formed due to the aggregation of particles. For the three hydrotalcite-supported BiOCl catalysts, compared to pure hydrotalcite, a significant decrease was found in their surface area (see Table 1). Moreover, with the increase in the loading amount of BiOCl, the surface area of supported catalysts decreased continually, probably because of the increased pore lock phenomenon derived from the introduction of BiOCl. A similar trend also appeared in the three mixed-oxide-supported BiOCl catalysts, and the increased loading of BiOCl to Mn-containing mixed oxides also led to a significantly decreased surface area. Noticeably, the lowest surface area of 0.015BiOCl/MnMgAlO was found in comparison with the other two mixed-oxide-supported catalysts. This might be associated with the appearance of Bi_2O_3 with a larger molecular size than BiOCl, as described in XRD characterization.

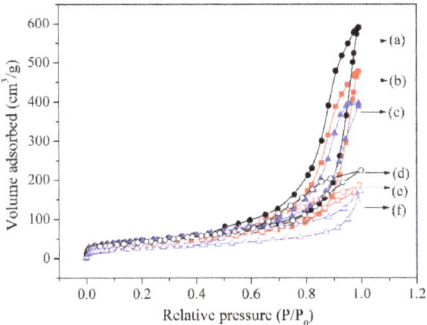

Figure 2. N_2 adsorption–desorption isotherms of (a) 0.005BiOCl/MnMgAlO, (b) 0.01BiOCl/MnMgAlO, (c) 0.015BiOCl/MnMgAlO, (d) 0.005BiOCl/MnMgAl-HT, (e) 0.01BiOCl/MnMgAl-HT and (f) 0.015BiOCl/MnMgAl-HT.

Table 1. Catalytic performance of samples in the removal reaction of formaldehyde [a].

Entry	Catalysts	Mn/Mg/Al Molar Ratio	SBET (m²/g)	Removal Efficiency of Formaldehyde [b] (%)
1	blank	— [c]	—	0
2	BiOCl	—	35	1.5
3	MnMgAl-HT	1.45/1.46/1	85	7.8
4	0.015BiOCl/MnMgAl-HT	1.46/1.46/1	73	51.4
5	0.012BiOCl/MnMgAl-HT	1.45/1.46/1	77	54.5
6	0.01BiOCl/MnMgAl-HT	1.45/1.47/1	78	54.9
7	0.005BiOCl/MnMgAl-HT	1.47/1.46/1	81	45.5
8	MnMgAlO	1.45/1.45/1	210	9.5
9	0.015BiOCl/MnMgAlO	1.46/1.47/1	197	80.6
10	0.012BiOCl/MnMgAlO	1.45/1.46/1	204	91.5
11	0.01BiOCl/MnMgAlO	1.45/1.47/1	205	91.0
12	0.005BiOCl/MnMgAlO	1.46/1.45/1	207	75.4

[a] Reaction conditions: catalyst 100 mg; formaldehyde 100 ppm; air/formaldehyde 100 mL min⁻¹; atmospheric pressure; temperature 60 °C; time 3 h. [b] Removal efficiency = (the outlet concentration of CO_2/the inlet concentration of formaldehyde) × 100. [c] "—" means no detection.

TGA curves of 0.01BiMnMgAl-HT were typically illustrated in Figure 3. Three major weight losses appeared at about 55 °C, 90 °C and 290 °C, respectively. The first two peaks might be related to the removal of physically adsorbed and interlayer water, respectively. The last peak could be attributed to the dehydroxylation of layers, which led to the complete collapse of the layered structure. Moreover, no significant weight loss was found after 450 °C, indicating that the mesoporous structure of mixed oxides had been formed by calcining their parent hydrotalcite-like compounds at the present calcination temperature (450 °C). Simultaneously, it was known that BiClO was stable before 600 °C; thus, the above results also further confirmed that the structure of BiClO remained unchanged as described in XRD characterization.

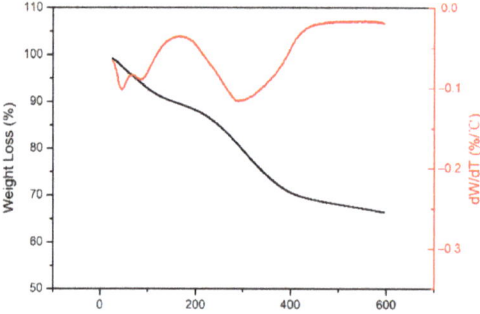

Figure 3. Thermogravimetric weight loss curve and derivative plots of 0.01BiMnMgAl−HT.

The H_2-TPR images of samples are shown in Figure 4. For the uncalcined xBiOCl/MnMgAl-HT, two hydrogen consumption peaks were found at about 300 °C and 450 °C, which could be attributed to the redox process of Mn. The former could be assigned to the hydrogen consumption peak of Mn^{4+} to Mn^{3+}, while the latter might be due to the hydrogen consumption peak of Mn^{3+} to Mn^{2+}. With regard to the peak area of the two characteristic peaks, the latter is much larger than the former. This indicated that Mn^{3+} and Mn^{4+} exist together in uncalcined xBiOCl/MnMgAl-HT, of which Mn^{3+} is the dominant state. In contrast, for the calcined xBiOCl/MnMgAlO, only one hydrogen consumption peak appeared at about 410 °C, suggesting the presence of single Mn^{3+} due to the redox of Mn^{4+} to Mn^{3+} during calcination. Moreover, xBiOCl/MnMgAlO showed a lower reduction temperature of Mn^{3+} than xBiOCl/MnMgAl-HT. It was accepted that the shifts of hydrogen consumption peaks to lower temperatures were ascribed to the formation of more active oxygen species [28,36]. It could be deduced that the relatively large number of surface oxygen species in xBiOCl/MnMgAlO could be produced by their more surface oxygen vacancies, which were generated to maintain the electrostatic balance due to the presence of the more reduced element state of Mn^{3+}. This also suggested that xBiOCl/MnMgAlO would show increased redox capacity in comparison with xBiOCl/MnMgAl-HT.

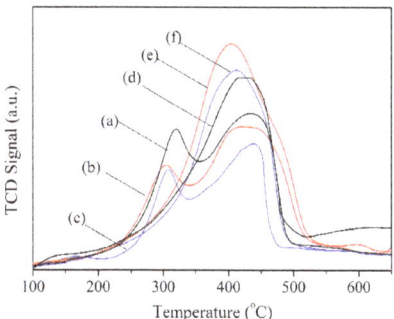

Figure 4. H_2-TPR images of (a) 0.005BiOCl/MnMgAlO, (b) 0.01BiOCl/MnMgAlO, (c) 0.015BiOCl/MnMgAlO, (d) 0.005BiOCl/MnMgAl-HT, (e) 0.01BiOCl/MnMgAl-HT and (f) 0.015BiOCl/MnMgAl-HT.

The O_2-TPD profiles of 0.005BiOCl/MnMgAlO, 0.01BiOCl/Mn MgAlO and 0.015BiO Cl/MnMgAlO are depicted in Figure 5. Generally, three oxygen desorption peaks were detected, including low temperature (378 °C), medium temperature (500~550 °C) and high temperature (>600 °C), which could be ascribed to the surface active oxygen (O_2^- and O^-), the surface lattice oxygen (O^{2-}) and the bulk lattice oxygen, respectively [20,21,29]. Moreover, with the rise in BiOCl loading, the low-temperature desorption peaks in three catalysts enhanced first and then weakened, indicating that the amount of active oxygen species increased first and then decreased. However, no significant shifts were found in the position of their low-temperature desorption peaks, suggesting that the relative number of surface active oxygen species did not change with the increased BiOCl loading. For the medium desorption peaks of the three supported catalysts, no significant differences were detected in the desorption peak intensities of 0.005BiOCl/MnMgAlO and 0.01BiOCl/MnMgAlO, while the peak position shifted to lower temperature with the increasing of BiOCl loading. This indicated that the surface lattice oxygen in the catalyst with high BiOCl loading was more active; however, the increased BiOCl loading did not influence the amount of surface lattice oxygen. With respect to 0.015BiOCl/MnMgAlO, compared to 0.01BiOCl/MnMgAlO, the peak intensities associated with low temperature and medium temperature desorption weakened, revealing that the amount of surface active oxygen and lattice oxygen both decreased, which might be attributed to the excessive loading of BiOCl. Noticeably, the desorption peak related to bulk lattice oxygen appeared in 0.015BiOCl/MnMgAlO, which might be generated to compensate for the decreased surface active oxygen and lattice oxygen originated from the introduction of excessive BiOCl.

The surface basicity of the obtained three mixed-oxide-supported catalysts was further investigated, and their CO_2-TPD profiles are shown in Figure 6. Three CO_2 desorption peaks were detected, which could be assigned to the weak basic sites (around 100 °C), moderate basic sites (around 235 °C) and strong basic sites (>390 °C) [37,38]. The weak basic sites could be ascribed to the abundant surface OH^- groups, the moderate basic sites might be related to Mg–O and Al–O pairs and the strong basic sites might be associated with coordinatively unsaturated lattice oxygen (O^{2-}) [39–41]. Furthermore, Figure 6 also illustrated that the characteristic peaks of weak basic sites and strong basic sites in 0.01BiOCl/MnMgAlO both shifted to high temperatures in comparison with those in 0.005BiOCl/MnMgAlO. A much higher CO_2 uptake was also found in 0.01BiOCl/MnMgAlO than in 0.005BiOCl/MnMgAlO (see Table 2). This indicated that the basicity of weak and strong basic sites increased with the rise in BiOCl loading. The increased basicity of weak basic sites could be attributed to the more surface OH⁻ groups in the catalyst with higher BiOCl loading than in the catalyst with lower BiOCl loading. The increased basicity of strong basic sites might be caused by the effect of the increased BiOCl loading on the mixed oxide support. However, with the further increase in the loading of BiOCl, slight shifts in the characteristic peaks ascribed to the weak and strong basic sites to low temperatures were found in the CO_2-TPD profiles of 0.015BiOCl/MnMgAlO. Simultaneously, a relatively decreased CO_2 uptake was also detected (see Table 2). The presence of a small amount of Bi_2O_3 in 0.015BiOCl/MnMgAlO, as confirmed by XRD characterization due to the excessive loading, might afford the above-mentioned slight decreased surface basicity.

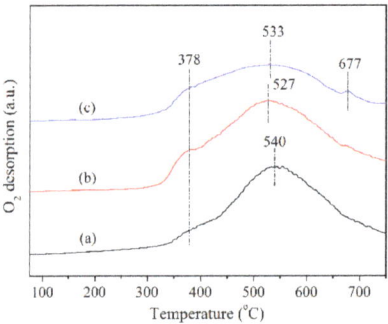

Figure 5. O_2-TPD images of (a) 0.005BiOCl/MnMgAlO, (b) 0.01BiOCl/MnMgAlO and (c) 0.015BiOCl/MnMgAlO.

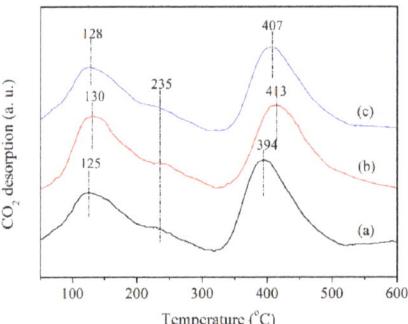

Figure 6. CO_2-TPD images of (a) 0.005BiOCl/MnMgAlO, (b) 0.01BiOCl/MnMgAlO and (c) 0.015BiOCl/MnMgAlO.

Table 2. Basicity of different catalysts.

Sample	CO_2 Uptake (μmol/g) [a]		
	Weak Basic Sites	**Moderate Basic Sites**	**Strong Basic Sites**
0.005BiOCl/MnMgAlO	32.5	6.7	45.3
0.01BiOCl/MnMgAlO	37.8	9.5	53.2
0.015BiOCl/MnMgAlO	34.6	8.6	48.6

[a] Relative amounts of the basic sites (%) calculated from the peak area.

XPS was performed to investigate the surface chemical state of catalysts. XPS spectra of Bi 4f of pure commercial BiOCl, 0.01BiOCl/MnMgAl-HT and 0.01BiOCl/MnMgAlO were typically illustrated in Figure 7A. It was found that two characteristic peaks at about

164.5 eV and 159.2 eV corresponding to Bi^{3+} appeared in the XPS spectra of pure BiOCl. In contrast, in the spectra of 0.01BiOCl/MnMgAl-HT and 0.01BiOCl/MnMgAlO, those two peaks showed small shifts to low binding energies, indicating the presence of the $+(3-x)$ oxidation state of Bi. This could be attributed to the formation of more oxygen vacancies on the surface of BiOCl in the supported catalysts in comparison with those on the surface of commercial BiOCl, which originated from the interaction of BiClO and supports during loading [31,32]. In view of the two supported catalysts, no significant differences were found in their XPS spectra, suggesting the almost identical surface chemical states of Bi in 0.01BiOCl/MnMgAl-HT and 0.01BiOCl/MnMgAlO.

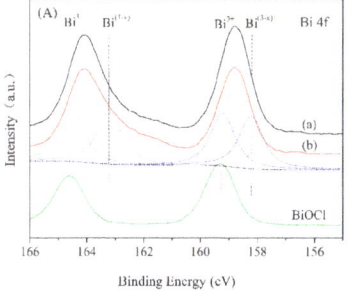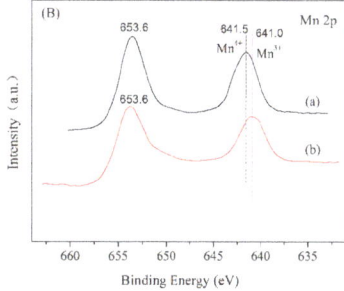

Figure 7. XPS spectra of (**A**) Bi 4f and (**B**) Mn 2p, (a) 0.01BiOCl/MnMgAl−HT and (b) 0.01BiOCl/MnMgAlO.

XPS spectra of Mn 2p of 0.01BiOCl/MnMgAl-HT and 0.01BiOCl/MnMgAlO were typically illustrated in Figure 7B. In the XPS spectra of 0.01BiOCl/MnMgAl-HT, two peaks at 641.5 eV and 653.6 eV could be attributed to Mn 2p2/3 and Mn 2p1/2. By the curve fitting analysis of Mn 2p2/3, Mn3+ was found to be the predominant oxidation state in 0.01BiOCl/MnMgAl-HT. When it comes to the XPS spectra of 0.01BiOCl/MnMgAlO, the peak assigned to Mn 2p2/3 shifted to 641.0 eV, indicating the absence of Mn4+ due to the redox of Mn^{4+} to Mn^{3+} during calcination. This fit well with the results of H_2-TPR and further demonstrated the different chemical states of Mn in the two supported catalysts.

SEM images of 0.01BiOCl/MnMgAl-HT and 0.01BiOCl/MnMgAlO are typically displayed in Figure 8A,B. 0.01BiOCl/MnMgAl-HT showed lamella structure with lateral sizes of platelets about 50~150 nm, indicating that the layered morphology of hydrotalcite-like compounds and BiOCl was not altered during loading. After calcination, two spherical or ellipsoidal particles with an average diameter of 300–400 and 20–100 nm were found in the SEM image of 0.01BiOCl/MnMgAlO, which could be assigned to mixed oxide support and BiOCl, respectively.

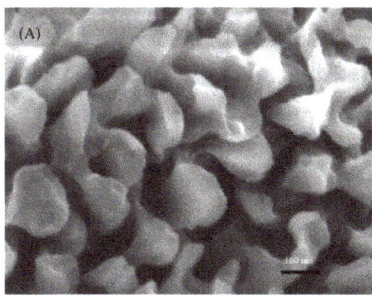

Figure 8. SEM images of (**A**) 0.01BiOCl/MnMgAl-HT and (**B**) 0.01BiOCl/MnMgAlO.

2.2. Catalytic Performance and Stability of Catalysts

In the present work, the obtained xBiOCl/MnMgAl-HT and xBiOCl/MnMgAlO catalysts were used in the oxidation removal of formaldehyde (see Table 1). The results revealed that pure BiOCl hardly showed effective formaldehyde removal efficiency (Entry 2). Though BiOCl has widely been accepted as a good photocatalyst for oxidation removal of pollutants, under the current UV-free reaction conditions, BiOCl itself did not seem to be active in the oxidation removal of formaldehyde. Similarly, the catalytic performance of the pure MnMgAl-HT or MnMgAlO was also poor, and the formaldehyde removal efficiency was no more than 10% (Entry 3 and 7). However, with the xBiOCl/MnMgAl-HT or xBiOCl/MnMgAlO used as catalyst, dramatically enhanced formaldehyde removal efficiency was found (Entry 4–6 and 8–10). Such an increased catalytic performance could be related to the generation of more oxygen vacancies on the surface of BiOCl in the supported catalysts than in the pure BiOCl, which caused more surface active oxygen species and thus facilitated the oxidation removal reaction. This also suggested that certain synergetic catalysis was present between BiOCl and the supports of MnMgAl-HT or MnMgAlO.

With regard to the catalytic performance of the supported BiOCl materials before and after calcination, xBiOCl/MnMgAlO showed much higher catalytic performance than xBiOCl/MnMgAl-HT. It was known that the supports of heterogeneous catalysts influenced mainly their catalytic performance by regulating the adsorption capacity to reactants. The dried formaldehyde itself is not ionized and thus is neutral. However, as the oxidation progressed, the metabolite water that appeared in the reaction system induced forming the dimeric of formaldehyde with weak acidity. For the supported catalysts, the surface basicity of mixed-oxide-supported BiOCl compounds was much higher than that of hydrotalcite-supported catalysts because the inherent basic sites of hydrotalcite-like compounds were fully explored during calcination. Therefore, more formaldehyde molecules could be adsorbed on the surface of mixed-oxide-supported catalysts, which had been verified by the following in situ FTIR experiments (see Figure 11). This increased the probability of formaldehyde molecules being further oxidized and could afford the high catalytic performance of xBiOCl/MnMgAlO to a certain degree. At the same time, H_2-TPR and XPS spectra of Mn 2p illustrated that the presence of more Mn^{3+} in xBiOCl/MnMgAlO also induced more oxygen vacancies on their support surface, which led to more surface active oxygen species and higher redox capacity in xBiOCl/MnMgAlO than that in xBiOCl/MnMgAl-HT. This might also contribute to the higher catalytic performance of xBiOCl/MnMgAlO.

Furthermore, the loading of BiOCl was also found to significantly influence the catalytic performance of supported catalysts. The hydrotalcite- and mixed-oxide-supported catalysts both exhibited an enhanced formaldehyde removal efficiency first with the increased loading of BiOCl to 0.01%, which could be attributed to the demand for enough catalytic active sites for the oxidation removal reaction. However, with the further increase in the loading of BiOCl, the formaldehyde removal efficiency then decreased. Such a decreased removal efficiency could be related to the decreased surface area and surface basicity of catalysts. In addition, for 0.015BiOCl/MnMgAlO, the presence of a small minority of Bi_2O_3 secondary phases with weak acidity could cause a decreased adsorption capacity of catalysts for formaldehyde, which could also afford the decrease in the formaldehyde removal efficiency.

The effect of the reaction time and temperature on the catalytic performance of six supported BiOCl catalysts was also investigated in detail (see Figure 9). Generally speaking, xBiOCl/MnMgAlO displayed significantly high catalytic performance in the formaldehyde removal reaction in comparison with xBiOCl/MnMgAl-HT. Under a sufficient reaction time and temperature, the formaldehyde removal efficiency could reach 100% over xBiOCl/MnMgAlO, while about 70% of formaldehyde removal efficiency was achieved over xBiOCl/MnMgAl-HT. Such a difference could be attributed to the different surface properties and chemical state of Mn in the two supported catalysts, which was discussed above. However, the performance trends of the two catalysts were found to be almost identical with the rise in reaction time and temperature. A sharp increase in the formaldehyde removal efficiency appeared with the increased reaction time and temperature, which might be related to the increased probability of adsorption and surface reactivity of reactant molecules at high temperatures and long times. In the presence of xBiOCl/MnMgAl-HT, the optimal reaction time and temperature were 240 min and 90 °C, respectively. In contrast, with xBiOCl/MnMgAlO as a catalyst, the complete formaldehyde removal efficiency could be achieved when the reaction was performed at 70 °C and 220 min, respectively. Such reaction results were rarely achieved over low-priced metal catalysts at such a low temperature and short time, so the Mn-containing mixed-oxides-supported BiOCl catalyst showed potential application value.

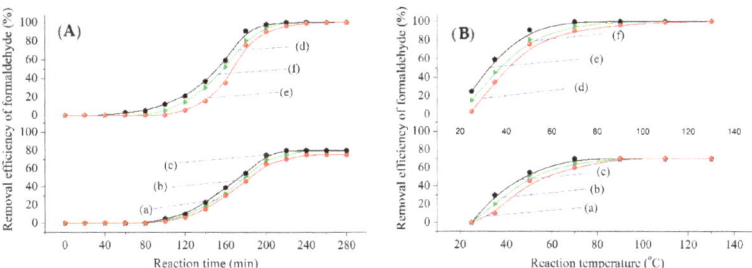

Figure 9. Effect of reaction time (**A**) and temperature (**B**) on the catalytic performance of catalysts. (a) 0.005BiOCl/MnMgAl-HT, (b) 0.01BiOCl/MnMgAl-HT, (c) 0.015BiOCl/MnMgAl-HT, (d) 0.005BiOCl/MnMgAlO, (e) 0.01BiOCl/MnMgAlO and (f) 0.015BiOCl/MnMgAlO.

Reaction conditions: Left part, catalyst 100 mg, formaldehyde 100 ppm, air/formaldehyde 100 mL min^{-1}, atmospheric pressure; temperature 60 °C; Right part, catalyst 100 mg, formaldehyde 100 ppm, air/formaldehyde 100 mL min^{-1}, atmospheric pressure, time 3 h.

0.01BiOCl/MnMgAlO, which exhibited the best catalytic performance in the oxidation removal of formaldehyde, was selected as the representative catalyst to carry out the stability test at 70 °C. The results in Figure 10 illustrated that the formaldehyde removal efficiency was maintained at more than 90% for 21 h. The slight decrease in catalytic performance could be attributed to the accumulation of a little carbonates on the surface of catalysts. This indicated that the Mn-containing mixed-oxide-supported BiOCl catalyst was stable in the present oxidation removal conditions.

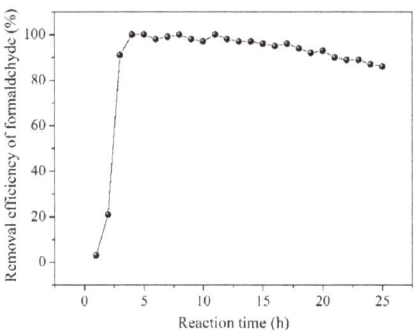

Figure 10. Stability of the representative catalyst 0.01BiOCl/MnMgAlO.

Reaction conditions: catalyst 100 mg, formaldehyde 100 ppm, air/formaldehyde 100 mL min^{-1}, atmospheric pressure; temperature 70 °C.

2.3. Structure Activity Relationship of Catalysts and Possible Reaction Mechanism

In situ ATR-IR spectra of 0.01BiOCl/MnMgAlO and 0.01BiOCl/MnMgAl-HT are typically illustrated in Figure 11. When the 0.01BiOCl/MnMgAlO sample was exposed to HCHO in air, the characteristic peaks of polyformaldehyde (929 cm^{-1}), dioxymethylene (1091 cm^{-1}), carbonate (1238 and 1280 cm^{-1}), formate (1340 and 1560 cm^{-1}), bicarbonate (1430 and 1603 cm^{-1}), water (1645 cm^{-1}) and CO_2 (2341 cm^{-1}) were observed [42,43] (see Figure 11A). Moreover, all the above peaks of intermediate or product species enhanced with the increasing exposure time from 10 to 30 min, indicating that more HCHO molecules were oxidized. However, with the further extension of the reaction time to 40 min, the characteristic peaks corresponding to polyformaldehyde, dioxymethylene, carbonate, formate and bicarbonate species weakened, and simultaneously, the CO_2 characteristic peak enhanced significantly. This suggested that more intermediate species were further oxidized into CO_2 with the prolonging of the reaction time.

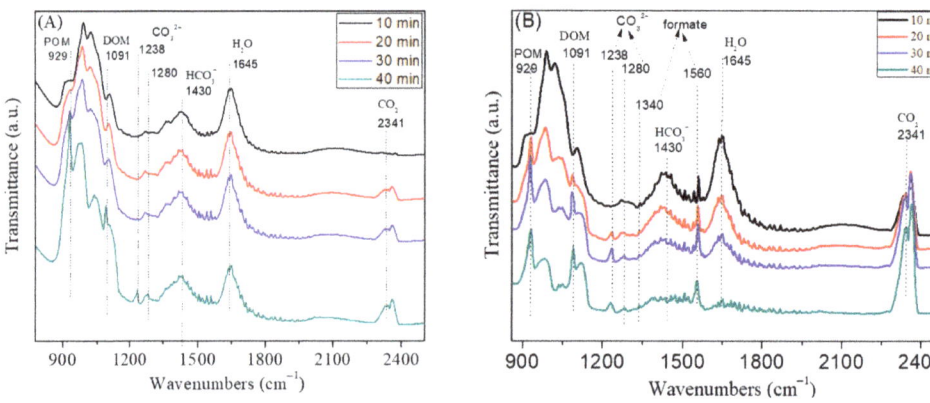

Figure 11. FTIR spectra of (**A**) 0.01BiOCl/MnMgAl−HT and (**B**) 0.01BiOCl/MnMgAlO.

In the case of the 0.01BiOCl/MnMgAl-HT, upon being exposed to HCHO and air, the intensity of the characteristic peaks ascribed to polyformaldehyde (929 cm^{-1}) and dioxymethylene (1091 cm^{-1}) increased continuously when the reaction time increased from 10 to 40 min (see Figure 11B). Moreover, no peaks ascribed to formate species were detected, and the enhancement of the CO_2 characteristic peak was not obvious compared to the peak in the spectra of 0.01BiOCl/MnMgAlO. This indicated that the intermediate product of polyformaldehyde and dioxymethylene were not oxidized effectively into formate and CO_2 over 0.01BiOCl/MnMgAl-HT, probably because of its relatively few surface oxygen vacancies and basicity. This also suggested that the conversion of polyformaldehyde and dioxymethylene into formate might be the rate-determining step for HCHO oxidation in the present catalytic systems. Furthermore, the following possible reaction mechanism was also proposed (see Scheme 1). According to the Mars-Van-Krevelen oxidation-reduction mechanism, O_2 and lattice oxygen O^{2-} can be transformed into each other. Thus, O_2 was firstly transformed into O^{2-}. Then, O^{2-} could be oxidized to O_2^- with a stronger oxidation capacity on the oxygen vacancies of the catalyst. Simultaneously, formaldehyde could be adsorbed by the basic surface -OH groups. The adsorbed formaldehyde molecules in the form of polyformaldehyde were oxidized into dioxymethylene, and then reacted with the active oxygen species O_2^- to produce formate and carbonate (rate-determining step). Subsequently, the formate and carbonate would further react with active oxygen species to obtain CO_2 and H_2O.

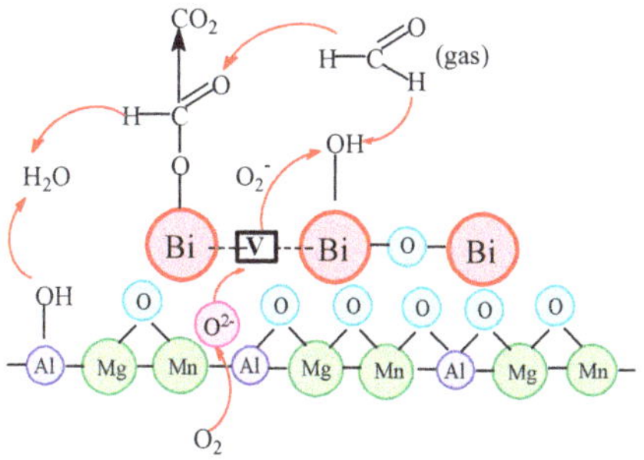

Scheme 1. Possible mechanism of the oxidation removal of formaldehyde over 0.01BiOCl/MnMgAlO.

3. Experimental

3.1. Catalyst Preparation

Typically, 0.04 mol of $Mn(NO_3)_2 \cdot 6H_2O$, 0.04 mol of $Mg(NO_3)_2 \cdot 6H_2O$ and 0.04 mol of $Al(NO_3)_3 \cdot 9H_2O$ were added concurrently to 120 mL of deionized water under vigorous stirring at room temperature to prepare the aqueous solution A. x mol of $BiCl_3$ (x = 0.005, 0.01, 0.012 and 0.015 mol) was added to 60 mL absolute alcohol under vigorous stirring at room temperature to prepare the alcohol solution B. Then, 0.06 mol Na_2CO_3 was added to 120 mL of deionized water with stirring to prepare the aqueous solution C. Then, solution A and B were simultaneously added dropwise to solution C with stirring at 80 °C for 1.0~1.5 h. In the process of dropping, the appropriate amount of NaOH was added into the mixture to maintain its pH value at 10.0 ± 0.5. The resulting gel-like slurry was further aged with stirring for 12 h at room temperature, and then was filtered and washed with deionized water until pH = 7. The filtrate was dried in air at 373 K for 12 h, and the obtained solid samples were designated as xBiOCl/MnMgAl-HT.

XBiOCl/MnMgAl-HT was further calcined at 450 °C under N_2 for 8 h, and the as-prepared solid samples were designated as xBiOCl/MnMgAlO (x = 0.005, 0.01 and 0.015 mol).

3.2. Characterization

The elemental composition of samples was detected using the inductively coupled plasma-optical (ICP) emission spectroscopy (PerkinElmer ICP OPTIMA–3000). Powder X-ray diffraction (XRD) experiments were run on a Rigaku Miniflex diffractometer with a Ni filter using a Cu target. The X-ray gun operated at 30 mA and 50 kV with a scan speed rate of 0.2°/min. N_2 adsorption–desorption isotherms of samples were measured using static adsorption procedures at 77 K on a Micromeritics ASAP-2000 instrument (Norcross, GA), and the surface areas of samples were obtained using the BET method. Thermogravimetric analysis (TGA) was performed on a Setaram TGA-92 thermal analyzer under nitrogen, and the samples were heated from 50 to 700 °C at 10 °C/min. X-ray photoelectron spectra (XPS) of samples were obtained using Perkin-Elmer PHI 5000C, and all binding energies were calibrated according to the reference of contaminant carbon (C1S = 284.6 eV). Scan electron microscope (SEM) experiments were performed on the JEOL JSM-7600F microscope.

The temperature-programmed reduction (H_2-TPR) equipment (Finesorb-3010) consisted of a thermal conductivity detector (TCD) connected to a flow-control system and a programmed heating unit. In each test, 0.1 g of the sample (40–60 mesh) was placed in a quartz reactor and reduced in10% H_2 in N_2 at a flow rate of 60 mL min^{-1} with a heating rate of 10 °C min^{-1} from 50 °C to 700 °C.

CO_2 temperature-programmed desorption (CO_2-TPD) and O_2 temperature-programmed desorption (O_2-TPD) tests were performed on the same instrument as H_2-TPR. Prior to each test, 0.1 g of the sample (40–60 mesh) was pre-treated in a Ar flow (60 mL min^{-1}) at 300 °C for 30 min with a heating rate of 10 °C min^{-1}, followed by cooling down to 50 °C in the same flow in order to yield clean surfaces. Afterward, a flow of CO_2 or 21% O_2/N_2 at a rate of 20 mL min^{-1} was passed through the sample for 1 h at 50 °C. Additionally, the physically adsorbed CO_2 or O_2 was purged off for 1 h. Finally, the sample was heated from 50 °C to 700 °C at a rate of 10 °C min^{-1} for the desorption of the previous adsorbed CO_2 or O_2 in a flow of Ar (60 mL min^{-1}).

In situ ATR-IR experiments were studied on a Fourier transform infrared spectroscopy (Bruker Tensor-27) from 400 to 4000 cm^{-1} equipped with an attenuated total reflection (ATR) accessory. Considering the low concentration of formaldehyde in our test, formaldehyde droplets (37 w.%) were dropped to the catalysts powder. The samples were exposed to air and identified at different times (10, 20, 30, 40 min).

3.3. Oxidation Removal Reaction of Formaldehyde

Typically, 100 mg catalyst was loaded in an affixed-bed quartz tube reactor (diameter = 4 mm). Under atmospheric pressure, a gas reactant mixture of 100 ppm formaldehyde and air was passed through the catalyst bed at 100 mL min^{-1}. The gaseous formaldehyde was produced by passing air flow over the formaldehyde solution at 0 °C. Outflow gases were analyzed using an online Agilent 8600 gas chromatograph equipped with an FID detector. The removal efficiency of formaldehyde was evaluated in terms of CO_2 concentration according to the carbon balance.

4. Conclusions

xBiOCl/MnMgAlO was a promising catalyst in the low-temperature oxidation removal of formaldehyde in air. The complete removal efficiency of formaldehyde could be achieved at 70 °C when the reaction was performed for 200 min. Such a high catalytic performance could be attributed to the synergetic catalysis of abundant surface oxygen vacancies and high surface basicity. The surface oxygen vacancies derived from the presence of the reduced element state of $Bi^{(3-x)+}$ on the surface of BiOCl and Mn^{3+} on the surface of MnMgAlO support, which induced the generation of more active oxygen species and thus benefited the removal reaction of formaldehyde. Simultaneously, xBiOCl/MnMgAlO showed increased surface basicity in comparison with their parent hydrotalcites due to the perfect exposure of base sites during calcination. This favored the adsorption of reactant formaldehyde molecules and further increased the probability of formaldehyde removal by oxidation, which also contributed to their efficient catalytic performance. Furthermore, xBiOCl/MnMgAlO showed excellent stability and potential application prospect.

Author Contributions: X.W., G.W. and Y.M. designed the experiments; X.W. and G.W. performed the experiments; X.W. analyzed the data and wrote the paper. All authors have read and agreed to the published version of the manuscript.

Funding: This research was funded by the Key Research and Development Program of Jiangsu Province [grant numbers BE2018718] and the Open Research fund of Key Laboratory for Organic Electronics and Information Displays.

Conflicts of Interest: No conflict of interest was declared here.

References

1. Salthammer, T.; Mentese, S.; Marutzky, R. Formaldehyde in the indoor environment. *Chem. Rev.* **2010**, *110*, 2536–2572. [CrossRef]
2. Hakim, M.; Broza, Y.Y.; Barash, O.; Peled, N.; Phillips, M.; Amann, A.; Haick, H. Volatile organic compounds of lung cancer and possible biochemical pathways. *Chem. Rev.* **2012**, *112*, 5949–5966. [CrossRef] [PubMed]
3. Wenger, O.S. Vapochromism in organometallic and coordination complexes: Chemical sensors for volatile organic compounds. *Chem. Rev.* **2013**, *113*, 3686–3733. [CrossRef]
4. Bai, B.Y.; Qiao, Q.; Li, J.H.; Hao, J.M. Progress in research on catalysts for catalytic oxidation of formaldehyde. *Chin. J. Catal.* **2016**, *37*, 102–122. [CrossRef]
5. Torres, J.Q.; Royer, S.; Bellat, J.P.; Giraudon, J.M.; Lamonier, J.F. Formaldehyde: Catalytic oxidation as a promising soft way of elimination. *ChemSusChem* **2013**, *6*, 578–592. [CrossRef] [PubMed]
6. Huang, H.B.; Xu, Y.; Feng, Q.Y.; Leung, D.Y.C. Low temperature catalytic oxidation of volatile organic compounds: A review. *Catal. Sci. Technol.* **2015**, *5*, 2649–2669. [CrossRef]

7. Nie, L.H.; Yu, J.G.; Jaroniec, M.; Tao, F.F. Room-temperature catalytic oxidation of formaldehyde on catalysts. *Catal. Sci. Technol.* **2016**, *6*, 3649–3669. [CrossRef]

8. Zhang, Y.P.; Mo, J.H.; Li, Y.G.; Sundell, J.; Wargocki, P.; Zhang, J.; Little, J.C.; Corsi, R.; Deng, Q.H.; Leung, M.H.K.; et al. Can commonly-used fan-driven air cleaning technologies improve indoor air quality? *A literature review. Atmos. Environ.* **2011**, *45*, 4329–4343. [CrossRef]

9. Chen, M.H.; Qin, Y.P.; Wang, W.Z.; Li, X.Y.; Wang, J.J.; Wen, H.; Yang, Z.Q.; Wang, P. Engineering oxygen vacancies via amorphization in conjunction with W-doping as an approach to boosting catalytic properties of Pt/Fe-W-O for formaldehyde oxidation. *J. Hazar. Mater.* **2021**, *416*, 126224. [CrossRef]

10. Yang, T.F.; Huo, Y.; Liu, Y.; Rui, Z.B.; Ji, H.B. Efficient formaldehyde oxidation over nickel hydroxide promoted Pt/γ-Al₂O₃ with a low Pt content. *Appl. Catal. B Environ.* **2017**, *200*, 543–551. [CrossRef]

11. Tan, H.Y.; Wang, J.; Yu, S.Z.; Zhou, K.B. Support morphology-dependent catalytic activity of Pd/CeO₂ for formaldehyde oxidation. *Environ. Sci. Technol.* **2015**, *49*, 8675–8682. [CrossRef] [PubMed]

12. Ye, J.W.; Cheng, B.; Wageh, S.; Al-Ghamdib, A.A.; Yu, J.G. Flexible Mg–Al layered double hydroxide supported Pt on Al foil for use in room-temperature catalytic decomposition of formaldehyde. *RSC Adv.* **2016**, *6*, 34280–34287. [CrossRef]

13. Chen, Y.X.; Huang, Z.W.; Zhou, M.J.; Hu, P.P.; Du, C.T.; Kong, L.D.; Chen, J.M.; Tang, X.F. The active sites of supported silver particle catalysts in formaldehyde oxidation. *Chem. Commun.* **2016**, *52*, 9996–9999. [CrossRef] [PubMed]

14. Xu, Z.H.; Yu, J.G.; Jaroniec, M. Efficient catalytic removal of formaldehyde at room temperature using AlOOH nanoflakes with deposited P. *Appl. Catal. B Environ.* **2015**, *163*, 306–312. [CrossRef]

15. Li, Y.B.; Wang, C.Y.; Zhang, C.B.; He, H. Formaldehyde oixidation on Pd/TiO₂ catalysts at room temperature: The effects of surface oxygen vacancies. *Top. Catal.* **2020**, *63*, 810–816. [CrossRef]

16. Lin, M.Y.; Yu, X.L.; Yang, X.Q.; Ma, X.Y.; Ge, M.F. Exploration of the active phase of the hydrotalcite-derived cobalt catalyst for HCHO oxidation. *Chin. J. Catal.* **2019**, *40*, 703–712. [CrossRef]

17. Qi, Y.Q.P.; Zhang, W.R.; Zhang, Y.S.; Bai, G.M.; Wang, S.W.; Liang, P. Formaldehyde oxidation at room temperature over layered MnO₂. *Catal. Commun.* **2021**, *153*, 106293. [CrossRef]

18. Ma, C.; Sun, S.; Lu, H.; Hao, Z.; Yang, C.G.; Wang, B.; Chen, C.; Song, M.Y. Remarkable MnO₂ structure-dependent H₂O promoting effect in HCHO oxidation at room temperature. *J. Hazar. Mater.* **2021**, *414*, 125542. [CrossRef]

19. Tao, Y.; Li, R.; Huang, A.B.; Ma, Y.N.; Ji, S.D.; Jin, P.; Luo, H.J. High catalytic activity for formaldehyde oxidation of an interconnected network structure composed of d-MnO₂ nanosheets and c-MnOOH nanowires. *Adv. Manuf.* **2020**, *8*, 429–439. [CrossRef]

20. Wang, Z.; Wang, W.Z.; Zhang, L.; Jiang, D. Surface oxygen vacancies on Co₃O₄ mediated catalytic formaldehyde oxidation at room temperature. *Catal. Sci. Technol.* **2016**, *6*, 3845–3853. [CrossRef]

21. Lu, S.H.; Li, K.L.; Huang, F.L.; Chen, C.C.; Sun, B. Efficient MnOₓ-Co₃O₄-CeO₂ catalysts for formaldehyde elimination. *Appl. Surf. Sci.* **2017**, *400*, 277–282. [CrossRef]

22. Bai, L.; Wyrwalski, F.; Safariamin, M.; Bleta, R.; Lamonier, J.F.; Przybylski, C.; Monflier, E.; Ponchel, A. Cyclodextrin-cobalt (II) molecule-ion pairs as precursors to active Co₃O₄/ZrO₂ catalysts for the complete oxidation of formaldehyde: Influence of the cobalt source. *J. Catal.* **2016**, *341*, 191–204. [CrossRef]

23. Huang, Y.C.; Fan, W.J.; Long, B.; Li, H.B.; Qiu, W.T.; Zhao, F.Y.; Tong, Y.X.; Ji, H.B. Alkali-modified non-precious metal 3D-NiCo₂O₄ nanosheets for efficient formaldehyde oxidation at low temperature. *J. Mater. Chem. A* **2016**, *4*, 3648–3654. [CrossRef]

24. Dai, Z.J.; Yu, X.W.; Huang, C.; Li, M.; Su, J.F.; Guo, Y.P.; Xu, H.; Ke, Q.F. Nanocrystalline MnO₂ on an activated carbon fiber for catalytic formaldehyde removal. *RSC Adv.* **2016**, *6*, 97022–97029. [CrossRef]

25. Rong, S.P.; Zhang, P.Y.; Wang, J.L.; Liu, F.; Yang, Y.J.; Yang, G.L.; Liu, S. Ultrathin manganese dioxide nanosheets for formaldehyde removal and regeneration performance. *Chem. Eng. J.* **2016**, *306*, 1172–1179. [CrossRef]

26. Bai, B.Y.; Qiao, Q.; Li, J.H.; Ha, J.M. Synthesis of three-dimensional ordered mesoporous MnO₂ and its catalytic performance in formaldehyde oxidation. *Chin. J. Catal.* **2016**, *37*, 27–31. [CrossRef]

27. Bai, B.Y.; Qiao, Q.; Arandiyan, H.; Li, J.H.; Hao, J.M. Three-dimensional ordered mesoporous MnO₂ supported Ag nanoparticles for catalytic removal of formaldehyde. *Environ. Sci. Technol.* **2016**, *50*, 2635–2640. [CrossRef]

28. Zhu, L.; Wang, J.L.; Rong, S.P.; Wang, H.Y.; Zhang, P.Y. Cerium modified birnessite-type MnO₂ for gaseous formaldehyde oxidation at low temperature. *Appl. Catal. B Environ.* **2017**, *211*, 212–221. [CrossRef]

29. Huang, Y.X.; Ye, K.H.; Li, H.B.; Fan, W.J.; Zhang, Y.M.; Ji, H.B. A highly durable catalysts based on CoₓMn₃₋ₓO₄ nanosheets for low-temperature formaldehyde oxidation. *Nano Res.* **2016**, *9*, 3881–3892. [CrossRef]

30. Peña O'Shea, V.A.; Álvarez-Galván, M.C.; Fierro, J.L.G.; Arias, P.L. Influence of feed composition on the activity of Mn and PdMn/Al₂O₃ catalysts for combustion of formaldehyde/methanol. *Appl. Catal. B Environ.* **2005**, *57*, 191–199. [CrossRef]

31. Li, H.; Qin, F.; Yang, Z.; Cui, X.; Wang, J.; Zhang, L. New reaction pathway induced by plasmon for selective benzyl alcohol oxidation on BiOCl possessing oxygen vacancies. *J. Am. Chem. Soc.* **2017**, *139*, 3513–3521. [CrossRef] [PubMed]

32. Ning, X.; Li, Y.; Yu, H.; Peng, F.; Wang, H.; Yang, Y. Promoting role of bismuth and antimony on Pt catalysts for the selective oxidation of glycerol to dihydroxyacetone. *J. Catal.* **2016**, *335*, 95–104. [CrossRef]

33. Wang, X.L.; Wu, G.D.; Zhang, X.L.; Wang, D.F.; Lan, J.Y.; Li, J.Y. Selective oxidation of glycerol to glyceraldehyde with H₂O₂ catalyzed by CuNiAl hydrotalcites supported BiOCl in Neutral Media. *Catal. Lett.* **2019**, *149*, 1046–1056. [CrossRef]

34. Cavani, F.; Trifiro, F.; Vaccari, A. Hydrotalcite-type anionic clays: Preparation, properties and applications. *Catal. Today* **1991**, *11*, 173–301. [CrossRef]
35. Labajos, F.M.; Rives, V.; Ulibarri, M.A. Effect of hydrothermal and thermal treatments on the physicochemical properties of Mg-Al hydrotalcite-like materials. *J. Mater. Sci.* **1992**, *27*, 1546–1552. [CrossRef]
36. Yan, Z.X.; Xu, Z.H.; Cheng, B.; Jiang, C.J. Co_3O_4 nanorod-supported Pt with enhanced performance for catalytic HCHO oxidation at room temperature. *Appl. Surf. Sci.* **2017**, *404*, 426–434. [CrossRef]
37. Prescott, H.A.; Li, Z.J.; Kemnitz, E.; Trunschke, A.; Deutsch, J.; Lieske, H.; Auroux, A. Application of calcined Mg–Al hydrotalcites for Michael additions: An investigation of catalytic activity and acid–base properties. *J. Catal.* **2005**, *234*, 119–130. [CrossRef]
38. Wu, G.D.; Wang, X.L.; Chen, B.; Li, J.P.; Zhao, N.; Wei, W.; Sun, Y.H. Fluorine-modified mesoporous Mg–Al mixed oxides: Mild and stable base catalysts for O-methylation of phenol with dimethyl carbonate. *Appl. Catal. A Gen.* **2007**, *329*, 106–111. [CrossRef]
39. Li, H.; Li, J.; Jia, F.L.; Zhang, L.Z. Oxygen vacancy-mediated photocatalysis of BiOCl: Reactivity, selectivity, and perspectives. *Angew. Chem. Int. Ed.* **2017**, *56*, 2–19. [CrossRef]
40. Wu, G.D.; Wang, X.L.; Wei, W.; Sun, Y.H. Fluorine-modified mesoporous Mg–Al mixed oxides: A solid base with variable basic sites and tunable basicity. *Appl. Catal. A Gen.* **2010**, *377*, 107–113. [CrossRef]
41. Ishikawa, S.; Goto, Y.; Kawahara, Y.; Inukai, S.; Hiyoshi, N.; Dummer, N.F.; Murayama, T.; Yoshida, A.; Sadakane, M.; Ueda, W. Synthesis of crystalline microporous Mo−V−Bi oxide for selective oxidation of light alkanes. *Chem. Mater.* **2017**, *29*, 2939–2950. [CrossRef]
42. Xia, Y.T.; Wang, J.L.; Gu, C.Q.; Ling, Y.; Gao, Z.M. MnO_2/Al foil decorated air cleaner with self-driven property for the abatement of indoor formaldehyde. *Chem. Eng. J.* **2020**, *382*, 122872. [CrossRef]
43. Zou, N.; Nie, Q.; Zhang, X.R.; Zhang, G.K.; Wang, J.L.; Zhang, P.Y. Electrothermal regeneration by Joule heat effect on carbon cloth based MnO_2 catalyst for long-term formaldehyde removal. *Chem. Eng. J.* **2019**, *357*, 1–10. [CrossRef]

MDPI

Article

Alkali-Free Hydrothermally Reconstructed NiAl Layered Double Hydroxides for Catalytic Transesterification

Nazrizawati A. Tajuddin [1], Jinesh C. Manayil [2], Adam F. Lee [3] and Karen Wilson [3,*]

[1] School of Chemistry and Environment, Faculty of Applied Sciences, University of Technology MARA, Shah Alam 40450, Malaysia; nazriza@uitm.edu.my
[2] Energy & Bioproducts Research Institute (EBRI), Aston University, Birmingham B4 7ET, UK; j.manayil@aston.ac.uk
[3] Centre for Applied Materials & Industrial Chemistry (CAMIC), School of Science, RMIT University, 124 La Trobe Street, Melbourne, VIC 3000, Australia; adam.lee2@rmit.edu.au
* Correspondence: karen.wilson2@rmit.edu.au

Abstract: NiAl layered double hydroxides (LDHs) are promising bifunctional catalysts comprising tunable redox and Lewis acidic sites. However, most studies of NiAl LDH employ alkali hydroxide carbonate precipitants which may contaminate the final LDH catalyst and leach into reaction media. Here, we report an alkali-free route to prepare Ni_xAl LDHs with a composition range x = 1.7 to 4.1 using $(NH_4)_2CO_3$ and NH_4OH as precipitants. Activation of LDHs by calcination–rehydration protocols reveal Ni_xAl LDHs can be reconstructed under mild hydrothermal treatment (110 °C for 12 h), with the degree of reconstruction increasing with Ni content. Catalyst activity for tributyrin transesterification with methanol was found to increase with Ni content and corresponding base site loadings; TOFs also increased, suggesting that base sites in the reconstructed LDH are more effective for transesterification. Hydrothermally reconstructed $Ni_{4.1}Al$ LDH was active for the transesterification of C_4–C_{12} triglycerides with methanol and was stable towards leaching during transesterification.

Keywords: biodiesel; layered double hydroxides; transesterification; hydrothermal; nickel; aluminum; solid base

Citation: Tajuddin, N.A.; Manayil, J.C.; Lee, A.F.; Wilson, K. Alkali-Free Hydrothermally Reconstructed NiAl Layered Double Hydroxides for Catalytic Transesterification. *Catalysts* **2022**, *12*, 286. https://doi.org/10.3390/catal12030286

Academic Editors: Ioan-Cezar Marcu and Octavian Dumitru Pavel

Received: 11 January 2022
Accepted: 20 February 2022
Published: 3 March 2022

Publisher's Note: MDPI stays neutral with regard to jurisdictional claims in published maps and institutional affiliations.

1. Introduction

Layered double hydroxides (LDH) of general formula $[M^{2+}_{(1-x)}Al^{3+}_x(OH)_2]^{x+}$ $(A^{n-})_{x/n}\cdot yH_2O$ are interesting materials for photo-, electro-, and thermocatalysis [1,2] and electronic applications [3–5]. They typically adopt three-dimensional, lamellar structures of mixed metal oxyhydroxide platelets, with the interlayer voids providing nanoporous chemical reactors [6]. The catalytic versatility of LDHs arises from their structural flexibility, wherein a range of divalent M^{2+} and trivalent M^{3+} cations can be introduced into the platelets and diverse charge-balancing anions A^{n-} (e.g., CO_3^{2-}, SO_4^{2-}, Cl^-, OH^-) within the interlayers [7,8]. Consequently, LDHs offer tunable redox and electronic properties and morphologies amenable to delamination into individual two-dimensional platelets [9] or assembly into complex architectures such as sand roses [10,11]. The morphology of LDHs is largely dictated by their synthesis; low temperature co-precipitation favors random lamellar arrangements in which precursor impurities are retained [9], whereas moderate temperature calcination to remove impurities necessitates subsequent reconstruction [12,13] of mixed metal oxide crystallites by hydrothermal treatment to recover lamellar structures (calcination–rehydration) [14–18].

LDHs featuring transition metals such as Co, Fe, or Ni as divalent cations have been the focus of electrocatalytic applications [19] due to their good conductivity [20], low cost, and ease of generating host structures with high proton mobility [19]. NiAl LDH materials also find widespread application as precursors to Ni/Al_2O_3 catalysts [21] for hydrogenation/hydrodeoxygenation [19,22], H_2 production by steam reforming [23], and lignin depolymerization/hydrogenolysis [24].

LDH materials are typically synthesized by the coprecipitation of soluble M^{2+}/M^{3+} ionic solutions using NaOH or KOH, in the presence of the corresponding counter ion, typically CO_3^{2-} [8,25]. The use of ammonia or urea as a precipitating agent is preferable for applications where intercalated alkali cations from the residual precipitating agent could impair performance of the resulting LDH [26–28]. While routes such as hydrothermal urea hydrolysis [29] have been employed to prepare NiAl LDHs, the use of NH_4CO_3/NH_4OH precipitants has not been explored, to our knowledge, and may offer milder synthesis conditions.

Activation of LDH materials is often required to increase their catalytic utility [16]. This commonly involves calcination to decompose the precipitated LDH by removing charge compensating anions (accompanied by dehydroxylation) to form an amorphous mixed oxide. Reconstruction of these mixed oxides into lamellar structures is observed on subsequent hydration, termed the 'memory effect', which also introduces vacancies into the metal oxyhydroxide layers, due to the replacement of interlayer CO_3^{2-} with OH^- [30,31]. Calcined MgAl LDH materials can be readily reconstructed by vapor phase rehydration [12,32], but relies on moderate temperature (<550 °C) calcination to avoid significant formation of tetrahedral Al^{3+} and spinel structures [33]. Indeed, not all calcined LDHs can be fully reconstructed by such treatments, as the partial oxidation of the M^{2+} cation or the formation of thermodynamically stable mixed oxides are also possible [34]. A consequence is that calcined ZnAl LDH requires more forcing hydrothermal treatments to partially regenerate the initial lamella structure, due to the stability of Al(OH) and ZnO formed on calcination [28]. NiAl LDH materials are also challenging to reconstruct due to the formation of Al-doped NiO and spinel-like phases at elevated temperatures [21,35,36]. The influence of the Ni:Al ratio on the hydrothermal reconstruction of calcined NiAl LDH prepared by NH_4OH/NH_4CO_3 precipitation has not been previously reported. However, as alkali-free precipitants typically yield smaller LDH crystallites than their alkali analogues [26], these may be more susceptible toward reconstruction.

NiAl LDH materials are expected to exhibit mild basic properties (albeit weaker than for MgAl LDH, due to the lower electropositivity of the transition metal) [36,37], and hence may be suitable for base-catalyzed reactions. Here, we explore the hydrothermal reconstruction and catalytic activity of a family of NiAl LDHs for the transesterification of model (C4-C12) triglycerides to fatty acid methyl esters (FAME).

2. Results and Discussion

2.1. Catalyst Characterisation

The successful synthesis of alkali-free NiAl LDH materials was verified by XRD, SEM, and EDS. XRD reflections of the as-precipitated LDHs (Figure 1a) were in good agreement with reference patterns for the takovite structure (JSPDS 15-0087), while EDS confirmed that Ni:Al bulk atomic ratios were similar (albeit slightly higher) to their nominal values, varying between 1.7:1 and 4.1:1 across the LDH family (Table 1). Corresponding BET surface areas were also in accordance with literature values for NiAl LDHs, spanning 260–60 $m^2 \cdot g^{-1}$ [8]. Surface areas of LDH materials are a complex convolution of crystallite size and density; however, for the same morphology, increased Ni content is expected to lead to a decrease in the specific surface area. NiAl LDH compositions were also estimated from TGA by quantifying weight losses attributed to the desorption of intercalated H_2O (~150 °C) and CO_3^{2-} decomposition (261–355 °C) (Table S1 and Figure S2) and are in good agreement with expectations [38,39]. Higher Ni loadings are expected to expand the interlayers, thereby increasing water intercalation.

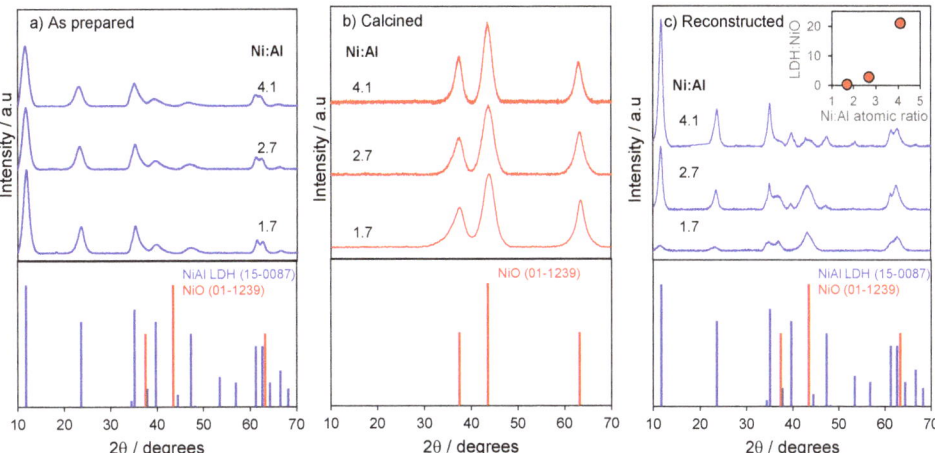

Figure 1. Powder XRD patterns of (**a**) As-prepared; (**b**) Calcined; and (**c**) Hydrothermally reconstructed, alkali-free NiAl LDH materials.

Table 1. Physicochemical properties of as-prepared and reconstructed NiAl LDH materials.

	As-Prepared				Reconstructed		
Nominal Ni:Al Atomic Ratio	Bulk Ni:Al Atomic Ratio [a]	Surface Ni:Al Atomic Ratio [b]	NiAl LDH Formula [c]	BET Surface Area [d]/m²·g⁻¹	BET Surface Area [d]/m²·g⁻¹	Base Site Loading [e] /mmol·g⁻¹	Base Site Loading /Molecules·g⁻¹
1.5:1	1.7:1	1.0	$[Ni_{0.6}Al_{0.4}(OH)_2].(CO_3)_{0.18}0.56H_2O$	149	183	0.062	3.7×10^{19}
3:1	2.7:1	2.6	$[Ni_{0.7}Al_{0.3}(OH)_2].(CO_3)_{0.14}0.58H_2O$	236	207	0.072	4.4×10^{19}
4:1	4.1:1	4.1	$[Ni_{0.8}Al_{0.2}(OH)_2].(CO_3)_{0.10}0.72H_2O$	77	134	0.099	6.0×10^{19}

[a] from EDX (error ±0.13), [b] from XPS, [c] CO_3^{2-}, and H_2O quantified by TGA, [d] Standard error ±30 m²·g⁻¹, [e] from CO_2 pulse chemisorption.

A comparison of surface and bulk Ni compositions from XPS and EDS, respectively, confirm an expected linear relationship for a pure phase LDH (Figure S1a) and, hence, a negligible surface segregation of Ni or Al pure phases. Ni 2p XP spectra of the NiAl LDHs (Figure S1b) reveal a well-resolved spin–orbit split doublet with $2p_{3/2}$ and $2p_{1/2}$ peaks at 854 eV and 872 eV binding energies, respectively, and shake-up satellites at 860 eV and 879 eV consistent with Ni^{2+} [40]. Corresponding Al 2p XP spectra reveal a similar spin–orbit split doublet with a $2p_{3/2}$ peak at 75.2 eV consistent with Al^{3+} in an LDH. Activation of NiAl LDHs by calcination and hydrothermal reconstruction was subsequently assessed. A 350 °C calcination temperature was selected as this achieved sufficient dehydroxylation/decarbonylation to destruct the ordered LDH phase in all materials (Figure S2) while also minimizing sintering. All calcined NiAl LDHs exhibited reflections at 37.4°, 43.9°, and 63.3° (Figure 1b), consistent with the formation of NiO (JSPDS 01–1239) [41]. The genesis of a crystalline $NiAl_2O_4$ spinel phase was not observed, and requires higher temperatures (~900 °C) [42]. Calcination at 350 °C thus generates NiO crystallites and amorphous alumina, similar to previous reports [38,43,44]. Higher temperature calcination of an alkali-precipitated Ni_2Al LDH reports the formation of an Al-containing NiO-like oxide at 450 °C alongside an Al-rich amorphous component. Previous Rietveld refinement also revealed a smaller lattice parameter for the former phase than observed for pure NiO (lattice parameter of 0.418 nm) [43], which increased with Ni:Al content of the parent LDH. A slight increase in the lattice parameter (determined from the d(200) reflection) of the NiO phase with Ni:Al ratio was also observed for our NiAl LDHs (from 0.413 to 0.416 nm); this indicates that trace Al must likewise be incorporated into alkali-free NiO crystallites obtained after 350 °C calcination.

The reconstruction of 350 °C calcined NiAl LDHs was subsequently attempted at 110 °C in water. Note that it was not possible to reconstruct calcined NiAl LDHs using water vapor (Figure S3) in accordance with previous reports [21,35,36]. The complete reconstruction of calcined LDHs requires rehydration of mixed metal oxides and dissolution of any segregated M^{2+} oxide, which for Ni^{2+} requires more forcing conditions owing to the formation of stable (NiAl)O mixed oxides on calcination [13].

Hydrothermal treatments have proven more effective than water vapor in reconstructing the lamellar structure of calcined MgAl [45] and ZnAl [28] LDHs. In the present work, recovery of an LDH phase was strongly dependent on the Ni:Al ratio in the parent material (Figure 1c), with $Ni_{1.7}Al$ and $Ni_{2.7}Al$ LDHs retaining significant NiO. In contrast, the $Ni_{4.1}Al$ material exhibited negligible NiO and a majority reconstructed LDH phase, with the lattice parameter for the c axis determined to be 2.25 nm (calculated from the d(006) reflection at 23.6°). This is almost identical to that for the parent material (2.26 nm), and corresponds to an interlayer spacing (d) of 0.75 nm (note lattice parameter c = 3d [46]). This observation is significant since it is generally regarded that reconstruction of calcined NiAl LDHs by rehydration alone is not possible and requires harsh conditions (e.g., 160 °C in NH_4OH) [36]. Successful hydrothermal reconstruction in this work may reflect enhanced Al incorporation into the NiO phase obtained by lower temperature (350 °C) calcination versus literature reports, and concomitantly less phase-separated tetrahedral Al^{3+} (difficult to reincorporate during LDH reconstruction [33]).

The morphology of the $Ni_{4.1}Al$ LDH was studied by SEM following calcination and hydrothermal reconstruction (Figure 2). The precipitated NiAl LDH exhibited a characteristic sand-rose structure comprising assemblies of fused platelets. These structures evolve due to initially fast crystallisation of LDH seeds which preferentially grow along their (001) planes, disfavouring nanoparticle aggregation [47]. HRTEM (Figure S4) reveals the lattice spacing of LDH crystallites within precipitated platelets as 0.33 nm, corresponding to the d(110) plane, which is slightly larger than that determined from the corresponding XRD reflection but within the range expected for LDH lattice fringes. Calcination transforms the parent LDH into spherical agglomerates (Figure 2b), which are converted back to a fused lamellar structure by hydrothermal treatment (Figure 2c). Lattice fringes of the reconstructed $Ni_{4.1}Al$ LDH were determined by XRD and HRTEM as 0.30 and 0.31 nm, respectively. Slight contractions in the interlayer spacing relative to the parent material are attributed to the replacement of interlayer CO_3^{2-} by OH^- anions. NiAl LDHs are expected to exhibit basic character, originating from undercoordinated oxygen sites or Me–OH surface groups formed to charge balanced Al^{3+} ions replaced by Ni^{2+} in LDH layers as the Ni:Al ratio increases [48]. The quantification of base site loadings by CO_2 pulse chemisorption confirmed that basicity increases with Ni:Al ratio (Table 1).

2.2. Transesterification Activity

The impact of Ni:Al ratio on the activity of reconstructed NiAl LDHs was subsequently evaluated for the base-catalyzed transesterification of tributyrin with methanol (Figures 3 and S5).

TOFs increased from 60 to 180 h^{-1} with rising Ni:Al ratio, mirroring the trends in conversion (which ranged from 15% to 30% over 6 h reaction) (see Figure S5), indicative of increased accessibility or base strength of active sites, the former being consistent with a greater extent of hydrothermal reconstruction observed by XRD (Figure 1c inset). These TOFs are in line with literature values for tributyrin transesterification over related ZnAl (180–290 h^{-1}) [28] and Mg_2Al (329 h^{-1}) [45,49] LDHs, considering their relative basicity (NiAl ≈ ZnAl < MgAl [16,50]), which are comparable to values for nanocrystalline MgO (220 h^{-1}) [51] and Mg-ZrO_2 (100 h^{-1}) [52]. Selectivity to methyl butyrate was >95–80%, with a slight decrease observed with Ni:Al ratio attributed to increased conversion across the series, and reduced transesterification activity of diglycerides (see discussion later) [53].

The most active $Ni_{4.1}Al$ LDH catalyst was selected for the transesterification of longer-chain C_8-C_{12} TAGs (Figure 4 and Figure S6). $Ni_{4.1}Al$ LDH was active for all three TAGs; however, conversions and TOFs fell with increasing alkyl chain length, likely reflecting poorer base site accessibility for the bulkier TAGs [45]. FAME selectivities also fell with increasing chain length, reaching ~40% for trilaurin after 24 reaction, suggesting poor mass transport [45,54] of bulky diglyceride (DAG) and monoglyceride (MAG) reactively-formed intermediates, and that their competitive adsorption with TAG hinders subsequent conversion under batch conditions [55,56]. Slow DAG conversion has previously been accounted for, owing to activation barriers for transesterification being higher than for their corresponding triglycerides. The removal of the first fatty acid chain leads to increased electron density at the -CH_2-OH center of the glyceride backbone, which hinders subsequent attack by adsorbed MeO- at the remaining ester groups [53].

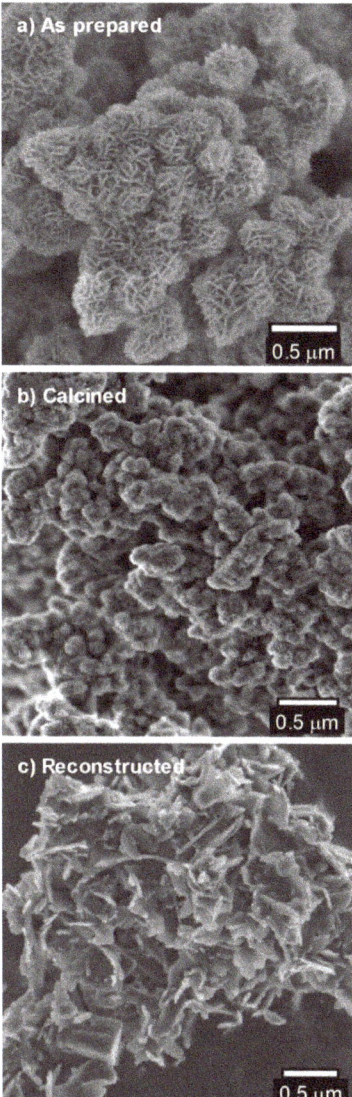

Figure 2. SEM images of (**a**) As-prepared; (**b**) Calcined; and (**c**) Hydrothermally reconstructed, alkali-free Ni4.1Al LDH materials.

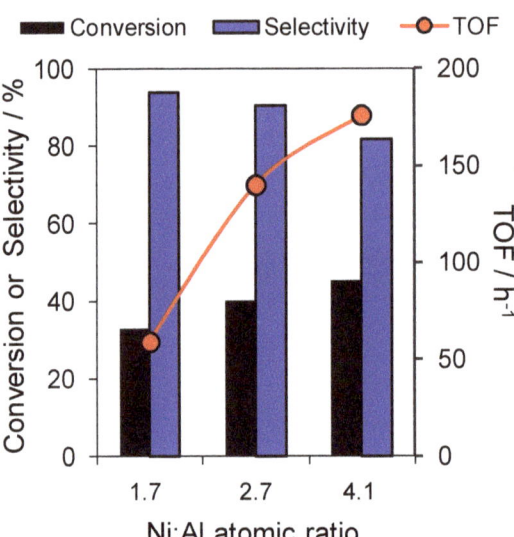

Figure 3. Activity of reconstructed NiAl LDH for transesterification of tributyrin with methanol as a function of NiAl ratio; conversion and selectivity after 24 h, TOF from initial rate over 1 h reaction. Reaction conditions: 100 mg catalyst, 110 °C, 10 mmol TAG, 30:1 methanol:TAG molar ratio, 650 rpm stirring.

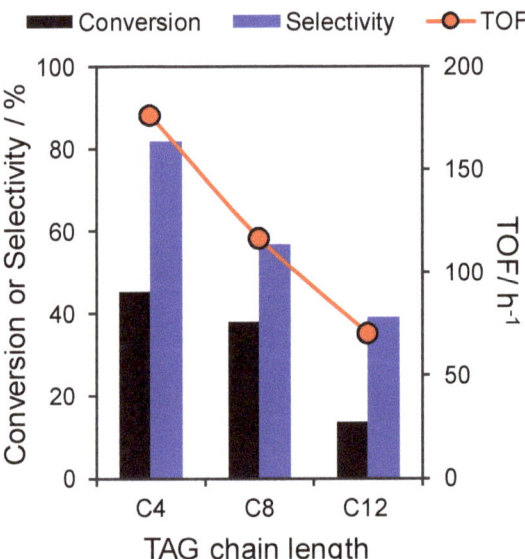

Figure 4. Catalytic performance of reconstructed $Ni_{4.1}Al$ LDH for transesterification of C_4-C_{12} TAGs with methanol; conversion and selectivity values after 24 h, TOF calculated from initial rate over 1 h reaction. Reaction conditions: 100 mg catalyst, 110 °C, 10 mmol TAG, 30:1 methanol:TAG molar ratio (20 wt% 1-butanol for C_8–C_{12} TAG), 650 rpm stirring.

The stability of the $Ni_{4.1}Al$ LDH catalyst was investigated by a recycling study for tributyrin transesterification, which confirmed the heterogenous nature of the catalysis (Figure S7). The same catalyst could be recycled, retaining ~60% of activity for tributyrin transesterification after three cycles. ICP-OES revealed a negligible decrease in Ni:Al (Table S2) after the third recycle, suggesting that loss of activity was attributed to deactivation by poisoning of base sites or pore blockage by reaction intermediates [45].

There are few previous studies of NiAl LDH or NiO-derived catalysts for TAG transesterification, which hinders performance benchmarking, and of these all employed alkali precipitation syntheses and, hence, homogenous contributions to catalytic activity cannot be discounted [57,58]. Calcined NiAl and ZnAl LDHs are reported as inactive for sunflower oil transesterification with methanol at 65 °C [50]. Our reconstructed $Ni_{4.1}Al$ LDH, which affords 14% trilaurin conversion (albeit at a higher temperature), is promising, despite their essentially microporous nature arising from a lamellar structure [59] with layer spacing of 0.3 nm observed in HRTEM (Figure S4), which hinders access of bulky TAGs to interlayers and in-pore base sites. Hence, most TAG activation is expected to occur over the external surface of the 2D NiAl LDH lamellar sheets in the sand rose structure observed in Figure 2c. The introduction of meso- and macroporosity is expected to significantly enhance TAG accessibility; hence, alkali-free, hydrothermally activated NiAl LDHs show potential as heterogeneous solid base catalysts for transesterification.

3. Experimental

3.1. Catalyst Synthesis

All reagents were of analytical grade and were used without further purification. Ni_xAl LDHs were synthesized in a 500 mL Radley's reactor-ready stirred vessel. Aqueous solutions of $Ni(NO_3)_2·6H_2O$ (1.5 M) and $Al(NO_3)_3·9H_2O$ (1.5 M) were mixed together in desired proportions to yield Ni:Al molar ratios spanning 1.5 to 4, with a final solution volume of 100 cm^3. Ammonium carbonate (2 M, 100 cm^3) was used as the precipitating agent and was added by a syringe pump (1 $mL·min^{-1}$) to the metal nitrate mixture while being stirred. Once the addition of metal nitrate and ammonium carbonate solutions was completed, ammonia solution was added dropwise to maintain a constant pH of 9.5. The mixture was aged while stirring at 65 °C for 18 h, after which the resulting precipitate was filtered and washed repeatedly with deionised water until the washings were pH 7. The resulting solid powder was then calcined at 350 °C (ramp rate of 1 $°C·min^{-1}$) for 5 h under flowing O_2 (20 $mL·min^{-1}$). The calcined solid was subjected to hydrothermal treatment in 50 mL of water, in a stirred Teflon-lined autoclave at 110 °C under autogenous pressure for 12 h. The treated solid was recovered by centrifugation and oven-dried at 80 °C before being stored in a vacuum desiccator prior to use.

3.2. Catalyst Characterisation

Powder X-ray diffraction was performed on a Bruker D8 ADVANCE diffractometer (Coventry, UK) using Cu K_α X-ray radiation (0.15418 nm). The diffraction patterns were recorded between $2\theta = 10–80°$. Crystallite sizes were determined by application of the Scherrer equation. The ratio of LDH:NiO following reconstruction was determined from the relative heights of the most intense reflections for NiAl LDH (13.8°) and NiO (43.6°). X-ray photoelectron spectroscopy was performed on a Kratos Axis HSi spectrometer (Manchester, UK) fitted with a charge neutralizer and magnetic focusing lens employing monochromated Al K_α radiation at 90 W. Spectral fitting was performed using CasaXPS version 2.3.16, with binding energies referenced to the C 1s Peak at 284.5 eV. N_2 porosimetry was undertaken on a Quantachrome Nova 4000 porosimeter (Hook, UK) on samples degassed at 120 °C for 3 h. Surface areas were calculated by the Brunauer–Emmett–Teller (BET) method from the desorption isotherm for $P/P_0 < 0.2$. Base site densities were determined by CO_2 pulse titration using a Quantachrome ChemBET 3000 chemisorption analyzer (Hook, UK) with a thermal conductivity detector. Then, 50 mg of the sample was placed in a quartz cell, outgassed for 1 h under flowing He at 120 °C, cooled to 40 °C, and then titrated with 50 µL CO_2 pulses at room temperature until uptake saturated. Thermogravimetric analysis with online mass spectrometry (TGA-MS) was performed on a Mettler Toledo TGA/DSC2 Star system (Leicester, UK) under flowing nitrogen during sample heating to

800 °C at 10 °C·min^{-1}. TGA was performed under N_2 to avoid any complications from oxidation of defects formed when a loss of H_2O/CO_2 would complicate the calculation of absolute mass loss during the TGA experiment. Scanning electron microscopy (SEM) was performed on a JEOL JSM-7000F microscope fitted with an energy dispersive X-ray spectroscopy detector (EDS) (Welwyn Garden City, UK) using an accelerating voltage of 20 kV. Transmission electron microscopy (TEM) analysis was carried out using a JEOL 2100F FEG STEM microscope operating at 200 keV. Samples were dispersed in methanol and drop cast onto copper grids coated with a holey carbon film (Agar Scientific Ltd., Essex, UK).

3.3. Catalytic Activity

Transesterification was performed using Radley's Starfish reactor, equipped with ACE^{TM} pressure flasks (Vineland, NJ, USA) modified with a dip tube to enable aliquots to be periodically withdrawn. Reactions were performed at 110 °C, using 10 mmol of triglyceride (TAG) and 308 mmol (12.5 mL) methanol, with 0.0025 of mol (0.59 cm^3) dihexyl ether as the internal standard; an additional 20 wt% 1-butanol was introduced for longer chain C_8-C_{12} TAGs (tricaprylin and trilaurin, 99%, Alfa Aesar (Lancashire, UK) to improve their miscibility. Aliquots were periodically withdrawn during reaction. They were then filtered and diluted with dichloromethane, and analyzed by off-line gas chromatography (GC) using a Varian 450-GC (Crawley, UK) fitted with a Phenomenex (California, USA) ZB-5HT Inferno capillary column (15 m × 0.32 mm × 0.1 μm) for C_4–C_8 TAGs. Heavier TAGs were analyzed using a temperature-programmed, on-column injector and a Phenomenex ZB-1HT Inferno wide-bore capillary column (15 m × 0.53 mm × 0.1 μm). Turnover frequencies (TOFs) were calculated by normalizing initial rates of TAG conversion derived from the linear portion of reaction profiles (<20% conversion) to base site loadings obtained from CO_2 chemisorption. Triglyceride conversion and selectivity to fatty acid methyl ester (FAME) were calculated as follows:

$$\text{Conversion} = \frac{\text{Concentration of TAG at T} = 0(\text{mol/L}) - \text{Concentration of TAG at T} = t(\text{mol/L})}{\text{Concentration of TAG at T} = 0(\text{mol/L})} \times 100 \quad (1)$$

$$\text{Selectivity} = \frac{\text{Concentration of product formed (mol/L)}}{\text{Total product concentration (mol/L)}} \times 100 \quad (2)$$

Leaching studies were conducted by hot filtration and recycle tests to establish the catalyst stability during tributyrin transesterification in methanol at 110 °C. The recovered catalyst was filtered between recycles, repeatedly washed with methanol to remove weakly bound residues, and then dried at 80 °C. After the third cycle, the catalyst was analyzed by ICP-OES to determine the final Ni and Al content.

4. Conclusions

NiAl LDHs with Ni:Al molar ratios spanning 1.5–4.1 were synthesized via an alkali-free route using NH_4OH and NH_4CO_3 as pH regulator and precipitant, respectively. The influence of activation protocol on 350 °C calcined Ni_xAl LDH was explored, with hydrothermal reconstruction at 110 °C proving to be the most effective for regenerating the parent LDH crystalline lamellar structure and maximizing base site loading. The extent of hydrothermal regeneration increased with Ni:Al ratio, attributed to suppressed NiO and AlO_x phase separation during calcination for higher Ni content. Base site loading increased with Ni content, attributed to undercoordinated oxygen sites as Al^{3+} is replaced by Ni^{2+}, and/or increased surface hydroxyls on lamellae. The initial rates of tributyrin transesterification with methanol (and final conversion) increased with Ni loading and, hence, the base site density. However, TOFs also increased with Ni:Al ratio, indicative of a change in base site strength or accessibility across the family (the latter consistent with enhanced reconstruction). The most active $Ni_{4.1}Al$ LDH catalyst was moderately stable towards site-blocking, and effective for the transesterification of C_4–C_{12} triglycerides to FAMEs. Future work will explore the versatility of these catalysts for transesterification of real oil feedstocks and the development of macroporous architectures in order to improve the mass transport of bulky TAGs. The somewhat weaker basicity of NiAl versus MgAl LDH counterparts may be advantageous with respect to poisoning by fatty acid impurities or adventitious CO_2.

Supplementary Materials: The following are available online: https://www.mdpi.com/article/10.3390/catal12030286/s1, Bulk and surface Ni content by XPS and EDS; XPS, TGA, XRD, HRTEM, and transesterification reaction profiles, conversion and selectivity following catalyst recycle tests, and elemental analysis of recycled catalyst. Figure S1. (a) Ni surface and bulk loading of NiAl LDHs determined by XPS and EDS respectively; (b) High-resolution XP spectra of NiAl LDHs with nominal Ni:Al atomic ratios of (i) 1.5, (ii) 3 and (iii) 4; Figure S2. Thermogravimetric analysis of NiAl LDHs with nominal Ni:Al atomic ratios of (A) 1.5, (B) 3 and (C) 4; Figure S3. Powder XRD patterns of Ni4Al LDH after different hydrothermal treatments; Figure S4. HRTEM images of (top) precipitated and (bottom) hydrothermally reconstructed Ni4.1Al LDH. Hydrothermal treatment involved 350 °C calcination and subsequent 110 °C water treatment; Figure S5. Reaction profiles for C4 TAG transesterification with methanol by NiAl LDH as a function of Ni:Al atomic ratio. Reaction conditions: 100 mg catalyst, 10 mmol of triglyceride (TAG), 308 mmol (12.5 mL) methanol with 0.0025 mol (0.59 cm^3) dihexylether as internal standard. Reaction was done at 650 rpm, 110 °C for 24 h; Figure S6. Reaction profiles for C4-C12 TAG transesterification with methanol by reconstructed Ni4.1Al LDH. Reaction conditions: 100 mg catalyst, 10 mmol of triglyceride (TAG), 308 mmol (12.5 mL) methanol with 0.0025 mol (0.59 cm^3) dihexylether as internal standard. 20 wt% 1-butanol was introduced for longer chain C8-C12 TAGs (tricaprylin and trilaurin) to improve their miscibility. Reaction was performed at 110 °C, 650 rpm, for 24 hours; Figure S7. Recycling of hydrothermally reconstructed Ni4Al LDH in the transesterification of tributyrin with methanol. Reactions conditions: 100 mg catalyst, 110 °C, 10 mmol tributyrin (TAG), 30:1 methanol:TAG molar ratio, 650 rpm stirring, 24 h reaction; Table S1. Weight losses from TGA; Table S2. ICP-OES analysis on reconstructed Ni4Al LDH after 3 recycles.

Author Contributions: Conceptualization, K.W. and A.F.L.; Data curation, N.A.T.; Formal analysis, N.A.T.; Investigation, N.A.T.; Methodology, J.C.M.; Supervision, J.C.M., A.F.L. and K.W.; Writing—original draft, N.A.T.; Writing—review and editing, A.F.L. and K.W. All authors have read and agreed to the published version of the manuscript.

Funding: This research was funded through a grant to N.A.T. from the University of Technology MARA (UiTM) under the Young Talent Research Grant (600-RMC/YTR/5/3 (017/2020)). K.W. and A.F.L. acknowledge the Australian Research Council for funding under DP200100204 and DP200100313.

Data Availability Statement: Data is available from the corresponding author on request.

Acknowledgments: We also acknowledge Lee Durndell (University of Plymouth) and Christopher Parlett (University of Manchester and Diamond Light Source) for assistance with HRTEM and Mark Issacs (University College London and HarwellXPS—EPSRC National Facility for XPS) for XPS measurements.

Conflicts of Interest: The authors declare no conflict of interest.

References

1. Yan, K.; Liu, Y.; Lu, Y.; Chai, J.; Sun, L. Catalytic application of layered double hydroxide-derived catalysts for the conversion of biomass-derived molecules. *Catal. Sci. Technol.* **2017**, *7*, 1622–1645. [CrossRef]
2. Song, Y.; Beaumont, S.K.; Zhang, X.; Wilson, K.; Lee, A.F. Catalytic applications of layered double hydroxides in biomass valorisation. *Curr. Opin. Green Sustain. Chem.* **2020**, *22*, 29–38. [CrossRef]
3. Jing, C.; Zhang, Q.; Liu, X.; Chen, Y.; Wang, X.; Xia, L.; Zeng, H.; Wang, D.; Zhang, W.; Dong, F. Design and fabrication of hydrotalcite-like ternary NiMgAl layered double hydroxide nanosheets as battery-type electrodes for high-performance supercapacitors. *RSC Adv.* **2019**, *9*, 9604–9612. [CrossRef]
4. Li, R.; Hu, Z.; Shao, X.; Cheng, P.; Li, S.; Yu, W.; Lin, W.; Yuan, D. Large Scale Synthesis of NiCo Layered Double Hydroxides for Superior Asymmetric Electrochemical Capacitor. *Sci. Rep.* **2016**, *6*, 18737. [CrossRef] [PubMed]
5. Szilagyi, I. Layered Double Hydroxide-Based Nanomaterials—From Fundamentals to Applications. *Nanomaterials* **2019**, *9*, 1174. [CrossRef]

6. Tokudome, Y.; Tarutani, N.; Nakanishi, K.; Takahashi, M. Layered double hydroxide (LDH)-based monolith with interconnected hierarchical channels: Enhanced sorption affinity for anionic species. *J. Mater. Chem. A* **2013**, *1*, 7702–7708. [CrossRef]
7. Wang, Y.; Yan, D.; El Hankari, S.; Zou, Y.; Wang, S. Recent Progress on Layered Double Hydroxides and Their Derivatives for Electrocatalytic Water Splitting. *Adv. Sci.* **2018**, *5*, 1800064. [CrossRef]
8. Cavani, F.; Trifirò, F.; Vaccari, A. Hydrotalcite-type anionic clays: Preparation, properties and applications. *Catal. Today* **1991**, *11*, 173–301. [CrossRef]
9. Wang, Q.; O'Hare, D. Recent Advances in the Synthesis and Application of Layered Double Hydroxide (LDH) Nanosheets. *Chem. Rev.* **2012**, *112*, 4124–4155. [CrossRef]
10. Taviot-Guého, C.; Prévot, V.; Forano, C.; Renaudin, G.; Mousty, C.; Leroux, F. Tailoring Hybrid Layered Double Hydroxides for the Development of Innovative Applications. *Adv. Funct. Mater.* **2018**, *28*, 1703868. [CrossRef]
11. Tonelli, D.; Gualandi, I.; Musella, E.; Scavetta, E. Synthesis and Characterization of Layered Double Hydroxides as Materials for Electrocatalytic Applications. *Nanomaterials* **2021**, *11*, 725. [CrossRef] [PubMed]
12. Valente, J.S.; Lima, E.; Toledo-Antonio, J.A.; Cortes-Jacome, M.A.; Lartundo-Rojas, L.; Montiel, R.; Prince, J. Comprehending the Thermal Decomposition and Reconstruction Process of Sol−Gel MgAl Layered Double Hydroxides. *J. Phys. Chem. C* **2010**, *114*, 2089–2099. [CrossRef]
13. Mascolo, G.; Mascolo, M.C. On the synthesis of layered double hydroxides (LDHs) by reconstruction method based on the "memory effect". *Microporous Mesoporous Mater.* **2015**, *214*, 246–248. [CrossRef]
14. Cao, Z.; Li, B.; Sun, L.; Li, L.; Xu, Z.P.; Gu, Z. 2D Layered Double Hydroxide Nanoparticles: Recent Progress toward Preclinical/Clinical Nanomedicine. *Small Methods* **2020**, *4*, 1900343. [CrossRef]
15. Dewangan, N.; Hui, W.M.; Jayaprakash, S.; Bawah, A.-R.; Poerjoto, A.J.; Jie, T.; Jangam, A.; Hidajat, K.; Kawi, S. Recent progress on layered double hydroxide (LDH) derived metal-based catalysts for CO2 conversion to valuable chemicals. *Catal. Today* **2020**, *356*, 490–513. [CrossRef]
16. Takehira, K. Recent development of layered double hydroxide-derived catalysts—Rehydration, reconstitution, and supporting, aiming at commercial application. *Appl. Clay Sci.* **2017**, *136*, 112–141. [CrossRef]
17. Tang, S.; Yao, Y.; Chen, T.; Kong, D.; Shen, W.; Lee, H.K. Recent advances in the application of layered double hydroxides in analytical chemistry: A review. *Anal. Chim. Acta* **2020**, *1103*, 32–48. [CrossRef]
18. Yang, Z.-Z.; Zhang, C.; Zeng, G.-M.; Tan, X.-F.; Wang, H.; Huang, D.-L.; Yang, K.-H.; Wei, J.-J.; Ma, C.; Nie, K. Design and engineering of layered double hydroxide based catalysts for water depollution by advanced oxidation processes: A review. *J. Mater. Chem. A* **2020**, *8*, 4141–4173. [CrossRef]
19. Birjega, R.; Vlad, A.; Matei, A.; Ion, V.; Luculescu, C.; Dinescu, M.; Zavoianu, R. Growth and characterization of ternary Ni, Mg–Al and Ni–Al layered double hydroxides thin films deposited by pulsed laser deposition. *Thin Solid Film.* **2016**, *614*, 36–41. [CrossRef]
20. Stevens, M.B.; Enman, L.J.; Korkus, E.H.; Zaffran, J.; Trang, C.D.; Asbury, J.; Kast, M.G.; Toroker, M.C.; Boettcher, S.W. Ternary Ni-Co-Fe oxyhydroxide oxygen evolution catalysts: Intrinsic activity trends, electrical conductivity, and electronic band structure. *Nano Res.* **2019**, *12*, 2288–2295. [CrossRef]
21. Clause, O.; Rebours, B.; Merlen, E.; Trifiró, F.; Vaccari, A. Preparation and characterization of nickel-aluminum mixed oxides obtained by thermal decomposition of hydrotalcite-type precursors. *J. Catal.* **1992**, *133*, 231–246. [CrossRef]
22. Yue, X.; Zhang, L.; Sun, L.; Gao, S.; Gao, W.; Cheng, X.; Shang, N.; Gao, Y.; Wang, C. Highly efficient hydrodeoxygenation of lignin-derivatives over Ni-based catalyst. *Appl. Catal. B Environ.* **2021**, *293*, 120243. [CrossRef]
23. Li, D.; Lu, M.; Aragaki, K.; Koike, M.; Nakagawa, Y.; Tomishige, K. Characterization and catalytic performance of hydrotalcite-derived Ni-Cu alloy nanoparticles catalysts for steam reforming of 1-methylnaphthalene. *Appl. Catal. B Environ.* **2016**, *192*, 171–181. [CrossRef]
24. Sturgeon, M.R.; O'Brien, M.H.; Ciesielski, P.N.; Katahira, R.; Kruger, J.S.; Chmely, S.C.; Hamlin, J.; Lawrence, K.; Hunsinger, G.B.; Foust, T.D.; et al. Lignin depolymerisation by nickel supported layered-double hydroxide catalysts. *Green Chem.* **2014**, *16*, 824–835. [CrossRef]
25. Theiss, F.L.; Ayoko, G.A.; Frost, R.L. Synthesis of layered double hydroxides containing Mg^{2+}, Zn^{2+}, Ca^{2+} and Al^{3+} layer cations by co-precipitation methods—A review. *Appl. Surf. Sci.* **2016**, *383*, 200–213. [CrossRef]
26. Olfs, H.W.; Torres-Dorante, L.O.; Eckelt, R.; Kosslick, H. Comparison of different synthesis routes for Mg–Al layered double hydroxides (LDH): Characterization of the structural phases and anion exchange properties. *Appl. Clay Sci.* **2009**, *43*, 459–464. [CrossRef]
27. Cantrell, D.G.; Gillie, L.J.; Lee, A.F.; Wilson, K. Structure-reactivity correlations in MgAl hydrotalcite catalysts for biodiesel synthesis. *Appl. Catal. A: Gen.* **2005**, *287*, 183–190. [CrossRef]
28. Tajuddin, N.A.; Manayil, J.C.; Isaacs, M.A.; Parlett, C.M.A.; Lee, A.F.; Wilson, K. Alkali-Free Zn–Al Layered Double Hydroxide Catalysts for Triglyceride Transesterification. *Catalysts* **2018**, *8*, 667. [CrossRef]
29. Jiang, L.; Guo, H.; Li, C.; Zhou, P.; Zhang, Z. Selective cleavage of lignin and lignin model compounds without external hydrogen, catalyzed by heterogeneous nickel catalysts. *Chem. Sci.* **2019**, *10*, 4458–4468. [CrossRef] [PubMed]
30. Ye, H.; Liu, S.; Yu, D.; Zhou, X.; Qin, L.; Lai, C.; Qin, F.; Zhang, M.; Chen, W.; Chen, W.; et al. Regeneration mechanism, modification strategy, and environment application of layered double hydroxides: Insights based on memory effect. *Coord. Chem. Rev.* **2022**, *450*, 214253. [CrossRef]

31. Yuan, Z.; Bak, S.-M.; Li, P.; Jia, Y.; Zheng, L.; Zhou, Y.; Bai, L.; Hu, E.; Yang, X.-Q.; Cai, Z.; et al. Activating Layered Double Hydroxide with Multivacancies by Memory Effect for Energy-Efficient Hydrogen Production at Neutral pH. *ACS Energy Lett.* **2019**, *4*, 1412–1418. [CrossRef]

32. Kikhtyanin, O.; Lesnik, E.; Kubička, D. The occurrence of Cannizzaro reaction over Mg-Al hydrotalcites. *Appl. Catal. A Gen.* **2016**, *525*, 215–225. [CrossRef]

33. Rocha, J.; del Arco, M.; Rives, V.; Ulibarri, A.M. Reconstruction of layered double hydroxides from calcined precursors: A powder XRD and 27Al MAS NMR study. *J. Mater. Chem.* **1999**, *9*, 2499–2503. [CrossRef]

34. Pérez-Ramírez, J.; Mul, G.; Kapteijn, F.; Moulijn, J.A. On the stability of the thermally decomposed Co-Al hydrotalcite against retrotopotactic transformation. *Mater. Res. Bull.* **2001**, *36*, 1767–1775. [CrossRef]

35. Sato, T.; Fujita, H.; Endo, T.; Shimada, M.; Tsunashima, A. Synthesis of hydrotalcite-like compounds and their physico-chemical properties. *React. Solids* **1988**, *5*, 219–228. [CrossRef]

36. Prinetto, F.; Ghiotti, G.; Graffin, P.; Tichit, D. Synthesis and characterization of sol–gel Mg/Al and Ni/Al layered double hydroxides and comparison with co-precipitated samples. *Microporous Mesoporous Mater.* **2000**, *39*, 229–247. [CrossRef]

37. Belskaya, O.; Leont'eva, N.; Gulyaeva, T.; Cherepanova, S.; Talzi, V.; Drozdov, V.; Likholobov, V. Influence of a doubly charged cation nature on the formation and properties of mixed oxides MAlO x (M= Mg^{2+}, Zn^{2+}, Ni^{2+}) obtained from the layered hydroxide precursors. *Russ. Chem. Bull.* **2013**, *62*, 2349–2361. [CrossRef]

38. Cherepanova, S.V.; Leont'eva, N.N.; Arbuzov, A.B.; Drozdov, V.A.; Belskaya, O.B.; Antonicheva, N.V. Structure of oxides prepared by decomposition of layered double Mg–Al and Ni–Al hydroxides. *J. Solid State Chem.* **2015**, *225*, 417–426. [CrossRef]

39. Claydon, R.; Wood, J. A Mechanistic Study of Layered-Double Hydroxide (LDH)-Derived Nickel-Enriched Mixed Oxide (Ni-MMO) in Ultradispersed Catalytic Pyrolysis of Heavy Oil and Related Petroleum Coke Formation. *Energy Fuels* **2019**, *33*, 10820–10832. [CrossRef]

40. Naumkin, A.V.; Gaarenstroom, S.W.; Cedric, J. NIST Standard Reference Database 20, Version 4.1. Available online: https://srdata.nist.gov/xps (accessed on 22 November 2021).

41. Abelló, S.; Bolshak, E.; Gispert-Guirado, F.; Farriol, X.; Montané, D. Ternary Ni–Al–Fe catalysts for ethanol steam reforming. *Catal. Sci. Technol.* **2014**, *4*, 1111–1122. [CrossRef]

42. Li, C.; Wang, L.; Wei, M.; Evans, D.G.; Duan, X. Large oriented mesoporous self-supporting Ni–Al oxide films derived from layered double hydroxide precursors. *J. Mater. Chem.* **2008**, *18*, 2666–2672. [CrossRef]

43. Kovanda, F.; Rojka, T.; Bezdička, P.; Jirátová, K.; Obalová, L.; Pacultová, K.; Bastl, Z.; Grygar, T. Effect of hydrothermal treatment on properties of Ni–Al layered double hydroxides and related mixed oxides. *J. Solid State Chem.* **2009**, *182*, 27–36. [CrossRef]

44. Trifirò, F.; Vaccari, A.; Clause, O. Nature and properties of nickel-containing mixed oxides obtained from hydrotalcite-type anionic clays. *Catal. Today* **1994**, *21*, 185–195. [CrossRef]

45. Woodford, J.J.; Dacquin, J.-P.; Wilson, K.; Lee, A.F. Better by design: Nanoengineered macroporous hydrotalcites for enhanced catalytic biodiesel production. *Energy Environ. Sci.* **2012**, *5*, 6145–6150. [CrossRef]

46. Prestopino, G.; Arrabito, G.; Generosi, A.; Mattoccia, A.; Paci, B.; Perez, G.; Verona-Rinati, G.; Medaglia, P.G. Emerging switchable ultraviolet photoluminescence in dehydrated Zn/Al layered double hydroxide nanoplatelets. *Sci. Rep.* **2019**, *9*, 11498. [CrossRef]

47. Nishimura, S.; Takagaki, A.; Ebitani, K. Characterization, synthesis and catalysis of hydrotalcite-related materials for highly efficient materials transformations. *Green Chem.* **2013**, *15*, 2026–2042. [CrossRef]

48. Gonçalves, A.A.S.; Costa, M.J.F.; Zhang, L.; Ciesielczyk, F.; Jaroniec, M. One-Pot Synthesis of MeAl2O4 (Me = Ni, Co, or Cu) Supported on γ-Al₂O₃ with Ultralarge Mesopores: Enhancing Interfacial Defects in γ-Al₂O₃ to Facilitate the Formation of Spinel Structures at Lower Temperatures. *Chem. Mater.* **2018**, *30*, 436–446. [CrossRef]

49. Creasey, J.J.; Chieregato, A.; Manayil, J.C.; Parlett, C.M.A.; Wilson, K.; Lee, A.F. Alkali- and nitrate-free synthesis of highly active Mg–Al hydrotalcite-coated alumina for FAME production. *Catal. Sci. Technol.* **2014**, *4*, 861–870. [CrossRef]

50. Sankaranarayanan, S.; Antonyraj, C.A.; Kannan, S. Transesterification of edible, non-edible and used cooking oils for biodiesel production using calcined layered double hydroxides as reusable base catalysts. *Bioresour. Technol.* **2012**, *109*, 57–62. [CrossRef]

51. Montero, J.M.; Brown, D.R.; Gai, P.L.; Lee, A.F.; Wilson, K. In situ studies of structure–reactivity relations in biodiesel synthesis over nanocrystalline MgO. *Chem. Eng. J.* **2010**, *161*, 332–339. [CrossRef]

52. Rabee, A.; Manayil, J.; Isaacs, M.; Parlett, C.; Durndell, L.; Zaki, M.; Lee, A.; Wilson, K. On the Impact of the Preparation Method on the Surface Basicity of Mg–Zr Mixed Oxide Catalysts for Tributyrin Transesterification. *Catalysts* **2018**, *8*, 228. [CrossRef]

53. Asakuma, Y.; Maeda, K.; Kuramochi, H.; Fukui, K. Theoretical study of the transesterification of triglycerides to biodiesel fuel. *Fuel* **2009**, *88*, 786–791. [CrossRef]

54. Kaneda, K.; Mizugaki, T. Design of high-performance heterogeneous catalysts using hydrotalcite for selective organic transformations. *Green Chem.* **2019**, *21*, 1361–1389. [CrossRef]

55. Chantrasa, A.; Phlernjai, N.; Goodwin, J.G. Kinetics of hydrotalcite catalyzed transesterification of tricaprylin and methanol for biodiesel synthesis. *Chem. Eng. J.* **2011**, *168*, 333–340. [CrossRef]

56. Woodford, J.J.; Parlett, C.M.A.; Dacquin, J.-P.; Cibin, G.; Dent, A.; Montero, J.; Wilson, K.; Lee, A.F. Identifying the active phase in Cs-promoted MgO nanocatalysts for triglyceride transesterification. *J. Chem. Technol. Biotechnol.* **2014**, *89*, 73–80. [CrossRef]

57. Bin Jumah, M.N.; Ibrahim, S.M.; Al-Huqail, A.A.; Bin-Murdhi, N.S.; Allam, A.A.; Abu-Taweel, G.M.; Altoom, N.; Al-Anazi, K.M.; Abukhadra, M.R. Enhancing the Catalytic Performance of NiO during the Transesterification of Waste Cooking Oil Using a Diatomite Carrier and an Integrated Ni0 Metal: Response Surface Studies. *ACS Omega* **2021**, *6*, 12318–12330. [CrossRef]

58. Cross, H.E.; Brown, D.R. Entrained sodium in mixed metal oxide catalysts derived from layered double hydroxides. *Catal. Commun.* **2010**, *12*, 243–245. [CrossRef]
59. Gonzalez Rodriguez, P.; de Ruiter, M.; Wijnands, T.; ten Elshof, J.E. Porous Layered Double Hydroxides Synthesized using Oxygen Generated by Decomposition of Hydrogen Peroxide. *Sci. Rep.* **2017**, *7*, 481. [CrossRef] [PubMed]

Article

Electrodeposition of a Li-Al Layered Double Hydroxide (LDH) on a Ball-like Aluminum Lathe Waste Strips in Structured Catalytic Applications: Preparation and Characterization of Ni-Based LDH Catalysts for Hydrogen Evolution

Song-Hui Huang [1], Yu-Jia Chen [1], Wen-Fu Huang [1] and Jun-Yen Uan [1,2,*]

[1] Department of Materials Science and Engineering, National Chung Hsing University, 145 Xingda Rd., Taichung 40227, Taiwan; d099066006@mail.nchu.edu.tw (S.-H.H.); t820207@gmail.com (Y.-J.C.); g106066012@mail.nchu.edu.tw (W.-F.H.)
[2] Innovation and Development Center of Sustainable Agriculture (IDCSA), National Chung Hsing University, 145 Xingda Rd., Taichung 40227, Taiwan
* Correspondence: jyuan@dragon.nchu.edu.tw; Tel.: +886-422-840-500-401

Citation: Huang, S.-H.; Chen, Y.-J.; Huang, W.-F.; Uan, J.-Y. Electrodeposition of a Li-Al Layered Double Hydroxide (LDH) on a Ball-like Aluminum Lathe Waste Strips in Structured Catalytic Applications: Preparation and Characterization of Ni-Based LDH Catalysts for Hydrogen Evolution. *Catalysts* 2022, 12, 520. https://doi.org/10.3390/catal12050520

Academic Editors: Ioan-Cezar Marcu and Octavian Dumitru Pavel

Received: 10 March 2022
Accepted: 2 May 2022
Published: 5 May 2022

Publisher's Note: MDPI stays neutral with regard to jurisdictional claims in published maps and institutional affiliations.

Abstract: A functionally structured catalyst was explored for ethanol steam reforming (ESR) to generate H_2. Aluminum lathe waste strips were employed as the structured catalytic framework. The mixed metal oxide (Li-Al-O) was formed on the surface of Al lathe waste strips through calcination of the Li-Al-CO_3 layered double hydroxide (LDH), working as the support for the formation of Ni catalyst nanoparticles. NaOH and $NaHCO_3$ titration solutions were, respectively, used for adjusting the pH of the $NiCl_2$ aqueous solutions at 50 °C when developing the precursors of the Ni-based catalysts forming in-situ on the Li-Al-O oxide support. The Ni precursor on the Al structured framework was reduced in a H_2 atmosphere at 500 °C for 3 h, changing the hydroxide precursor into Ni nanoparticles. The titration agent (NaOH or $NaHCO_3$) effectively affected the physical and chemical characterizations of the catalyst obtained by the different titrations. The ESR reaction catalyzed by the structured catalysts at a relatively low temperature of 500 °C was studied. The catalyst using $NaHCO_3$ titration presented good stability for generating H_2 during ESR, achieving a high rate of H_2 volume of about 122.9 L/(g_{cat}·h). It also had a relatively low acidity on the surface of the Li-Al-O oxide support, leading to low activity for the dehydration of ethanol and high activity to H_2 yield. The interactions of catalysts between the Ni precursors and the Li-Al-O oxide supports were discussed in the processes of the H_2 reduction and the ESR reaction. Mechanisms of carbon formation during the ESR were proposed by the catalysts using NaOH and $NaHCO_3$ titration agents.

Keywords: structured catalyst; ethanol steam reforming; aluminum lathe waste strips; layered double hydroxide; Ni nanoparticle

1. Introduction

Renewable energy is sought to replace fossil fuels, which have caused environmental pollution, such as through the emission of greenhouse gases. Hydrogen is a clean and renewable energy carrier that can be used as a feedstock in fuel cell systems [1,2] and transportation [3]. Clean techniques for producing hydrogen are being developed, including the electrolysis of water [4], the photocatalysis of water [5,6], metal hydrolysis [7], the use of biomass [8], hydrocarbon reforming [2,9,10], and others. In recent decades, ethanol steam reforming (ESR) for hydrogen production has attracted interesting investigation due to its environmental benefits. Ethanol is a renewable raw material that can be produced in the fermentation of biomass, including sugar cane, corns, and in starch-rich and lignocellulosic materials [2,11,12]. The production of hydrogen by ESR is not only environmentally friendly but also highly efficient, unlike steam reforming for producing other fuels [13,14].

The overall steam reforming reaction of ethanol is given by Equation (1) [1,2,13–18].

$$C_2H_5OH + 3H_2O \rightarrow 6H_2 + 2CO_2 \ (\Delta H^0 = + 173.5 \ kJ/mol) \tag{1}$$

However, several side reactions may occur during the ESR process, depending on the species of catalyst used [2,13,19]. Catalysts play a crucial role in the ESR process. In addition to increasing the reaction rate, they activate ethanol conversion and hydrogen selection [2,13,18,19]. The use of noble metal catalysts (such as Rh, Ru, Pt, Pd, and Ir) and non-noble metal catalysts (such as Ni, Co, Cu, and Fe) in the ESR process has frequently been studied [2,13,16,18,19]. Noble metal catalysts have a high cost, limiting their widespread use in industry. A Ni-based catalyst has a lower cost and is effective in breaking C-C bonds and O-H bonds, as well as in CH_4 reforming [2,16,18–21]. Ni also favors the adsorption of hydrogen atoms on the catalyst surface to form molecular H_2 [2]. Accordingly, a Ni-based catalyst is used herein. Catalyst support selection is closely associated with the activity and stability of the catalyst [2,14,18]. Al_2O_3, working as catalyst support, is widely used for a Ni-based catalyst in the ESR process, owing to its good mechanical properties and thermal stability under reaction conditions [18,21,22]. However, at the acidic sites on Al_2O_3, ethanol can be dehydrated to ethylene, according to Equation (2) [2,13,14,16–21]:

$$C_2H_5OH \rightarrow C_2H_4 + H_2O \tag{2}$$

Finally, the polymerization of ethylene (C_2H_4) forms coke, causing the gradual deactivation of the catalyst. Furthermore, ethylene can be decomposed into carbon species, as described by Equation (3) [22]:

$$nC_2H_4 \rightarrow C_nH_{2n} + nH_2 \rightarrow nC + 2nH_2 \tag{3}$$

Some studies have found that basic additives or promoters, such as an alkali metal (such as Li, Na, and K) or an alkaline earth metal (such as Mg, Ca, and Sr) neutralize the acidic sites on Al_2O_3 [2,14,18]. Promoters favor H_2O adsorption and OH^- mobility on the Al_2O_3 surface, simultaneously accelerating carbon oxidation and reducing the rate of coke formation [21]. However, promoters do not disperse easily in the Al_2O_3 crystal structure or on the Al_2O_3 surface. Calcination of hydrotalcites i.e., layered double hydroxides (LDHs), at a high temperature can yield mixed metal oxides (MMO) [23–25]. The metal oxides are candidates for providing the support of catalyst particles in ESR. Favorable characteristics, such as a large surface area, thermal stability, and high metal dispersion, favor the use of LDHs in the ESR process [23,24,26–29]. LDHs are large classes of layered anionic clays. They have positively charged layers with divalent or trivalent metal cations at the octahedral sites of the hydroxyl slabs. Anions and water molecules intercalate the interlayer spaces with charge compensation [30–32]. Under special conditions, monovalent metal cations such as Li^+ occupy the vacancies in an aluminum octahedral sheet ($Al(OH)_3$) to form a positively charged layer [30–32]. For instance, Li-Al LDHs have the chemical formula $[LiAl_2(OH)_6]^+A^- \cdot mH_2O$, where A^- is an interlayered anion.

In this study, 500 °C-calcined Li-Al-CO_3 LDH is used as a support in the ESR process. Li-Al-CO_3 LDH after calcination at 500 °C may transform to Li-contained γ-Al_2O_3 [33,34]. Notably, Li^+ ions that are well dispersed in the γ-Al_2O_3 structure neutralize the acidic sites on the γ-Al_2O_3 surface. Hence, in this study, calcining LDHs at 500 °C to form Li-Al mixed metal oxides is expected to resolve some issues, such as the difficulty of dispersing promoters in oxide supports. In conventional fixed-bed reactors, the catalyst used for ESR is commonly in the form of powder or a powder-pressed pellet. A small-scale reactor not only has limited catalytic space, but also can be blocked by a large volume of carbon deposits during ESR [35,36]. Catalyst pellets, during ESR in a long-term reaction, may have coking problems, and the difficulties of thermal, chemical, and mechanical stresses, therefore, causing their disintegration and functional failure [35]. Developing stable catalysts with low carbon deposition when used in ESR is a challenge. Recently, structured catalysts and reactors have attracted attention in the field of catalyst research owing to their rapid heat and mass transfer and lower pressure drop [36–40]. Classes of structured catalysts include monolithic catalysts, foam catalysts, membrane catalysts, and three-dimensional micro-fibrous entrapped catalysts (3D-MFEC) [36–40]. Among these, the framework of 3D-MFECs can be metallic or ceramic. A metal framework is better than a ceramic one because it has higher heat and mass transfer [39]. However, structured catalytic frameworks must be prepared precisely, and the process is complicated, resulting in a high cost. Therefore, a lower-cost and stable structured catalyst must urgently be developed. In this study, three-dimensional ball-like aluminum alloy lathe waste strips with a dense and disordered structure are used as a structured catalytic framework. Aluminum alloy lathe waste strips are not economically recycled in industry, because of the oxide, filth, and oil that adheres to their surfaces. Herein, the use of aluminum alloy lathe waste strips as a structured catalytic framework reduces not only the cost of the framework material but also the pollution of the environment by such wastes.

In this work, aqueous solutions that mainly contained Li^+ and Al^{3+} ions were used for the formation of Li-Al-CO_3 LDH thin film on Al lathe waste strips. An important discovery was that, following the electrochemical deposition, the LDH was calcined at 400 °C to increase the hydrophilicity, easily leading to the in-situ growth of Ni precursors on the calcined LDH platelets in $NiCl_2$ aqueous solutions at 50 °C by using NaOH or $NaHCO_3$ titration. The Ni precursors were activated in an H_2 atmosphere at 500 °C for 3 h. The structured catalysts prepared using NaOH and $NaHCO_3$ titration methods were tested in the ESR process at 500 °C. After the ESR reactions, H_2 yields and the interactions of nickel loading with calcined LDH support of the catalysts on product selectivity were analyzed. In addition, catalytic stabilities, acidic and basic properties on the surface of the catalysts, and the mechanisms of carbon deposition in prepared samples were discussed in detail.

2. Results and Discussion

2.1. Electrodeposition and Calcination of Li-Al-CO_3 LDH Thin Film/Characterization of Ni-Based Catalyst on Calcined LDH Support

Figure 1 plots the curve of current against electrolysis time during the deposition process. The initial current decayed rapidly, indicating the occurrence of the Li-Al-CO_3 LDH deposition on the surfaces of the Al lathe waste strips. Subsequently, the current approached a plateau at about 2000 to 7200 sec, implying that the non-conducted Li-Al-CO_3 LDH film had steadily grown as a function of time. The thin film that was formed using the electrochemical technique was examined using an X-ray diffractometer, as shown in Figure S1. Characteristic peaks of Li-Al-CO_3 LDH (hydrotalcite) (JCPDS card no. 42-729) and the Al substrate were observed. The X-ray diffraction (XRD) pattern suggests that the electrodeposition method successfully deposits the Li-Al-CO_3 LDH thin film on to the ball-like aluminum lathe waste strips. Wetting with water is an important procedure that can help the catalyst precursor to grow on the LDH platelets' surface. Figure S2 presents the water contact angle on the surfaces of the Al substrate, Li-Al LDH film @ Al substrate (denoted as L@Al), and the L@Al samples after calcination at 100–500 °C. The calcined Li-Al LDH film @ Al substrate had a lower water contact angle than the original Al substrate and Li-Al LDH film @ Al substrate. In particular, the 400 °C-calcined Li-Al LDH film @ Al substrate had the lowest water contact angle (the highest wettability with water). Therefore, calcination at 400 °C was performed before the preparation of Ni precursor on the LDH @ Al substrate. Figure 2a presents the SEM surface morphology of the electrodeposited Li-Al LDH thin film on the Al lathe substrate. Figure 2b zooms in on part of Figure 2a. The SEM images show a typical nanoplate's microstructure on the Al lathe substrate. Figure 2c shows the SEM image of the 400 °C-calcined Li-Al LDH thin film on Al lathe substrate. The SEM image demonstrates that calcination at 400 °C does not change the LDH surface microstructure.

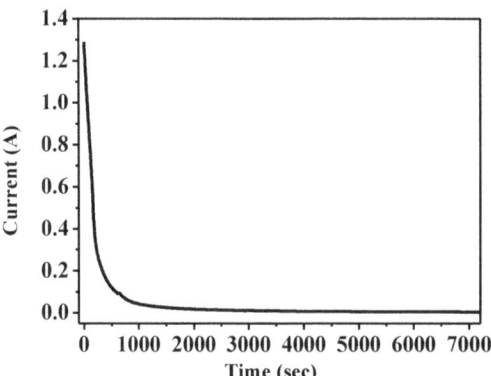

Figure 1. Current versus time curve during the electrochemical deposition of Li-Al-CO₃ LDH thin film in the Li⁺/Al³⁺ aqueous solution at a constant potential. A DC voltage of 5 V was applied in the electrochemical cell for 2 h at room temperature.

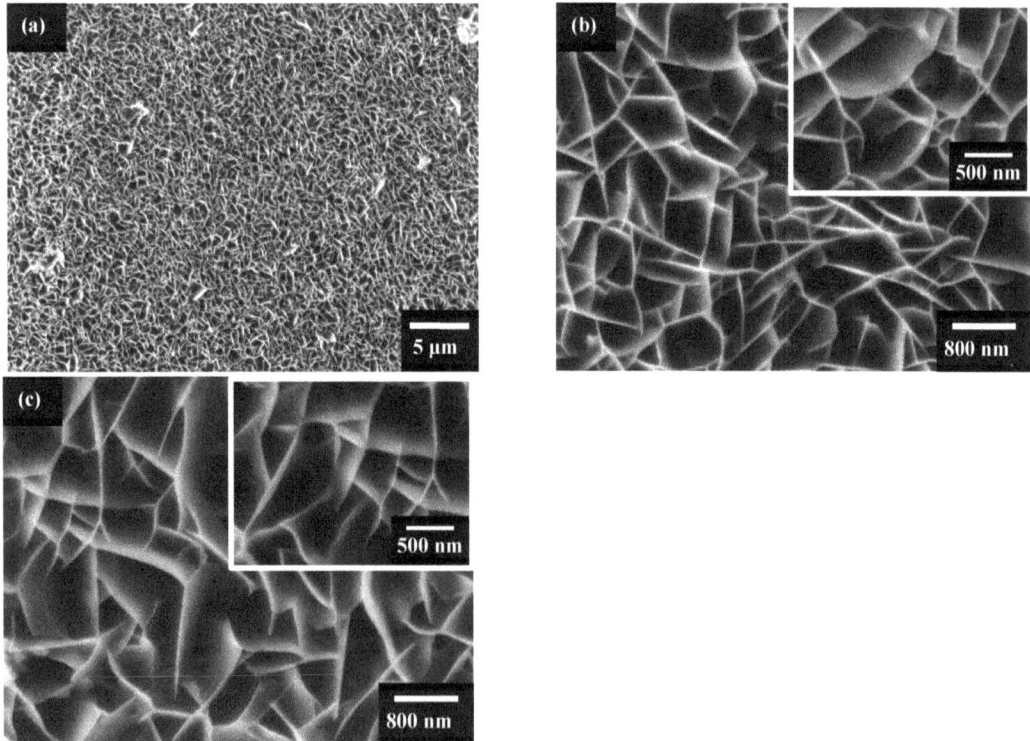

Figure 2. SEM surface morphologies: (**a**) Li-Al-CO₃ LDH thin film on Al substrate (A6061 lathe waste strip) and its high-magnification image in (**b**); (**c**) Li-Al-CO₃ LDH thin film calcined at 400 °C for 1 h.

Figure 3 shows the XRD patterns of Ni precursors (fresh catalysts) on the 400 °C-calcined LDH@Al substrates of the NaOH and NaHCO₃ prepared samples (denoted as NaOH_T and NaHCO₃_T). The two samples, NaOH and NaHCO₃, yield the same diffraction peaks at 2θ of 11°, 22.2°, 34.8°, 36.5°, 46°, 61.2°, which are characteristic of α-type nickel hydroxide (Ni catalyst precursor; Ni(OH)₂·xH₂O; JCPDS card no. 38-715). Sample

NaHCO$_3$_T yielded weak and broad peaks of Ni(OH)$_2$, while the sample NaOH_T yielded intensive peaks of Ni(OH)$_2$. The results suggest that the Ni(OH)$_2$ (Ni catalyst precursor) for sample NaHCO$_3$_T probably had poor crystallinity with nano-size structure, while the Ni(OH)$_2$ on NaOH_T had good crystallinity. The patterns of both samples also include diffraction peaks of the Al substrate. However, the calcined LDH phase was not observed in the patterns of either sample. In our previous study [25], calcining LDH at 300–400 °C formed an amorphous-like mixed metal oxide (MMO). Accordingly, intensive diffraction peaks of the calcined LDH phase were not present in the patterns of the NaOH_T and NaHCO$_3$_T samples, owing to their amorphous-like structure.

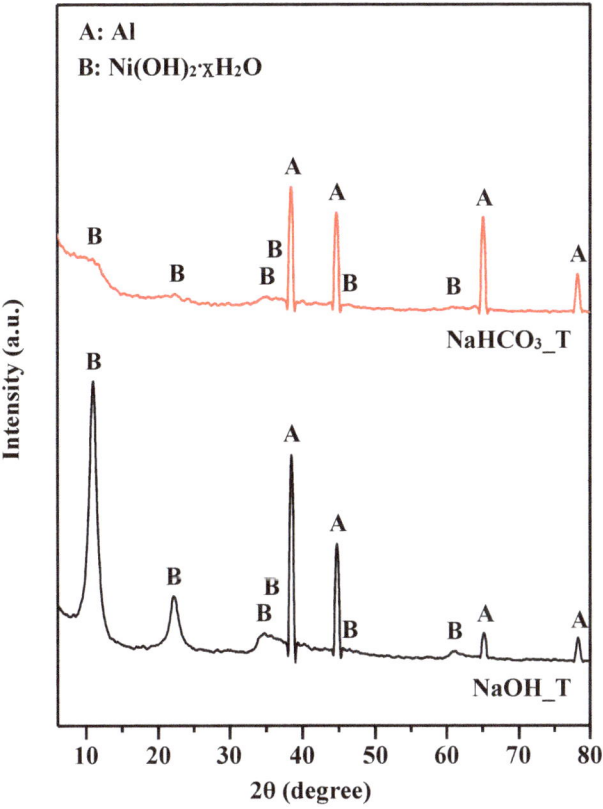

Figure 3. XRD patterns of Ni precursors (fresh catalysts) on the 400 °C calcined LDH@Al substrates for the sample NaOH_T (bottom pattern) and NaHCO$_3$_T (top pattern), respectively.

As shown in Figure 4a,b for the SEM microstructures of NaOH_T (fresh catalyst) and NaOH_T + R (reduced catalyst), Figure 4a displays the SEM surface observation on NaOH_T. Figure 4b shows the SEM surface microstructure of the NaOH_T + R. A few nanoparticles are presented in the calcined LDH platelets. As shown in Figure 4c,d, Figure 4c displays the SEM surface microstructure of the NaHCO$_3$_T, showing the cross-linked meshwork structures on platelets. Figure 4d presents the SEM surface microstructure of the sample NaHCO$_3$_T + R after H$_2$ reduction. Many nanoparticles are dispersed uniformly on the edges and platelets of the calcined LDH.

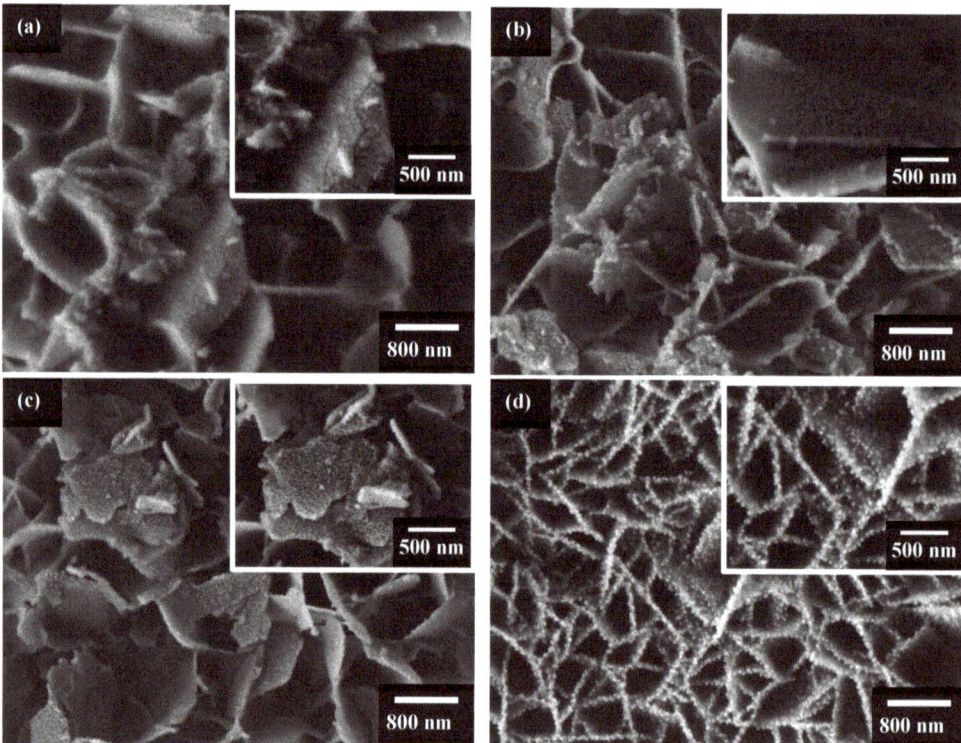

Figure 4. SEM surface morphologies: (**a**) NaOH_T fresh catalyst; (**b**) NaOH_T + R reduced catalyst; (**c**) NaHCO$_3$_T fresh catalyst; (**d**) NaHCO$_3$_T + R reduced catalyst.

2.2. Use of the Structured Catalyst in Ethanol Steam Reforming (ESR) Reaction

Figure 5a plots the H$_2$ yields as a function of time during ESR reactions. The H$_2$ yield (Y$_{H2}$) can be expressed below:

$$Y_{H_2} = \text{mole of } H_2 \text{ output/mole of ethanol input} \tag{4}$$

The ESR reaction catalyzed by the NaOH_T + R or by the NaHCO$_3$_T + R catalyst produced different H$_2$ yields. The NaHCO$_3$_T + R catalyst gave a stable H$_2$ yield of about 55%, higher than that obtained using the NaOH_T + R catalyst. The H$_2$ yield of NaOH_T + R catalyst decreased with time and rapidly dropped after 4.3 h. Figure 5b plots the cumulative volume of H$_2$ produced per gram of the catalyst by using the NaOH_T + R or NaHCO$_3$_T + R catalyst during the ESR reaction for 5 h. The cumulative volume of H$_2$ produced by the NaHCO$_3$_T + R catalyst increased linearly with time from the beginning to the end of the ESR reaction, indicating a constant rate of H$_2$ generation. The cumulative volume of H$_2$ was 614.5 L/g$_{cat}$ after 5 h of ESR. Restated, the cumulative volume of H$_2$ per hour was about 122.9 L/(g$_{cat}$·h). The cumulative volume of H$_2$ that was obtained using NaOH_T + R catalyst was around 435.9 L/g$_{cat}$, less than that produced by the NaHCO$_3$_T + R catalyst. The cumulative volume of H$_2$ per hour was about 87.2 L/(g$_{cat}$·h). Figure 5c,d present the compositions of the gases that were produced using the NaOH_T + R and NaHCO$_3$_T + R catalysts in ESR reactions for 5 h. The main gaseous productions were H$_2$, CO$_2$, CO, and CH$_4$. As shown in Figure 5c, the H$_2$ produced (Vol.%) by the NaOH_T + R catalyst generally fell with time, and rapidly dropped after 4.3 h. This result is consistent with the H$_2$ yield in Figure 5a. As shown in Figure 5d, the gas productions (H$_2$, CO$_2$, CO, and CH$_4$) had a stable output from the beginning to the end of the ESR reaction.

Importantly, the gas production remained about 70 Vol.% of H_2 until the reaction stopped. Frusteri et al., [41] reported that a doping with Li increased the stability of the Ni/MgO catalyst for ESR, mainly by reducing the sintering of the Ni catalyst particles. Herein, the NaHCO$_3$_T + R catalyst with the LiAlO mixed metal support powerfully and stably activated ESR for H_2 generation. However, with the NaOH_T + R catalyst, H_2 production rapidly declined after 4.3 h; the possible reason will be discussed in the following section.

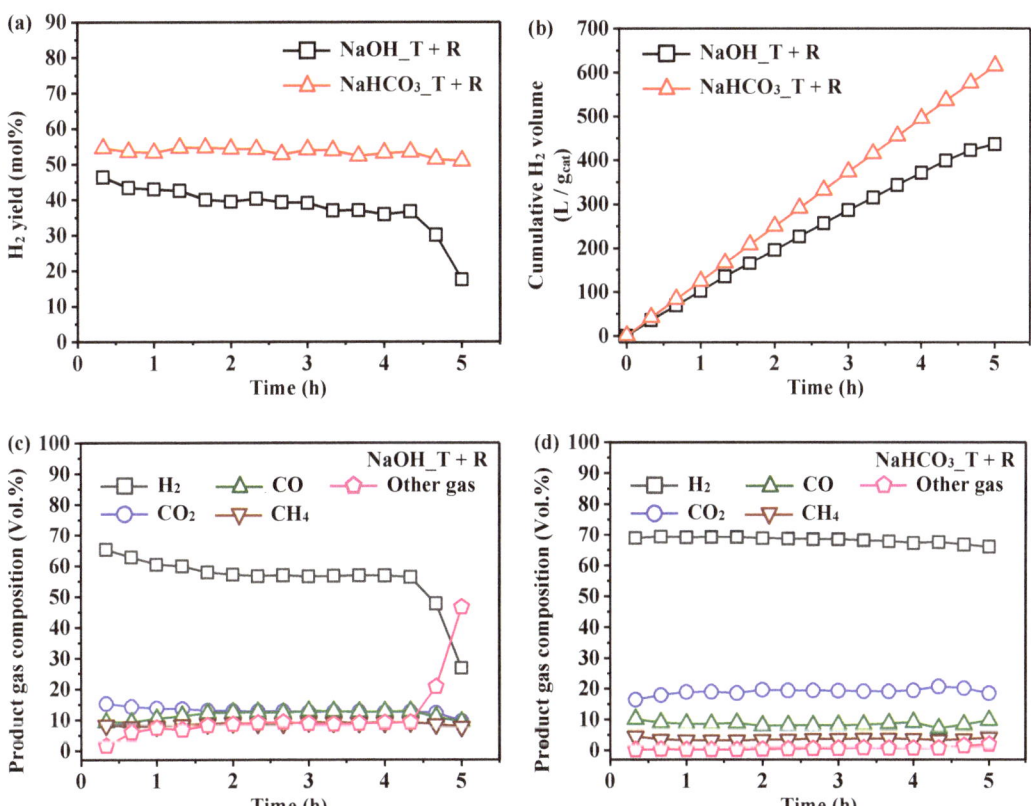

Figure 5. (**a**) H_2 yields and (**b**) cumulative H_2 volumes of the NaOH_T + R and NaHCO$_3$_T + R catalysts in the ESR reactions; Volume percentages of product gas composition of the NaOH_T + R catalyst (**c**) and NaHCO$_3$_T + R catalyst (**d**) in the ESR reactions.

The ethanol conversions (X_{EtOH}) and selectivity (S) of the productions after ESR reactions catalyzed by the NaOH_T + R and NaHCO$_3$_T + R catalysts were calculated using Equations (5)–(7) and were provided in Table 1.

$$X_{EtOH} \ (\%) \ = \ \frac{F_{ethanol, \ in} - F_{ethanol,out}}{F_{ethanol,in}} \tag{5}$$

$$S_{H_2} \ (\%) \ = \ \frac{F_{H_2 \ produced}}{3 \times (F_{ethanol,in} - F_{ethanol,out}) + (F_{water,in} - F_{water,out})} \times 100 \tag{6}$$

$$S_{i \ carbon-containing \ product} \ (\%) \ = \ \frac{F_{i \ carbon-containing \ product}}{n_i \times (F_{ethanol,in} - F_{ethanol, \ out})} \times 100 \tag{7}$$

where $F_{i, in/out}$ is the molar flow rate of the i specie at the inlet/outlet of the reactor, respectively, and n_i is the stoichiometric factor between the carbon-containing products and ethanol.

Table 1. Activations of the NaOH_T + R and NaHCO$_3$_T + R catalysts in the ESR reactions for 5 h: the averages of the ethanol conversion, selectivity of the production, and deposition weight of carbon species.

Catalyst	Conversion (%), Selectivity (mol%) and Deposition of Carbon Species (g/(g$_{cat}$·h))									
	X_{EtOH}	H_2	CO_2	CO	CH_4	C_2H_4	$(C_2H_5)_2O$	CH_3CHO	CH_3COCH_3	C
NaOH_T + R	30.35	37.97	13.27	11.59	8.82	20.93	7.40	40.58	1.14	0.943
NaHCO$_3$_T + R	21.94	74.04	30.84	13.91	5.80	11.01	2.57	43.21	0.77	0.869

In the analyses of the selectivity of the production, the chemical reaction pathways associated with ESR can be presumed be reducible to the following individual reactions [1,2,13–18].

$$C_2H_5OH + H_2O \rightarrow CH_4 + CO_2 + 2H_2 \tag{8}$$

$$C_2H_5OH \rightarrow CH_4 + CO + H_2 \tag{9}$$

$$2C_2H_5OH \rightarrow CH_3COCH_3 + CO + 3H_2 \tag{10}$$

$$2C_2H_5OH \rightarrow (C_2H_5)_2O + H_2O \tag{11}$$

$$C_2H_5OH \rightarrow CH_3CHO + H_2 \tag{12}$$

$$CH_3CHO \rightarrow CH_4 + CO \tag{13}$$

$$CH_3CHO + H_2O \rightarrow 2CO + 3H_2 \tag{14}$$

$$CH_4 + H_2O \rightarrow CO + 3H_2 \tag{15}$$

$$CO + H_2O \rightarrow CO_2 + H_2 \tag{16}$$

Table 1 shows the ethanol conversions after ESR reactions, and that the NaOH_T + R catalyst had a higher conversion (~30.35%) than the NaHCO$_3$_T + R catalyst (~21.94%). However, the H$_2$ selectivity (~37.97 mol%) by the NaOH_T + R catalyst was lower than that of the H$_2$ selectivity (~74.04 mol%) of the reaction catalyzed by NaHCO$_3$_T + R. Owing to the low H$_2$ selectivity, the application of the NaOH_T + R catalyst resulted in relatively low values of the selectivity of CO$_2$, CO, and CH$_4$, as shown in Table 1; they are 13.27 mol% of CO$_2$, 11.59 mol% of CO, and 8.82 mol% of CH$_4$, respectively. For comparison, the CO$_2$ selectivity was high up to 30.84 mol% when the NaHCO$_3$_T + R was employed, and this value was much higher than that of the CO$_2$ selectivity of 13.27 mol% catalyzed by the NaOH_T + R. Fundamentally, the high selectivity to CO$_2$ mainly represents that the ethanol steam reforming, as shown in Equation (8), and the water gas shift reaction, as shown in Equation (16) must proceed to produce CO$_2$. That is, the ethanol steam reforming of Equation (8), and water gas shift reaction of Equation (16) were not so well catalyzed while using the sample NaOH_T + R as the catalyst. In addition, the fact that CO$_2$ production exceeds CO production, especially when using the NaHCO$_3$_T + R catalyst (see Table 1), suggests a strong water gas shift reaction, as shown in Equation (16). Moreover, the high acetaldehyde selectivity (~40% up) of the two samples were detected in the ESR residual solutions. It implies that the reduced catalysts in both samples had successfully activated the ethanol dehydrogenation to acetaldehyde, as seen in Equation (12). Moreover, CH$_4$ can be easily produced in the ESR reaction process, as seen in Equations (8), (9) and (13). The methane steam reforming, as in Equation (15), was one of the main subreactions during ESR to consume CH$_4$ and produce H$_2$ (with one mole of CH$_4$ producing three moles of H$_2$). As shown in Table 1, the selectivity to CH$_4$ by the NaOH_T + R catalyst was 8.82 mol%, while the selectivity to CH$_4$ by the NaHCO$_3$_T + R catalyst was only 5.80 mol%. Therefore, it suggests that the NaHCO$_3$_T + R catalyst promoted the methane steam reforming, as seen in Equation (15), to consume more CH$_4$ and to produce a larger amount of H$_2$ than those in the case of the NaOH_T + R catalyst. Furthermore, the ethanol dehydrated to produce the non-desirable C$_2$H$_4$, as seen in Equation (2), and (C$_2$H$_5$)$_2$O, as seen in Equation (11) was also detected in Table 1. The high C$_2$H$_4$ selectivity (~20.93 mol%) and (C$_2$H$_5$)$_2$O selectivity (~7.40 mol%) were produced by the NaOH_T + R catalyst. When compared to the NaOH_T + R catalyst, the NaHCO$_3$_T + R catalyst had less C$_2$H$_4$ selectivity (~11.01 mol%) and (C$_2$H$_5$)$_2$O selectivity (~2.57 mol%). The polymerization of ethylene caused the deposition of carbon species on the catalyst's surface, as described by Equation (3). Correspondingly, the NaOH_T + R catalyst had the higher carbon formation of 0.943 g/(g$_{cat}$·h) than that catalyzed by the NaHCO$_3$_T + R catalyst (~0.869 g/(g$_{cat}$·h)). In summary of the above description, the NaHCO$_3$_T + R catalyst, due to the relatively high selectivity to H$_2$, the high selectivity of CO$_2$, and the low selectivity of CH$_4$, C$_2$H$_4$, and (C$_2$H$_5$)$_2$O, it is explained that the NaHCO$_3$_T + R catalyst exhibited better catalysis performance in the ESR reaction than the NaOH_T + R catalyst.

2.3. Ni-Based Catalysts on Calcined LDH Supports: Comparison between NaOH and NaHCO$_3$ Titration in Preparation of the Catalysts

Figure 6 presents the FT-IR spectra of the samples NaOH_T and NaHCO$_3$_T. The intense broadband at around 3425 cm^{-1} is consistent with the stretching vibration of hydroxyl groups that are hydrogen-boned to H$_2$O [42–47]. A band near 1632 cm^{-1} is attributed to the H$_2$O bending vibration [42,44–47]. The tiny absorption band at 2403 cm^{-1} is attributed to the CO$_2$ adsorbed on the sample surface, because the prepared samples were grown in the atmosphere [44,45]. Another tiny band, at about 2125 cm^{-1}, reflected the presence of intercalated C–O species in the precursors [47]. The narrow band at 1376 cm^{-1} is assigned to the asymmetric stretching vibration of CO$_3^{2-}$ [42,44,46,47]. A band around 847 cm^{-1} corresponded to an asymmetric stretching vibration of intercalated C–O species [44,46]. Two bands at about 612 cm^{-1} and 550 cm^{-1} are attributed to the in-plane deformations δ_{OH}, which are associated with the stretching vibration of Ni-OH [42,43,45,47]. Importantly, the FT-IR spectra revealed the vibrational bands of the OH group (~3425 cm^{-1}) and the Ni-OH group (~612 cm^{-1}), confirmed that the precursors of both samples were α-Ni(OH)$_2$ [43].

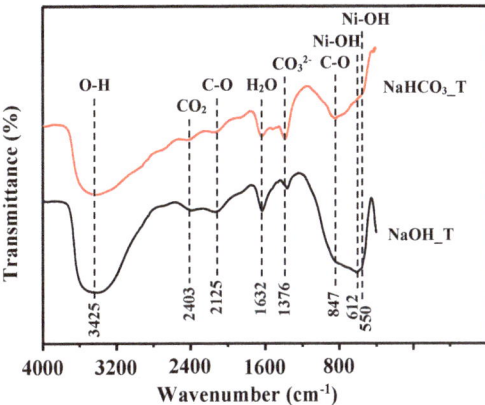

Figure 6. FT-IR spectra of Ni precursors (fresh catalysts) on the 400 °C-calcined LDH platelets of the samples NaOH_T and NaHCO$_3$_T.

Figure 7a,b present the SEM images and EDS compositional mappings of the samples NaOH_T and NaHCO$_3$_T. Ni and O were dispersed uniformly in the maps, indicating that the deposited Ni(OH)$_2$ were dispersed on the calcined LDH support. Figure 7c displays statistics concerning the Ni, Al, and O element contents of the samples NaOH_T and NaHCO$_3$_T. The average Ni content in NaOH_T was more than double that in NaHCO$_3$_T. The high Ni content in NaOH_T is also clearly evident in the FT-IR spectra (Figure 6), which exhibits a higher O-H vibration band ($\sim$3425 cm^{-1}), H$_2$O vibration band (1632 cm^{-1}), and Ni-OH vibration bands (612 and 550 cm^{-1}), than those of the sample NaHCO$_3$_T.

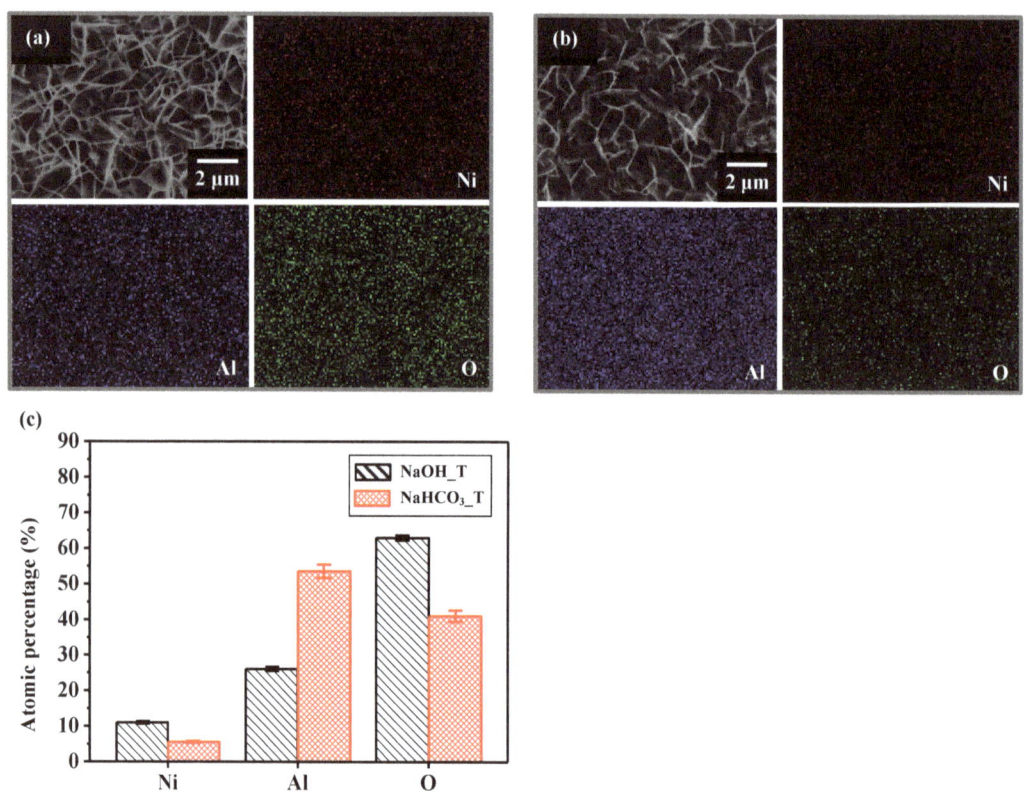

Figure 7. SEM surface morphologies of the sample NaOH_T and NaHCO$_3$_T with compositional maps for Ni, Al, and O are shown in (**a**,**b**), respectively; (**c**) EDS analyses of the elemental compositions and contents of the samples NaOH_T and NaHCO$_3$_T.

The above results suggest that sample NaOH_T had twice as much Ni(OH)$_2 \cdot$xH$_2$O on the calcined LDH support as sample NaHCO$_3$_T. Calcining Ni(OH)$_2 \cdot$xH$_2$O at 500 °C decomposed it to yield NiO [48,49]. The thermal decomposition reaction is as follows [48,49].

$$Ni(OH)_2 \cdot xH_2O_{(s)} \rightarrow Ni(OH)_{2(s)} + xH_2O_{(g)} \tag{17}$$

$$Ni(OH)_{2(s)} \rightarrow NiO_{(s)} + H_2O_{(g)} \tag{18}$$

where x is the amount of water in moles.

Equation (17) is the dehydration reaction of Ni(OH)$_2 \cdot$xH$_2$O, and Equation (18) is the decomposition of Ni(OH)$_2$ to NiO. Therefore, in this study, four times as much water vapor was produced from the nickel precursor of sample NaOH_T in the reduction process with an H$_2$ atmosphere at 500 °C than was produced from the nickel precursor of sample NaHCO$_3$_T. However, previous studies have reported [50–52] that a reduction reaction experiment for NiO@Al$_2$O$_3$ powder was carried out under an atmosphere with H$_2$ and water vapor, resulting in decreased amounts of the reduced Ni metallic particles. This is because, at high temperature, Al_2O_3 in the water vapor atmosphere easily forms aluminate, leading to the reduced nickel reacting with vicinal aluminate to form surface nickel aluminate ($NiAl_2O_4$) [50–52]. Zieliński claimed [52] that $NiAl_2O_4$ content depended on the concentration of water vapor in the environment. In particular, when $NiO@Al_2O_3$ powder was heated at 450 °C in a water vapor atmosphere in an H_2 reduction process, nickel aluminate was produced quickly. Nonstoichiometric $NiAl_2O_4$ that decorates the corners and edges of nickel crystallites retards NiO reduction and weakens catalyst performance. Previous studies [33,34] found that calcining Li-Al LDH at 500 °C transforms the crystal mainly into γ-Al_2O_3. In this study, the precursor ($Ni(OH)_2 \cdot xH_2O$) on a calcined LDH (γ-Al_2O_3) support of the sample NaOH_T may possess an excessive concentration of water vapor (in comparison to NaHCO$_3$_T) in the H_2 reduction process, inhibiting the reduction of $Ni(OH)_2 \cdot xH_2O$ to Ni particles. Hence, as shown in the SEM image of NaOH_T + R in Figure 4b, fewer Ni particles were found on the calcined LDH platelets than were found as shown in NaHCO$_3$_T + R (Figure 4d). Figure 8 presents a TEM dark-field image of sample NaOH_T + R, and the inset presents the TEM diffraction pattern from the dark-field image in Figure 8. The diffraction pattern exhibits the characteristic Bragg reflections of NiO and Ni, suggesting that NiO and Ni crystallites were presented together on the NaOH_T + R catalyst. The $NiAl_2O_4$ may not demonstrate much infraction, leading to no evident diffraction pattern.

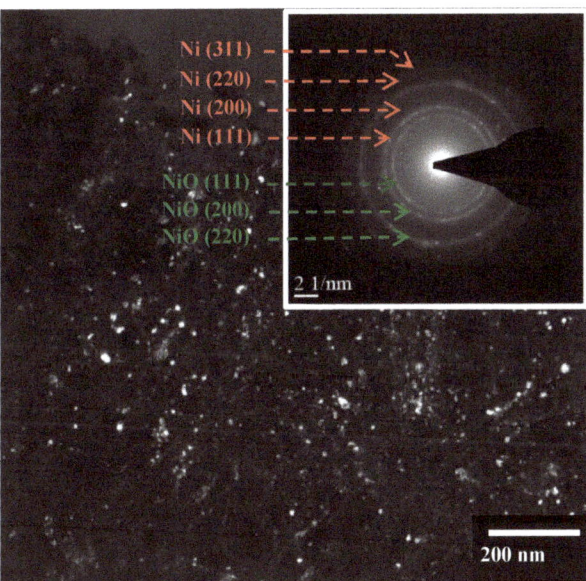

Figure 8. Dark-field image of the TEM cross-sectional microstructure of the NaOH_T + R catalyst. The inset shows the diffraction pattern of the bright particles from the TEM image.

Figure 9 exhibits the Ni $2p_{3/2}$ XPS spectra of the reduced NaOH_T + R and NaHCO$_3$_T + R catalysts. The binding energy of metallic Ni (Ni^0) is around 852.5 eV, and the accompanying shake-up satellite peak is observed at about 861.5 eV [21,53,54]. A peak at around 855.3 eV, which is characteristic of Ni^{2+}, is associated with NiO [53,54]. The NiO (Ni^{2+}) peaks from both reduced catalysts were higher than that of metallic Ni (Ni^0), perhaps because of the high oxygen affinity of nano-scale Ni particles [55]. Table 2 presents the surface compositions and quantitative elemental analyses of the reduced NaOH_T + R and NaHCO$_3$_T + R catalysts, as determined by XPS. The NaHCO$_3$_T + R catalyst had twice the Ni content (17.54 at.%) than that of NaOH_T + R catalyst (7.83 at.%). These results are consistent with the SEM images previously shown in Figure 4, and reveal that a relatively high amount of the reduced Ni particles were dispersed on $NaHCO_3\_T + R$.

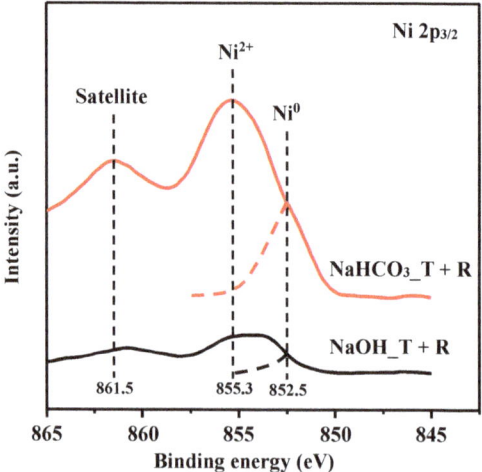

Figure 9. Ni $2p_{3/2}$ XPS spectra of the reduced $NaOH\_T + R$ and $NaHCO_3\_T + R$ catalysts.

Table 2. The quantitative elements and surface compositions of the reduced $NaOH\_T + R$ and $NaHCO_3\_T + R$ catalysts on the 500 °C calcined LDH thin films by XPS analysis.

Element	NaOH_T + R (at.%)	NaHCO₃_T + R (at.%)
Ni	7.83	17.54
Al	10.16	13.73
O	82.01	68.74

Figure 10a presents the TEM cross-sectional microstructure of the $NaHCO_3\_T + R$ catalyst; the image shows that Ni particles were dispersed uniformly on the platelet of the 500 °C-calcined LDH. Figure 10b shows the size distributions of the Ni particles on sample $NaHCO_3\_T + R$, as obtained from the TEM image. The mean particle size is 10.7 nm. For comparison, Figure 10c shows the size distributions of the Ni particles that were measured using the TEM image on sample $NaOH\_T + R$. The mean particle size is 14.6 nm. Figure S3 displays the TEM images of the Ni particles of sample $NaOH\_T + R$ and $NaHCO_3\_T + R$. The above results suggest that the reduced Ni particles in sample $NaHCO_3\_T + R$ were more dispersed, smaller, and more numerable than in sample $NaOH\_T + R$. The result may explain that the $NaHCO_3\_T + R$ catalyst for the ESR reaction resulted in a higher H_2 yield than the ESR results for the reaction catalyzed by the $NaOH\_T + R$ catalyst (Figure 5).

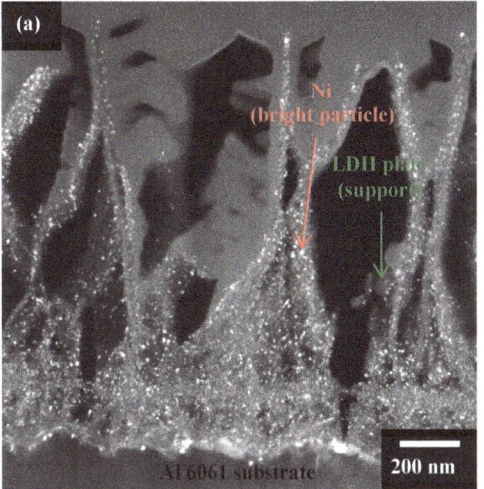

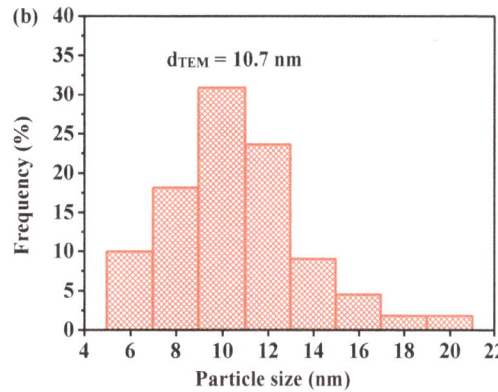

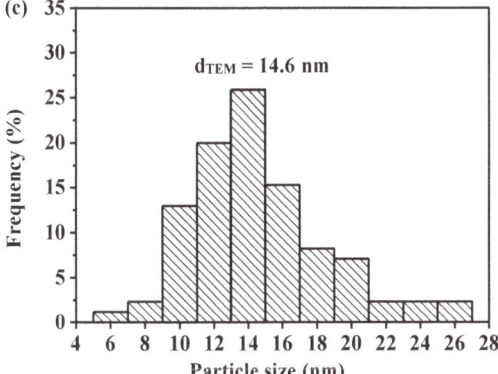

Figure 10. (**a**) TEM cross-sectional microstructure of the $NaHCO_3\_T + R$ catalyst; Particle size distributions of the $NaHCO_3\_T + R$ catalyst (**b**) and $NaOH\_T + R$ catalyst (**c**). The particle size distributions in (**b**,**c**) were collected from the TEM images in Figure S3.

2.4. Mechanism of Carbon (Coke) Formation in Ethanol Steam Reforming (ESR) Reaction

Figure 11 shows the SEM surface morphologies of the $NaOH\_T + R$ catalyst (Figure 11a) and $NaHCO_3\_T + R$ catalyst (Figure 11b), respectively, after ESR for 5 h. As shown, carbon filaments (filamentous cokes) were observed in both catalysts after the ESR reaction. The inset SEM micrographs in high magnification are also presented in Figure 11a,b, clearly indicating the carbon filaments and the Ni particles that were lifted at the top of the filaments. The size of the filaments on sample $NaOH\_T + R$ (Figure 11a) is much larger than that of the filaments on sample $NaHCO_3\_T + R$ (Figure 11b), As mentioned in Section 3.3, a higher Ni content on the $NaHCO_3\_T + R$ catalyst revealed a higher catalytic activity than in the $NaOH\_T + R$ catalyst. It is reasonable to interpret that fine carbon fiber filaments formed on the $NaHCO_3\_T + R$ catalyst (Figure 11b) due to the slower carbon growth rate. Inversely, the coarse carbon filaments formed on the $NaOH\_T + R$ catalyst due to a relatively high carbon growth rate. Figure 11c, taking the sample $NaHCO_3\_T + R$ as an example, displays the EDS spectrum for analyzing the surface elemental compositions of the bright particle on the top of the woven carbon filament after the ESR for 5 h. As shown, a high-intensity peak of Ni was found. The result proves that the Ni particles were lifted from the supports of calcined LDH at the tops of the carbon filaments.

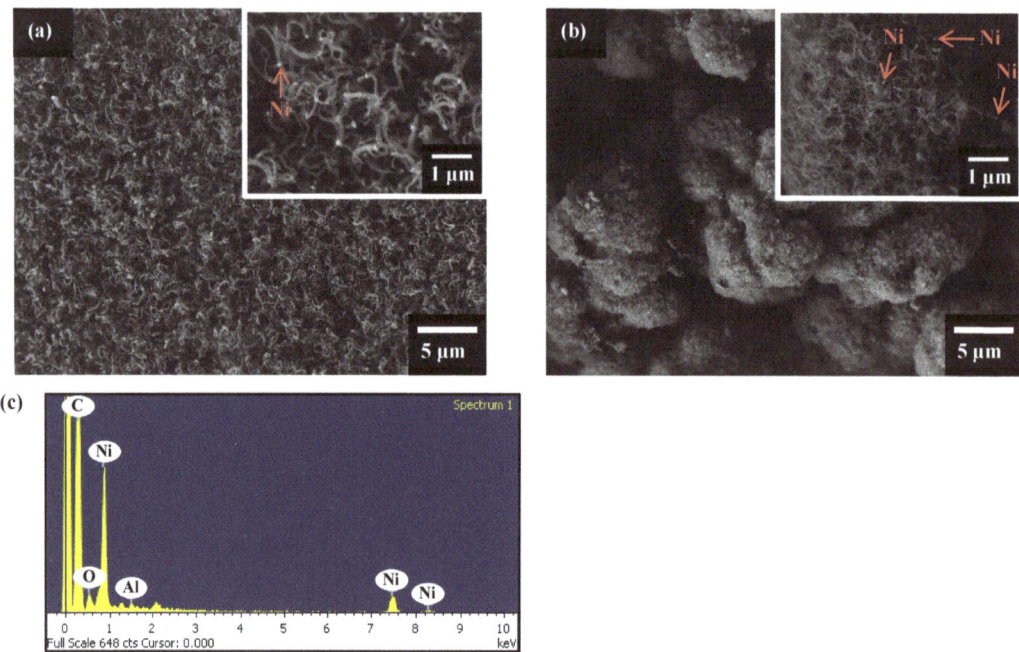

Figure 11. SEM surface morphologies of the catalysts after ESR reactions for 5 h: (**a**) NaOH_T + R catalyst; (**b**) NaHCO$_3$_T + R catalyst; (**c**) EDS spectrum presented as the high Ni contents were detected at bright particles on the tips of filamentous carbons in the SEM images (**a**,**b**).

An active catalyst can also exhibit the condition of carbon deposition and carbon interaction with supports after the catalyst has worked. The mechanisms of carbon deposition on the surfaces of catalysts (NaOH_T + R and NaHCO$_3$_T + R) and carbons interaction with Ni and calcined LDH supports during the ESR reactions are illustrated in Figure 12. The schematic models in Figure 12 (I)–(III) show this in atomic-scales, and Figure 12 (IV) shows this in nano-scale. Figure 12a (I) and 12b (I) are the schematic models of the crystal of the fresh catalysts (Ni precursors, NiO$_x$H$_z$) obtained, respectively, using the NaOH titration solution, as in Figure 12a (I), and the NaHCO$_3$ titration solution, as in Figure 12b (I). As schematically shown, the fresh catalysts (NaOH_T and NaHCO$_3$_T) were grown on the calcined LDH platelets. The precursor by NaOH titration, as in Figure 12a (I), has a much larger size than the precursor by NaHCO$_3$ titration, as in Figure 12b (I). In Figure 3 in Section 2.1, X-ray diffraction results of the precursor of sample NaHCO$_3$_T exhibited a much broader FWHM (full width at half maximum) (1.64°) than the FWHM (0.51°) of the strong peaks of the Ni(OH)$_2$ by sample NaOH_T, suggesting that the titration of the NaHCO$_3$ solution produced a much finer crystallite size of Ni(OH)$_2$ precursor than those produced by NaOH titration, according to the fundamental concept of the Scherrer equation to estimate nano-crystallite size. More Ni(OH)$_2$ precursors on the calcined LDH support in the NaOH_T catalyst suggest that the abundant water vapor would cause NiO$_x$H$_z$ interaction with calcined LDH (γ-Al$_2$O$_3$) supports to form NiAl$_2$O$_4$ on the NiO$_x$H$_z$ and NiO surface in the process of the H$_2$ reduction. As a result, Ni particles were not fully reduced after the H$_2$ reduction, as described in Section 2.3. Metallic Ni particles were reduced on the top of the surface of the support, and the dehydrated NiO particles, which were not reduced, were on the sub-surface near the support [56], as shown in Figure 12a (II). As compared with the NaOH_T catalyst, the fresh catalyst (NaHCO$_3$_T) using the NaHCO$_3$ titration method had fewer water molecules in the process of H$_2$ reduction because of the fine crystallite size (Figure 3) of NiO$_x$H$_z$ (Ni(OH)$_2$) on the calcined LDH support, and the low content of the Ni element, as illustrated in Figure 7. Hence, the NaHCO$_3$_T catalyst can fully contribute to the reduction of the metallic Ni within an H$_2$ atmosphere, as shown in Figure 12b (II). Figure 12a (III) shows the formation of carbon on the surface of the NaOH_T + R catalyst during the ESR reaction at the initial work condition of the catalyst. The mixed nickel compounds (NiO) that favor the interaction with calcined LDH and are located in the sub-surface region near the support can form relatively large particles [56]. The schematic model displayed in Figure 12a (II) shows that the top of the layer is the Ni particles, and the second layer is the NiO particles. It is a possible that the aggregation of NiO particles blocked the path or reduced the number for the diffusion of carbons to the support. As as result, some of the carbons were formed in the sub-surface near the support, and the others were formed through the Ni particles in the layer of the surface of NiO particles, as shown in Figure 12a (III). Figure 12b (III) shows the carbon formation of the NaHCO$_3$_T + R catalyst at the initial work condition during the ESR reaction. The carbons can easily diffuse through the smaller Ni clusters in the ESR reaction [35,57–60]. After the diffusion of the carbons, they interact with the calcined LDH support. The previous study [61,62] indicated that methane C–H bonds can be cleaved at the Lewis acid sites in γ-Al$_2$O$_3$. Thereafter, hydrogen is desorbed by proton-hydride recombination, and carbon remains on the surface of the support. Finally, the carbons were accumulated in the interface region between Ni clusters and the calcined LDH support, as shown in Figure 12b (III). Figure 12a,b (IV) show the schematic models of catalysts (NaOH_T + R and NaHCO$_3$_T + R), respectively, in nano-scales at a longer period of the ESR reactions. Figure 12a (IV) presents the carbon deposition continuously and then the growths of filamentous carbon on the surface of the NaOH_T + R catalyst at a longer period of the ESR reactions. Previous studies [56,60] have indicated that NiO$_x$H$_z$ (NiO) cluster particles are more stably bound on the γ-Al$_2$O$_3$ support. These cluster particles do not easily detach from the γ-Al$_2$O$_3$ support in the reaction process of the catalyst, owing to the low contact angle. Ultimately, the NiO and the mixed Ni cluster particles become buried by filamentous carbon, deactivating the catalyst. Figure 11a shows the SEM surface microstructure of the NaOH_T + R catalyst after the ESR reaction, and shows a few Ni particles on the tips of carbon filaments. The SEM image proves consistent with the model mechanism in Figure 12a. Figure 12b (IV) presents the carbons deposition and the growths of filamentous carbon on the surface of the NaHCO$_3$_T + R catalyst at a longer period of the ESR reactions. This is because Ni particles have an obtuse contact angle with the calcined LDH (γ-Al$_2$O$_3$) support and the binding energy between them is low [56,60]. The Ni particles can easily detach from the surface of the calcined LDH support. Hence, as the carbon filaments grow from the surface of the calcined LDH support, the Ni particles rose to the tips of the carbon filaments rather than being buried or encapsulated in the filaments. The model mechanism corresponds to the SEM image in Figure 11b, in which many of the Ni particles are on the tips of the carbon filaments. Cunha et al., [63] claimed that using the filamentous carbon as a support material with the Ni catalyst in steam and dry reforming of methane can increase the stability of the Ni particle, preventing surface migration and coalescence. In this study, the formation of filamentous carbons may act as a substitute, such as the LDH support that provided Ni particles on the surface and continued to activate with ethanol steam for generating H$_2$.

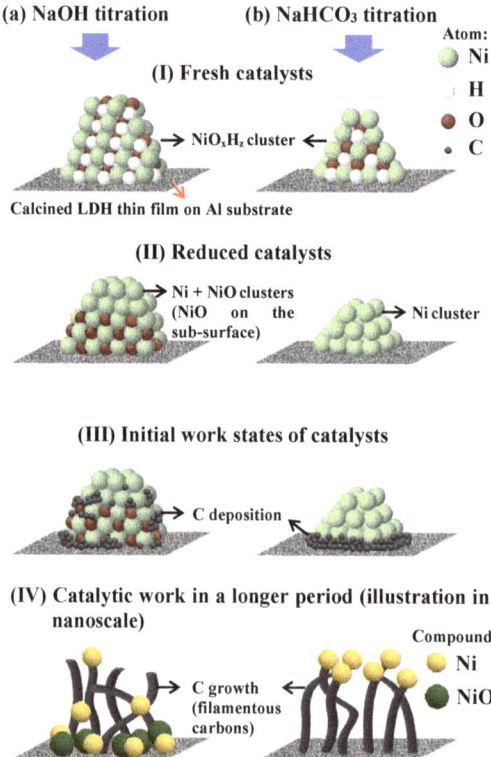

Figure 12. Schematic mechanisms illustrate the processes of carbon formation on the surfaces of the catalysts (Ni on calcined LDH) in the ESR reactions: (**a**) NaOH titrated catalysts and (**b**) NaHCO$_3$ titrated catalysts. The preparations of the catalysts and in the ESR reactions present in (I)–(IV): (I) Fresh catalysts; (II) Reduced catalysts; (III) Initial states of catalysts in the ESR reactions; (IV) NaOH titrated catalyst shows the formation of filamentous carbons burying gradually on the Ni and NiO particles; NaHCO$_3$ titrated catalyst shows that the Ni particles raise on the tips of the filamentous carbons. Note: The illustrations in (I)–(III) show processes in atomic scales, and (IV) shows in nanoscale.

Figure 13a depicts the temperature-programmed desorption (TPD) profiles of NH$_3$, which exhibits the information on the adsorption strength and density of acid sites on the surfaces of the NaOH_T + R and NaHCO$_3$_T + R catalysts. The adsorption peak in the low-temperature range from 50 to 200 °C is attributed to weak Lewis acid sites [64–66]. The weak acidity is reported in the previous study and indicates no or low catalytic activity [64]. The adsorption peaks in the intermediate to high temperature range (200–600 °C) are assigned to moderate to strong Lewis acid sites [64–66]. As shown in Figure 13a, the NaOH_T + R catalyst presented a broad desorption peak at about 200–600 °C, with the maximum intensity located at 338 °C. The NaHCO$_3$_T + R catalyst had three clear desorption peaks at about 298 °C, 386 °C, and 476 °C, respectively. The adsorption peak at 298 °C is probably attributed to the formation of NiAl$_2$O$_4$ species on the surface of the support (calcined LDH) [67,68]. The intensity of NH$_3$ adsorption at 298 °C on the NaOH_T + R catalyst was 1.5 times higher than with the NaHCO$_3$_T + R catalyst. Moreover, an asymmetric adsorption peak at about 430 °C on the NaOH_T + R catalyst may present the acid sites of NiO [69]. However, the NaHCO$_3$_T + R catalyst does not find any adsorption peak at 430 °C. It implies that more NiAl$_2$O$_4$, rather than reduced NiO, formed on the surface of the NaOH_T + R catalyst. The intensity of the NH$_3$-adsorption peak can be used to quantify the acidity on the surface of the catalyst. Therefore, the areas of TPD curve in the intermediate to high-temperature range (200–600 °C) for the NaOH_T + R and NaHCO$_3$_T + R catalysts were, respectively, integrated. The sum of the integrated function of the NaOH_T + R catalyst had about 1.2 times higher than the NaHCO$_3$_T + R catalyst. In other words, the acidity of the NaOH_T + R catalyst was higher than the NaHCO$_3$_T + R catalyst. The result corresponds to the mechanism in Figure 12 and discussion in Section 2.3 that the NiAl$_2$O$_4$, and not reduced NiO on the NaOH_T + R catalyst, led to the increase of the density of the acid sites. In addition, Fang et al. [68] suggested that the highly dispersed Ni species can drop the acid sites because the nano Ni particles cover some acid sites on the surface of the catalyst. Hence, the NaHCO$_3$_T + R catalyst had low acidity when compared to the NaOH_T + R catalyst. Herein, the superior NaHCO$_3$_T + R catalyst was selected to perform the CO$_2$-TPD experiment to understand the characterization of the Li-Al mixed oxide supported catalyst on the acidic and basic properties. Figure 13b shows the CO$_2$-TPD profile of the NaHCO$_3$_T + R catalyst, which shows the adsorption strength and density of base sites on the surface of the catalyst. For comparing the acidic and basic properties of the NaHCO$_3$_T + R catalyst, the NH$_3$-TPD curve as shown in Figure 13a was replotted in Figure 13b. For the CO$_2$-TPD curve, the broad and weak adsorption peak in the low-temperature range from 100 to 200 °C is ascribed to low Brønsted basicity (i.e., surface OH- groups) [70–72]. There are two clear adsorption peaks around the intermediate to high-temperature range (200–600 °C), associated with moderate to strong Lewis basicity [70–72]. A peak at around 200–400 °C with the intermediate-strength Lewis basicity is attributed to the Li$^+$-O^{2-} and Al^{3+}-O^{2-} acid-base pairs [70–72]. Another strong peak at about 400–600 °C with strong Lewis basicity is related to the presence of low-coordinated O^{2-} [70–72]. The peak areas of the NH$_3$/CO$_2$-TPD can be determined as the amounts of acid and base sites of different strengths [73]. Hence, the amounts of acid and base sites in Figure 13b were calculated by integrating the sums of NH$_3$/CO$_2$ in each desorption peak. The resulting value of the ratio of acid/base sites was 0.519. A previous study [69] described that the base strength distribution of the M-Al mixed oxide relies on the M^{n+} metal cation (such as Li$^+$) used for incorporation with Al^{3+} cation (Li$^+$/Al^{3+}). In our study, the Li$^+$ promoter in the Al$_2$O$_3$ structure could effectively provide the Li$^+$-O^{2-} acid-base pairs. Particularly, the substitution of Al^{3+} by Li$^+$ in the Al$_2$O$_3$ lattice could substantially increase the O^{2-} on the surface of the catalyst [69]. It could also find that the CO$_2$-TPD curve (Figure 13b) had a strong adsorption peak located at 400–600 °C. The result suggested that the Li$^+$ promoter could effectively neutralize the acid sites in the Li-Al mixed oxide supported the NaHCO$_3$_T + R catalyst, reducing the dehydration of ethanol in the ESR process. The acidic property of the catalyst favors activating the ethanol, and leads to the dehydration of ethanol, as described by Equation (2). The C$_2$H$_4$ gas is the main product. Figure 13c presents the volumes of the C$_2$H$_4$ production for the NaHCO$_3$_T + R and NaOH_T + R catalysts during the ESR reactions, respectively. At the initial ESR reaction for 1 h, the NaHCO$_3$_T + R catalyst had only 1 L of C$_2$H$_4$ production. The NaOH_T + R catalyst showed four times more C$_2$H$_4$ production than the NaHCO$_3$_T + R catalyst. After the ESR reaction for 5 h, the NaHCO$_3$_T + R catalyst showed about half of the volume of C$_2$H$_4$ production, as compared with the NaOH_T + R catalyst. Moreover, ethanol can also dehydrate to (C$_2$H$_5$)$_2$O [74,75], as described by Equation (11). The volumes of (C$_2$H$_5$)$_2$O production using the NaHCO$_3$_T + R and NaOH_T + R catalysts during the ESR reactions are presented in Figure 13d. At the initial ESR reaction for 1 h, the NaHCO$_3$_T + R catalyst had small volumes of (C$_2$H$_5$)$_2$O (0.17 L), while the NaOH_T + R catalyst had 0.89 L of (C$_2$H$_5$)$_2$O. The NaOH_T + R catalyst showed higher than five times the (C$_2$H$_5$)$_2$O product when compared with the NaHCO$_3$_T + R catalyst. After the ESR reaction for 5 h, about 0.72 L of (C$_2$H$_5$)$_2$O was produced by the NaHCO$_3$_T + R catalyst. The NaOH_T + R catalyst achieved nearly 2 L. The difference was about three times. The amounts of C$_2$H$_4$ and (C$_2$H$_5$)$_2$O production were consistent with the NH$_3$-TPD curves, suggesting that the lower acidity on NaHCO$_3$_T + R catalyst had high H$_2$ selectivity and less dehydration of ethanol to produce C$_2$H$_4$ and (C$_2$H$_5$)$_2$O. Although the ethanol conversion of the NaOH_T + R catalyst was higher than the $NaHCO_3\_T + R$ catalyst (Table 1), the high number of acid sites led to the activation of the dehydration of ethanol. As a result, the selectivity of the undesirable products, such as C_2H_4 and $(C_2H_5)_2O$, increased, while H_2 selectivity was reduced.

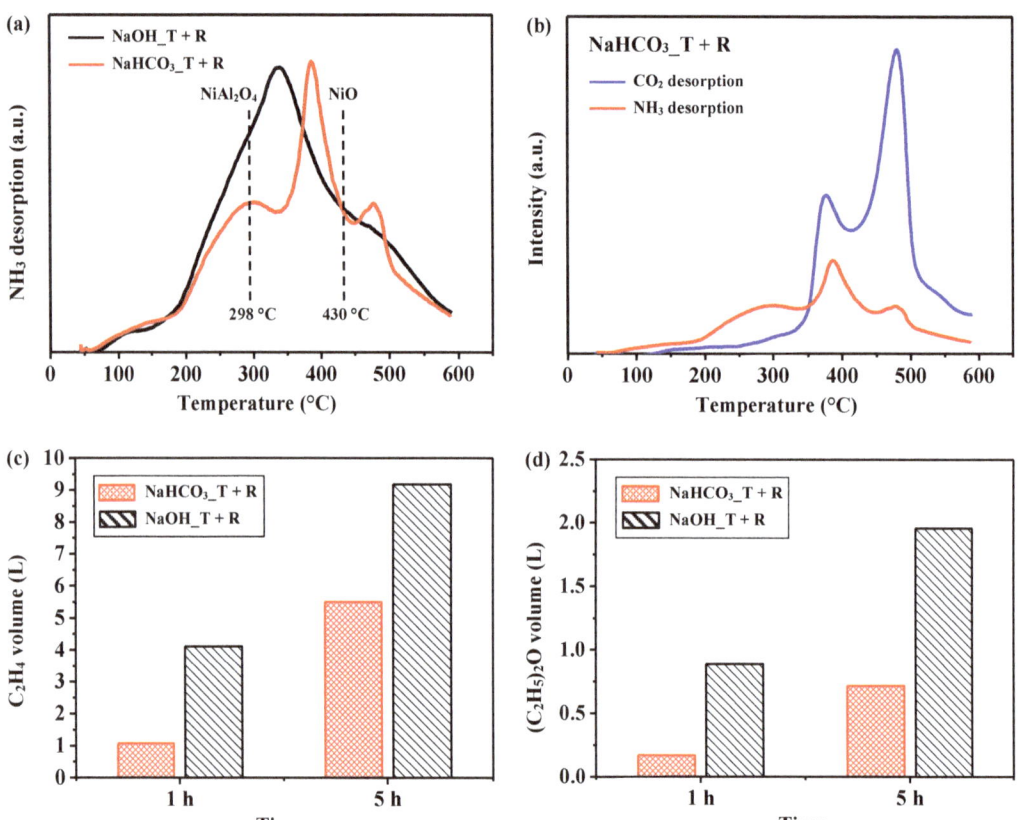

Figure 13. (**a**) NH_3-TPD profiles of the $NaOH\_T + R$ and $NaHCO_3\_T + R$ catalysts; (**b**) NH_3/CO_2-TPD profiles of the $NaHCO_3\_T + R$ catalyst; Gas volumes of the production of the $NaOH\_T + R$ and $NaHCO_3\_T + R$ catalysts at the initial and end states in the ESR reactions: (**c**) C_2H_4; (**d**) $(C_2H_5)_2O$.

3. Materials and Methods

3.1. Preparation and Calcination of Li-Al LDH Thin Film

The lathe strips from a 6061 aluminum alloy (Al-1 wt.% Mg–0.6 wt.% Si) rod were collected (C. S. Aluminium Co., Kaohsiung, Taiwan). The Al lathe waste strips were kneaded into a ball-like form with a diameter of approximately 50 mm. Before electrochemical deposition of the LDH, the ball-like Al lathe waste strip ball was ultrasonically cleaned in acetone and ethyl alcohol baths to remove debris and residual oil, and then were dried in air. Reagent-grade $LiOH·H_2O$ was dissolved in deionized (DI) water at 50 °C, and highly pure Ar gas was purged through the LiOH aqueous solution (Choneye Pure Chemicals, Taipei, Taiwan). Small pieces of Al foil (~1 × 5 mm) were added to the stirred LiOH aqueous solution (0.06 M) at 50 °C for 30 min, and then the solution was filtered to obtain an electrolyte solution. The Li/Al molar ratio is 1:2. In the electrochemical deposition, the anode was a platinum-coated titanium mesh, and the cathode was the ball-like Al lathe waste strip ball. A DC voltage of 5V was applied between the platinum-coated titanium mesh and the Al lathe waste strip ball for 2 h using a Princeton Applied Research Model

273 A Potentiostat/Galvanostat. The preparation of the electrochemical experiment was described in detail in our earlier study [25]. After electrodeposition, the LDH-coated Al lathe waste strip ball was cleaned in DI water and dried in the ambient atmosphere. Subsequently, the Li-Al-CO₃ LDH thin films were calcined in air at various temperatures from 100 °C to 500 °C for 1 h in order to find the optimum condition to increase the hydrophilicity of the calcined LDH surface. Water contact angles were measured on the LDH and calcined LDH thin film surfaces.

3.2. Preparation of Ni-Based Catalyst on Calcined LDH Thin Film

Reagent-grade $NiCl_2$ (Shimakyu's Pure Chemicals, Osaka, Japan), NaOH, and $NaHCO_3$ (Choneye Pure Chemicals, Taipei, Taiwan) were used in the preparation of the catalysts. Ni catalysts were prepared using the following two methods: NaOH titration and $NaHCO_3$ titration. Al lathe waste strip balls that were coated with 400 °C calcined LDH thin films were immersed in 0.15 M $NiCl_2$ aqueous solutions at 50 °C. Then, the 1.25 M NaOH and 1.14 M $NaHCO_3$ aqueous solutions were, respectively, titrated into two cups of 0.15 M $NiCl_2$ aqueous solutions that both contained the Al lathe waste strip balls with the calcined LDHs. When the pH value of the 0.15 M $NiCl_2$ aqueous solutions dropped to 6.7, the titrations of the NaOH and $NaHCO_3$ solutions were stopped, and the Al lathe waste strip balls with the calcined LDHs were continuously immersed in the solutions for hours in order to form Ni precursors on the calcined LDH platelets' surfaces. The Al lathe waste strip ball sample that was processed by titrating the NaOH solution was denoted as NaOH_T. The Al lathe waste strip ball sample that was processed by titrating the $NaHCO_3$ solution was denoted as $NaHCO_3$_T. The samples of NaOH_T and $NaHCO_3$_T were then reduced in highly pure H_2 gas (3 L/min) in a furnace at 500 °C for 3 h, and then furnace-cooled to room temperature. Upon H_2 reduction, samples NaOH_T and $NaHCO_3$_T became NaOH_T + R and $NaHCO_3$_T + R, respectively. According to weight change measurements between an Al lathe waste strip ball that was coated with 400 °C calcined LDH thin film and the ball with a catalyst (NaOH_T + R or $NaHCO_3$_T + R), there was about 55 mg of Ni catalyst in one Al lathe strip framework.

3.3. Tests of Catalysts in Ethanol Stream Reforming (ESR)

For each test of the ESR experiment, six pieces of ball-like Al lathe frameworks with catalysts (NaOH_T + R or $NaHCO_3$_T + R) were packed together in a stainless steel tubular reactor for ESR (Chi Huw Heating Co., Ltd., Taiwan). The tubular reactor had a size of 60 mm in diameter and 450 mm in length. In the catalyzed ESR process, the catalysts were tested under atmospheric pressure. Water/ethanol mixture with a molar ratio of 7.5 was fed in, flowing at 10 mL/min, using a diaphragm pump (SIMDOS 02, KNF Neuberger, Switzerland) into the preheating chamber at 250 °C. Highly pure N_2 gas was used as the carrier gas. The mixed solution was vaporized in this preheating chamber at 250 °C, and then the mixed steam (ethanol and water) was inputted into the catalyst reactor at 500 °C. The ESR reaction began in the catalyst reactor at 500 °C. During the ESR reaction, the exhaust gases from the catalyst reactor were cooled and dried, and the liquid productions were condensed in a cold trap. The gaseous and liquid productions were analyzed using a VARIO PLUS enhanced flue gas analyzer (MRU-VARIO PLUS, Germany) and gas chromatography-mass spectrometry (GC-MS; Agilent 7890CB, USA, and JEOL AccuTOF-GCX, Japan).

3.4. Characterization

The crystal structures of Li-Al-CO₃ LDH thin film and fresh catalysts were examined using glancing angle X-ray diffraction (GAXRD; Bruker, Darmstadt, Germany), with an incidence angle of 0.5°, and a Bruker D8 Advance ECO diffractometer with Cu K_α ($\lambda = 1.5406$ Å) radiation at 40 kV and 25 mA. Fourier transform infrared (FT-IR) spectra of the fresh catalysts were obtained using a Perkin Elmer Spectrum 65 FT-IR spectrometer in the range 4000–400 cm^{-1}, at a resolution of 4 cm^{-1} (Perkin Elmer, Norwalk, CT,

USA). Surface microstructure images of the supports, catalysts, and catalysts after ESR were obtained by field-emission scanning electron microscopy (FE-SEM; JEOL JSM-6700F, Tokyo, Japan) in the secondary electron imaging mode, with an accelerating voltage of 3 kV and an emission current of 10 μA. Energy-dispersive X-ray spectra (EDS) were obtained using a JEOL JSM-6700F FE-SEM with an acceleration voltage of 15 kV to determine the elemental contents and their distributions of fresh catalysts and catalysts after the ESR reaction. The surface chemical compositions of reduced Ni catalysts were obtained by X-ray photoelectron spectroscopy (XPS; ULVAC-PHI 5000 VersaProbe/Scanning ESCA Microprobe, Kanagawa, Japan/USA) with monochromatic Al K_α radiation (1486.6 eV). The X-ray source operated at 15 kV (25 W). Before XPS analysis, a sputtering area of 2×2 mm^2 was firstly pre-cleaned by sputtering at 2 kV with Ar ions for 0.3 min to remove surface-adsorbed molecules and contamination. A focused ion beam (FIB; Hitachi NX2000, Tokyo, Japan) was used to prepare a specimen for cross-sectional transmission electron microscopic (TEM) observation. The cross-sectional microstructure images and diffraction pattern of the reduced catalysts were obtained with a field-emission transmission electron microscope (FE-TEM; FEI EO Tecnai F20 G2 MAT S-TWIN Field Emission Gun, Hillsboro, OR, USA) at an accelerating voltage of 200 kV. Temperature programmed desorption (TPD) profiles of NH_3 and CO_2 of the reduced catalysts were obtained using a Quantachrome Autosorb iQ TPX instrument (Anton Paar, Graz, Austria). The mixed NH_3/He or CO_2/He gases flowed through the sample for 0.5 h at a room temperature of 25 °C. Subsequently, a steady flow of He gas was passed through the sample for 0.5 h to remove the weakly adsorbed NH_3/CO_2 on the surface. Finally, the sample was heated from 25–600 °C with a heating rate of 10 °C/min in an He-flowing atmosphere.

4. Conclusions

Recycled aluminum lathe waste strips were able to be a framework, used in structured catalysts for the ethanol steam reforming reaction. Li-Al-CO$_3$ LDH (hydrotalcite) nanoplatelets were successfully electrodeposited on the surface of the Al lathe waste framework. The LDH nanoplatelets with an interplatelet space of about 400 nm could highly increase the specific surface area of the Al lathe waste strips. The LDH after calcination could raise the hydrophilicity of the surface, aiding the adsorption with the NiCl$_2$ aqueous solution in the preparation of the catalyst. The Ni precursors were easily prepared by in situ growth in the NiCl$_2$ aqueous solutions at 50 °C, with the NaOH and NaHCO$_3$ titration solutions used to adjust the pH at 6.7, respectively. The Ni precursor using the NaOH or NaHCO$_3$ titration method could form the same crystal of α-type nickel hydroxide (Ni(OH)$_2$·xH$_2$O). However, the amount of Ni precursor using the NaOH titration method was about two times higher than with the NaHCO$_3$ titration method. This is because the Ni(OH)$_2$·xH$_2$O could dehydrate to NiO at 500 °C, simultaneously releasing the water vapor. The Ni precursor (Ni(OH)$_2$·xH$_2$O) on a calcined LDH (γ-Al$_2$O$_3$) support of the NaOH titrated sample may possess an excessive concentration of water vapor (in comparison to NaHCO$_3$_T) in the H$_2$ reduction process at 500 °C, inhibiting the reduction of Ni(OH)$_2$·xH$_2$O to Ni particles. Hence, the Ni particles were not fully reduced in the H$_2$ reduction process. Nonstoichiometric NiAl$_2$O$_4$ that decorated the corners and edges of nickel crystallites retarded NiO reduction. NiO particles that were not reduced to Ni particles were likely formed in the sub-surface near the calcined LDH support. The result from the mechanism of carbon formation indicated that the mixed Ni cluster particles and NiO particles were buried by filamentous carbon during the ESR reaction. The Ni precursor of the NaHCO$_3$ titrated sample obtained a fill reduction of metallic Ni particles due to the small amount of water vapor in the H$_2$ reduction process. After the reduction in the H$_2$ atmosphere at 500 °C, Ni particles could uniformly disperse on the edges and platelets of the calcined LDH support. Because the Ni particles had an obtuse contact angle (low binding energy) with the calcined LDH support, the Ni particles could easily detach from the surface of the support to rise on the tips of the carbon filaments during the ESR reaction. The formation of filamentous carbons may act as a substitute, such as the LDH support

Catalysts **2022**, *12*, 520

that provides Ni particles on the surface and continued to activate with ethanol steam in generating H_2. The mean Ni particle size on the $NaHCO_3$ titrated sample was 10.7 nm, which was smaller than the NaOH titrated sample (14.6 nm). Tests of catalytic activation of the $NaHCO_3$ titrated sample in the ESR for 5 h showed a stable H_2 yield of about 55 mol.%, and the rate of cumulative H_2 volume achieved was about 122.9 L/(g_{cat}·h). The main gas productions (H_2, CO_2, CO, and CH_4) had a stable output from the beginning to the end of the ESR reaction. The dehydration of ethanol showed a low selectivity to C_2H_4 (11.01%) and $(C_2H_5)_2O$ (2.57%), and deposition of carbon species of about 0.869 (g/(g_{cat}·h)), which was less than the NaOH titrated sample at 0.943 (g/(g_{cat}·h)). Possibly, Li^+ ions that existed in the mixed metal Li-Al-O oxide support neutralized the acid sites on the surface, decreasing Ni particle sintering. Conversely, the NiO particles with the $NiAl_2O_4$ wall on the surface of the NaOH titrated sample increased the number of acid sites. Although the ethanol conversion of the NaOH titrated sample (30.35%) was higher than the $NaHCO_3$ titrated sample (21.94%), the high number of acid sites led to the activation of the dehydration of ethanol. The selectivity of the C_2H_4 (20.93%) and $(C_2H_5)_2O$ (7.40%) were higher than the $NaHCO_3$ titrated sample. For the NaOH titrated sample, the H_2 yield decreased with time, and the rate of the cumulative H_2 volume was only about 87.2 L/(g_{cat}·h).

Supplementary Materials: The following supporting information can be downloaded at: https://www.mdpi.com/article/10.3390/catal12050520/s1, Figure S1: XRD pattern of the Li-Al LDH thin film on Al (A6061) substrate by electrochemical deposition. Figure S2: Water contact angle on the Al substrate, Li-Al LDH thin film @ Al substrate (L@Al), and Li-Al LDH thin film @ Al substrate (L@Al) after calcination at various temperatures. Figure S3: TEM bright-field images showed the distributions of Ni particles on the calcined LDH supports for the samples: (a) NaOH_T + R and (b) $NaHCO_3$_T + R.

Author Contributions: Conceptualization, J.-Y.U. and S.-H.H.; methodology, S.-H.H.; validation, J.-Y.U., S.-H.H., Y.-J.C. and W.-F.H.; formal analysis, S.-H.H.; investigation, S.-H.H.; resources, J.-Y.U.; data curation, S.-H.H.; writing—original draft preparation, S.-H.H.; writing—review and editing, S.-H.H.; visualization, J.-Y.U.; supervision, J.-Y.U.; project administration, S.-H.H.; funding acquisition, S.-H.H. All authors have read and agreed to the published version of the manuscript.

Funding: This study is financially supported by Ministry of Science and Technology, Taiwan (MOST 108-2221-E-005-034-MY3 and MOST 111-2923-E-194-002-MY3). The authors are grateful for their supports.

Data Availability Statement: The data that support the findings of this study are available on request from the corresponding author.

Conflicts of Interest: The authors declare no conflict to interest.

References

1. Benito, M.; Sanz, J.L.; Isabel, R.; Padilla, R.; Arjona, R.; Daza, L. Bio-ethanol steam reforming: Insights on the mechanism for hydrogen production. *J. Power Sources* **2005**, *151*, 11–17. [CrossRef]
2. Ni, M.; Leung, D.Y.C.; Leung, M.K.H. A review on reforming bio-ethanol for hydrogen production. *Int. J. Hydrogen Energy* **2007**, *32*, 3238–3247. [CrossRef]
3. Chang, W.-R.; Hwang, J.-J.; Wu, W. Environmental impact and sustainability study on biofuels for transportation applications. *Renew. Sustain. Energy Rev.* **2017**, *67*, 277–288. [CrossRef]
4. Lin, M.-Y.; Hourng, L.-W. Effects of magnetic field and pulse potential on hydrogen production via water electrolysis. *Int. J. Energy Res.* **2014**, *38*, 106–116. [CrossRef]
5. Wang, X.; Maeda, K.; Thomas, A.; Takanabe, K.; Xin, G.; Carlsson, J.M.; Domen, K.; Antonietti, M. A metal-free polymeric photocatalyst for hydrogen production from water under visible light. *Nat. Mater.* **2009**, *8*, 76–80. [CrossRef]
6. Sun, Y.; Jin, D.; Sun, Y.; Meng, X.; Gao, Y.; Dall'Agnese, Y.; Chen, G.; Wang, X.-F. G-C_3N_4/Ti_3C_2Tx (Mxenes) composite with oxidized surface groups for efficient photocatalytic hydrogen evolution. *J. Mater. Chem. A* **2018**, *6*, 9124–9131. [CrossRef]
7. Lin, M.-C.; Uan, J.-Y.; Tsai, T.-C. Fabrication of AlLi and Al_2Li_3/Al_4Li_9 intermetallic compounds by molten salt electrolysis and their application for hydrogen generation from water. *Int. J. Hydrogen Energy* **2012**, *37*, 13731–13736. [CrossRef]
8. Hosseini, S.E.; Abdul Wahid, M.; Jamil, M.M.; Azli, A.A.M.; Misbah, M.F. A review on biomass-based hydrogen production for renewable energy supply. *Int. J. Energy Res.* **2015**, *39*, 1597–1615. [CrossRef]
9. Abd El-Hafiz, D.R.; Ebiad, M.A.; Elsalamony, R.A.; Mohamed, L.S. Highly stable nano Ce–La catalyst for hydrogen production from bio-ethanol. *RSC Adv.* **2015**, *5*, 4292–4303. [CrossRef]

10. Chen, G.; Tao, J.; Liu, C.; Yan, B.; Li, W.; Li, X. Steam reforming of acetic acid using Ni/Al$_2$O$_3$ catalyst: Influence of crystalline phase of Al$_2$O$_3$ support. *Int. J. Hydrogen Energy* **2017**, *42*, 20729–20738. [CrossRef]
11. Olsson, L.; Hahn-Hagerdal, B. Fermentation of lignocellulosic hydrolysates for ethanol production. *Enzyme. Microb. Technol.* **1996**, *18*, 312–331. [CrossRef]
12. Huber, G.W.; Iborra, S.; Corma, A. Synthesis of transportation fuels from biomass: chemistry, catalysts, and engineering. *Chem. Rev.* **2006**, *106*, 4044–4098. [CrossRef] [PubMed]
13. Haryanto, A.; Fernando, S.; Murali, N.; Adhikari, S. Current status of hydrogen production techniques by steam reforming of ethanol: A review. *Energy Fuel* **2005**, *19*, 2098–2106. [CrossRef]
14. Phung, T.K.; Pham, T.L.M.; Nguyen, A.-N.T.; Vu, K.B.; Giang, H.N.; Nguyen, T.-A.; Huynh, T.C.; Pham, H.D. Effect of supports and promoters on the performance of Ni-based catalysts in ethanol steam reforming. *Chem. Eng. Technol.* **2020**, *43*, 672–688. [CrossRef]
15. Comas, J.; Mariño, F.; Laborde, M.; Amadeo, N. Bio-ethanol steam reforming on Ni/Al$_2$O$_3$ catalyst. *Chem. Eng. J.* **2004**, *98*, 61–68. [CrossRef]
16. Vaidya, P.D.; Rodrigues, A.E. Insight into steam reforming of ethanol to produce hydrogen for fuel cells. *Chem. Eng. J.* **2006**, *117*, 39–49. [CrossRef]
17. Mattos, L.V.; Jacobs, G.; Davis, B.H.; Noronha, F.B. Production of hydrogen from ethanol: Review of reaction mechanism and catalyst deactivation. *Chem. Rev.* **2012**, *112*, 4094–4123. [CrossRef]
18. Hou, T.; Zhang, S.; Chen, Y.; Wang, D.; Cai, W. Hydrogen production from ethanol reforming: Catalysts and reaction mechanism. *Renew. Sustain. Energy Rev.* **2015**, *44*, 132–148. [CrossRef]
19. Contreras, J.L.; Salmones, J.; Colín-Luna, J.A.; Nuño, L.; Quintana, B.; Córdova, I.; Zeifert, B.; Tapia, C.; Fuentes, G.A. Catalysts for H$_2$ production using the ethanol steam reforming (a review). *Int. J. Hydrogen Energy* **2014**, *39*, 18835–18853. [CrossRef]
20. Zanchet, D.; Santos, J.B.O.; Damyanova, S.; Gallo, J.M.R.; Bueno, J.M.C. Toward understanding metal-catalyzed ethanol reforming. *ACS Catal.* **2015**, *5*, 3841–3863. [CrossRef]
21. Sanchez-Sanchez, M.C.; Navarro, R.M.; Fierro, J.L.G. Ethanol steam reforming over Ni/M$_x$O$_y$-Al$_2$O$_3$ (M=Ce, La, Zr and Mg) catalysts: Influence of support on the hydrogen production. *Int. J. Hydrogen Energy* **2007**, *32*, 1462–1471. [CrossRef]
22. Elias, K.F.M.; Lucrédio, A.F.; Assaf, E.M. Effect of CaO addition on acid properties of Ni–Ca/Al$_2$O$_3$ catalysts applied to ethanol steam reforming. *Int. J. Hydrogen Energy* **2013**, *38*, 4407–4417. [CrossRef]
23. Huang, L.; Wang, J.; Gao, Y.; Qiao, Y.; Zheng, Q.; Guo, Z.; Zhao, Y.; O'Hare, D.; Wang, Q. Synthesis of LiAl$_2$-layered double hydroxides for CO$_2$ capture over a wide temperature range. *J. Mater. Chem. A* **2014**, *2*, 18454–18462. [CrossRef]
24. Sikander, U.; Sufian, S.; Salam, M.A. A review of hydrotalcite based catalysts for hydrogen production systems. *Int. J. Hydrogen Energy* **2017**, *42*, 19851–19868. [CrossRef]
25. Huang, S.-H.; Liu, S.-J.; Uan, J.-Y. Controllable luminescence of a Li–Al layered double hydroxide used as a sensor for reversible sensing of carbonate. *J. Mater. Chem.C* **2019**, *7*, 11191–11206. [CrossRef]
26. Tichit, D.; Gérardin, C.; Durand, R.; Coq, B. Layered double hydroxides: Precursors for multifunctional catalysts. *Top. Catal.* **2006**, *39*, 89–96. [CrossRef]
27. Feng, J.-T.; Lin, Y.-J.; Evans, D.G.; Duan, X.; Li, D.-Q. Enhanced metal dispersion and hydrodechlorination properties of a Ni/Al$_2$O$_3$ catalyst derived from layered double hydroxides. *J. Catal.* **2009**, *266*, 351–358. [CrossRef]
28. Li, M.; Wang, X.; Li, S.; Wang, S.; Ma, X. Hydrogen production from ethanol steam reforming over nickel based catalyst derived from Ni/Mg/Al hydrotalcite-like compounds. *Int. J. Hydrogen Energy* **2010**, *35*, 6699–6708. [CrossRef]
29. Vizcaíno, A.J.; Lindo, M.; Carrero, A.; Calles, J.A. Hydrogen production by steam reforming of ethanol using Ni catalysts based on ternary mixed oxides prepared by coprecipitation. *Int. J. Hydrog. Energy* **2012**, *37*, 1985–1992. [CrossRef]
30. Serna, C.J.; Rendon, J.L.; Iglesias, J.E. Crystal-chemical study of layered [Al$_2$Li(OH)$_6$]+ x·· nH$_2$O. *Clays. Clay Miner.* **1982**, *30*, 180–184. [CrossRef]
31. Cavani, F.; Trifiro', F.; Vaccari, A. Hydrotalcite-type anionic clays: Preparation, properties and applications. *Catal. Today* **1991**, *11*, 173–301. [CrossRef]
32. Evans, D.G.; Slade, R.C.T. Structural aspects of layered double hydroxides. *Struct. Bond.* **2006**, *119*, 1–87.
33. Hernandez, M.J.; Ulibarri, M.A.; Cornejo, J.; Pena, M.J.; Serna, C.J. Thermal stability of aluminium hydroxycarbonates with monovalent cations. *Thermochim. Acta* **1985**, *94*, 257–266. [CrossRef]
34. Shumaker, J.L.; Crofcheck, C.; Tackett, S.A.; Santillan-Jimenez, E.; Crocker, M. Biodiesel production from soybean oil using calcined Li–Al layered double hydroxide catalysts. *Catal. Lett.* **2007**, *115*, 56–61. [CrossRef]
35. Bartholomew, C.H. Mechanisms of catalyst deactivation. *Appl. Catal. A Gen.* **2001**, *212*, 17–60. [CrossRef]
36. López, E.; Divins, N.J.; Anzola, A.; Schbib, S.; Borio, D.; Llorca, J. Ethanol steam reforming for hydrogen generation over structured catalysts. *Int. J. Hydrogen Energy* **2013**, *38*, 4418–4428. [CrossRef]
37. Cybulski, A.; Moulijn, J.A. *Structured Catalysts and Reactors*; CRC Press: Boca Raton, FL, USA, 2005.
38. Palma, V.; Ruocco, C.; Castaldo, F.; Ricca, A.; Boettge, D. Ethanol steam reforming over bimetallic coated ceramic foams: Effect of reactor configuration and catalytic support. *Int. J. Hydrogen Energy* **2015**, *40*, 12650–12662. [CrossRef]
39. Santander, J.A.; Tonetto, G.M.; Pedernera, M.N.; López, E. Ni/CeO2–MgO catalysts supported on stainless steel plates for ethanol steam reforming. *Int. J. Hydrogen Energy* **2017**, *42*, 9482–9492. [CrossRef]

40. Kapteijn, F.; Moulijn, J.A. Structured catalysts and reactors—Perspectives for demanding applications. *Catal. Today* **2020**, *383*, 5–14. [CrossRef]

41. Frusteri, F.; Freni, S.; Chiodo, V.; Spadaro, L.; Di Blasi, O.; Bonura, G.; Cavallaro, S. Steam reforming of bio-ethanol on alkali-doped Ni/MgO catalysts: Hydrogen production for mc fuel cell. *Appl. Catal. A Gen.* **2004**, *270*, 1–7. [CrossRef]

42. Oliva, P.; Leonardi, J.; Laurent, J.F.; Delmas, C.; Braconnier, J.J.; Figlarz, M.; Fievet, F.; de Guibert, A. Review of the structure and the electrochemistry of nickel hydroxides and oxy-hydroxides. *J. Power Sources* **1982**, *8*, 229–255. [CrossRef]

43. Kamath, P.V.; Subbanna, G.N. Electroless nickel hydroxide: Synthesis and characterization. *J. Appl. Electrochem.* **1992**, *22*, 478–482. [CrossRef]

44. Liu, X.-M.; Zhang, X.-G.; Fu, S.-Y. Preparation of urchinlike NiO nanostructures and their electrochemical capacitive behaviors. *Mater. Res. Bull.* **2006**, *41*, 620–627. [CrossRef]

45. Duan, G.; Cai, W.; Luo, Y.; Sun, F. A hierarchically structured $Ni(OH)_2$ monolayer hollow-sphere array and its tunable optical properties over a large region. *Adv. Funct. Mater.* **2007**, *17*, 644–650. [CrossRef]

46. Li, J.; Yan, R.; Xiao, B.; Liang, D.T.; Lee, D.H. Preparation of nano-NiO particles and evaluation of their catalytic activity in pyrolyzing biomass components. *Energ. Fuel.* **2008**, *22*, 16–23. [CrossRef]

47. Zhu, Z.; Wei, N.; Liu, H.; He, Z. Microwave-assisted hydrothermal synthesis of $Ni(OH)_2$ architectures and their in situ thermal convention to NiO. *Adv. Powder Technol.* **2011**, *22*, 422–426. [CrossRef]

48. Song, Q.S.; Li, Y.Y.; Chan, S.L.I. Physical and electrochemical characteristics of nanostructured nickel hydroxide powder. *J. Appl. Electrochem.* **2005**, *35*, 157–162. [CrossRef]

49. Shangguan, E.; Chang, Z.; Tang, H.; Yuan, X.-Z.; Wang, H. Synthesis and characterization of high-density non-spherical $Ni(OH)_2$ cathode material for Ni–MH batteries. *Int. J. Hydrogen Energy* **2010**, *35*, 9716–9724. [CrossRef]

50. Zieliński, J. The effect of water on the reduction of nickel/alumina catalysts. *Catal. Lett.* **1992**, *12*, 389–394. [CrossRef]

51. Zieliński, J. Effect of alumina on the reduction of surface nickel oxide; morphology of the surfaces of the surfaces of Ni/Al_2O_3 catalysts. *J. Mol. Catal.* **1993**, *83*, 197–206. [CrossRef]

52. Zieliński, J. Effect of water on the reduction of nickel-alumina catalysts catalyst characterization by temperature-programmed reduction. *J. Chem. Soc. Faraday Trans.* **1997**, *93*, 3577–3580. [CrossRef]

53. Shalvoy, R.B.; Davis, B.H.; Reucroft, P.J. Studies of the metal–support interaction in coprecipitated nickel on alumina methanation catalysts using x-ray photoelectron spectroscopy (xps). *Surf. Interface Anal.* **1980**, *2*, 11–16. [CrossRef]

54. Jr, L.S.; Makovsky, L.E.; Stencel, J.M.; Brown, F.R.; Hercules, D.M. Surface spectroscopic study of tungsten-alumina catalysts using x-ray photoelectron, ion scattering, and raman spectroscopies. *J. Phys. Chem.* **1981**, *85*, 3700–3707.

55. Meng, F.; Li, X.; Li, M.; Cui, X.; Li, Z. Catalytic performance of CO methanation over La-promoted Ni/Al_2O_3 catalyst in a slurry-bed reactor. *Chem. Eng. J.* **2017**, *313*, 1548–1555. [CrossRef]

56. Czekaj, I.; Loviat, F.; Raimondi, F.; Wambach, J.; Biollaz, S.; Wokaun, A. Characterization of surface processes at the Ni-based catalyst during the methanation of biomass-derived synthesis gas: X-ray photoelectron spectroscopy (XPS). *Appl. Catal. A Gen.* **2007**, *329*, 68–78. [CrossRef]

57. Baker, R.T.K. Catalytic growth of carbon filaments. *Carbon* **1989**, *27*, 315–323. [CrossRef]

58. Helveg, S.; Lo´pez-Cartes, C.; Sehested, J.; Hansen, P.L.; Clausen, B.S.; Rostrup-Nielsen, J.R.; Abild-Pedersen, F.; Nørskov, J.K. Atomic-scale imaging of carbon nanofiber growth. *Nature* **2004**, *427*, 426–429. [CrossRef]

59. Jeong, N.; Lee, J. Growth of filamentous carbon by decomposition of ethanol on nickel foam: Influence of synthesis conditions and catalytic nanoparticles on growth yield and mechanism. *J. Catal.* **2008**, *260*, 217–226. [CrossRef]

60. Zhou, L.; Li, L.; Wei, N.; Li, J.; Takanabe, K.; Basset, J.-M. Effect of $NiAl_2O_4$ formation on Ni/Al_2O_3 stability during dry reforming of methane. *ChemCatChem* **2015**, *7*, 2508–2516. [CrossRef]

61. Barbier, J. Deactivation of reforming catalysts by coking-a review. *Appl. Catal.* **1986**, *23*, 225–243. [CrossRef]

62. Bitter, J.H.; Seshan, K.; Lercher, J.A. Deactivation and coke accumulation during CO_2/CH_4 reforming over pt catalysts. *J. Catal.* **1999**, *183*, 336–343. [CrossRef]

63. Cunha, A.F.; Morales-Torres, S.; Pastrana-Martínez, L.M.; Martins, A.A.; Mata, T.M.; Caetano, N.S.; Loureiro, J.M. Syngas production by bi-reforming methane on an Ni–K-promoted catalyst using hydrotalcites and filamentous carbon as a support material. *RSC Adv.* **2020**, *10*, 21158–21173. [CrossRef]

64. Berteau, P.; Delmon, B. Modified aluminas: Relationship between activity in 1-butanol dehydration and acidity measured by NH_3 TPD. *Catal. Today* **1989**, *5*, 121–137. [CrossRef]

65. Navarro, R.M.; Sanchez-Sanchez, M.C.; Fierro, J.L.G. Structure and activity of Pt-Ni catalysts supported on modified Al_2O_3 for ethanol steam reforming. *J. Nanosci. Nanotechnol.* **2015**, *15*, 6592–6603. [CrossRef]

66. Yergaziyeva, G.Y.; Dossumov, K.; Mambetova, M.M.; Strizhak, P.Y.; Kurokawa, H.; Baizhomartov, B. Effect of Ni, La, and Ce oxides on a Cu/Al_2O_3 catalyst with low copper loading for ethanol non-oxidative dehydrogenation. *Chem. Eng. Technol.* **2021**, *44*, 1890–1899. [CrossRef]

67. Zheng, S.; Liu, B.S.; Wang, W.S.; Wang, F.; Zhang, Z.F. Mesoporous and macroporous alumina-supported nickel adsorbents for adsorptive desulphurization of commercial diesel. *Adsorp. Sci. Technol.* **2015**, *33*, 337–353. [CrossRef]

68. Fang, X.; Zhang, R.; Wang, Y.; Yang, M.; Guo, Y.; Wang, M.; Zhang, J.; Xu, J.; Xu, X.; Wang, X. Plasma assisted preparation of highly active $NiAl_2O_4$ catalysts for propane steam reforming. *Int. J. Hydrogen Energy* **2021**, *46*, 24931–24941. [CrossRef]

69. Nivangune, N.; Kelkar, A. Selective synthesis of dimethyl carbonate via transesterification of propylene carbonate with methanol catalyzed by bifunctional Li-Al nano-composite. *ChemistrySelect* **2019**, *4*, 8574–8583. [CrossRef]

70. Di Cosimo, J.I.; Diez, V.K.; Xu, M.; Iglesia, E.; Apesteguia, C.R. Structure and surface and catalytic properties of Mg-Al basic oxides. *J. Catal.* **1998**, *178*, 499–510. [CrossRef]

71. Coleman, L.J.I.; Epling, W.; Hudgins, R.R.; Croiset, E. Ni/Mg–Al mixed oxide catalyst for the steam reforming of ethanol. *Appl. Catal. A Gen.* **2009**, *363*, 52–63. [CrossRef]

72. Pavel, O.D.; Tichit, D.; Marcu, I.-C. Acido-basic and catalytic properties of transition-metal containing Mg–Al hydrotalcites and their corresponding mixed oxides. *Appl. Clay Sci.* **2012**, *61*, 52–58. [CrossRef]

73. Antoniak-Jurak, K.; Kowalik, P.; Michalska, K.; Próchniak, W.; Bicki, R. Zn-Al mixed oxides decorated with potassium as catalysts for HT-WGS: Preparation and properties. *Catalysts* **2020**, *10*, 1094. [CrossRef]

74. Marino, F.J.; Cerrella, E.G.; Duhalde, S.; Jobbagy, M.; Laborde, M.A. Hydrogen from steam reforming of ethanol. Characterization and performance of copper-nickel supported catalysts. *Int. J. Hydrog. Energy* **1998**, *23*, 1095–1101. [CrossRef]

75. Marino, F.; Boveri, M.; Baronetti, G.; Laborde, M. Hydrogen production from steam reforming of bioethanol using Cu/Ni/K/γ-Al2O3 catalysts. Effect of Ni. *Int. J. Hydrogen Energy* **2001**, *26*, 665–668. [CrossRef]

MDPI

St. Alban-Anlage 66

4052 Basel

Switzerland

Tel. +41 61 683 77 34

Fax +41 61 302 89 18

www.mdpi.com

Catalysts Editorial Office

E-mail: catalysts@mdpi.com

www.mdpi.com/journal/catalysts